Beginning Anomaly Detection Using Python-Based Deep Learning

Implement Anomaly Detection Applications with Keras and PyTorch

Second Edition

Suman Kalyan Adari
Sridhar Alla

Apress®

Beginning Anomaly Detection Using Python-Based Deep Learning: Implement Anomaly Detection Applications with Keras and PyTorch, Second Edition

Suman Kalyan Adari
Tampa, FL, USA

Sridhar Alla
Delran, NJ, USA

ISBN-13 (pbk): 979-8-8688-0007-8
https://doi.org/10.1007/979-8-8688-0008-5

ISBN-13 (electronic): 979-8-8688-0008-5

Managing Director, Apress Media LLC: Welmoed Spahr
Acquisitions Editor: Celestin Suresh John
Development Editor: James Markham
Coordinating Editor: Gryffin Winkler

Cover designed by eStudioCalamar

Cover image by Tony Litvyak@Unsplash.com

Distributed to the book trade worldwide by Apress Media, LLC, 1 New York Plaza, New York, NY 10004, U.S.A. Phone 1-800-SPRINGER, fax (201) 348-4505, e-mail orders-ny@springer-sbm.com, or visit www.springeronline.com. Apress Media, LLC is a California LLC and the sole member (owner) is Springer Science + Business Media Finance Inc (SSBM Finance Inc). SSBM Finance Inc is a **Delaware** corporation.

For information on translations, please e-mail booktranslations@springernature.com; for reprint, paperback, or audio rights, please e-mail bookpermissions@springernature.com.

Apress titles may be purchased in bulk for academic, corporate, or promotional use. eBook versions and licenses are also available for most titles. For more information, reference our Print and eBook Bulk Sales web page at http://www.apress.com/bulk-sales.

Any source code or other supplementary material referenced by the author in this book is available to readers on GitHub (https://github.com/Apress). For more detailed information, please visit https://www.apress.com/gp/services/source-code.

Paper in this product is recyclable

Table of Contents

About the Authors

Suman Kalyan Adari is currently a machine learning research engineer. He obtained a B.S. in computer science at the University of Florida and an M.S. in computer science, specializing in machine learning, at Columbia University. He has been conducting deep learning research in adversarial machine learning since his freshman year at the University of Florida and has presented at the IEEE Dependable Systems and Networks workshop on Dependable and Secure Machine Learning held in Portland, Oregon, USA in June 2019. Currently, he works on various anomaly detection tasks spanning behavioral tracking and geospatial trajectory modeling.

He is quite passionate about deep learning, and specializes in various fields ranging from video processing to generative modeling, object tracking, time-series modeling, and more.

Sridhar Alla is the co-founder and CTO of Bluewhale, which helps organizations big and small in building AI-driven big data solutions and analytics, as well as SAS2PY, a powerful tool to automate migration of SAS workloads to Python-based environments using Pandas or PySpark. He is a published author of books and an avid presenter at numerous Strata, Hadoop World, Spark Summit, and other conferences. He also has several patents filed with the US PTO on large-scale computing and distributed systems. He has extensive hands-on experience in several technologies, including Spark, Flink, Hadoop, AWS, Azure, TensorFlow, Cassandra, and others. He spoke on anomaly detection using deep learning at Strata SFO in March 2019 and at Strata London in October 2019. He was born in Hyderabad, India, and now lives in New Jersey with his wife, Rosie, his daughters, Evelyn and Madelyn, and his son, Jayson. When he is not busy writing code, he loves to spend time with his family and also training, coaching, and organizing meetups.

About the Technical Reviewers

Puneet Sinha has accumulated more than 12 years of work experience in developing and deploying end-to-end models in credit risk, multiple marketing optimization, A/B testing, demand forecasting and brand evaluation, profit and price analyses, anomaly and fraud detection, propensity modeling, recommender systems, upsell/cross-sell models, modeling response to incentives, price optimization, natural language processing, and OCR using ML/deep learning algorithms.

Shubho Mohanty is a product thinker and creator, bringing two decades of experience in the "concept-to-market" life cycle of some of the unique, innovative, and highly successful industry-first products and platforms in the data and security spaces.

Shubho holds 12+ US patents in data, analytics, and cloud security. He has also been awarded IDG CIO100, 2020 for strategizing and developing a technology innovation ecosystem.

He currently serves as the Chief Product Officer at Calibo, where he leads the product vision, strategy, innovation, and development of Calibo's enterprise PaaS. Prior to Calibo, Shubho was the Global VP of Product & Engineering at CDK Global (formerly, ADP Inc). He has also served in various product leadership roles in organizations like Symantec and Microsoft. He also co-founded Ganos, a B2B data start-up.

He received his B.Tech. in Electrical Engineering from National Institute of Technology (NIT), India. He is a mentor to many high-repute start-up programs where he guides young entrepreneurs to solve some of the most pressing challenges. He is also an influential speaker at leading technology and industry forums.

Acknowledgments

Suman Kalyan Adari

I would like to thank my parents, Krishna and Jyothi, my sister, Niha, and my loving dog, Pinky, for supporting me throughout the entire process of writing this book as well as my various other endeavors.

Sridhar Alla

I would like to thank my wonderful, loving wife, Rosie Sarkaria, and my beautiful, loving children, Evelyn, Madelyn, and Jayson, for all their love and patience during the many months I spent writing this book. I would also like to thank my parents, Ravi and Lakshmi Alla, for their blessings and all the support and encouragement they continue to bestow upon me.

Introduction

Congratulations on your decision to explore the exciting world of anomaly detection using deep learning!

Anomaly detection involves finding patterns that do not adhere to what is considered as normal or expected behavior. Businesses could lose millions of dollars due to abnormal events. Consumers could also lose millions of dollars. In fact, there are many situations every day where people's lives are at risk and where their property is at risk. If your bank account gets cleaned out, that's a problem. If your water line breaks, flooding your basement, that's a problem. If all flights at an airport get delayed due to a technical glitch in the traffic control system, that's a problem. If you have a health issue that is misdiagnosed or not diagnosed, that's a very big problem that directly impacts your well-being.

In this book, you will learn how anomaly detection can be used to solve business problems. You will explore how anomaly detection techniques can be used to address practical use cases and address real-life problems in the business landscape. Every business and use case is different, so while we cannot copy and paste code and build a successful model to detect anomalies in any dataset, this book will cover many use cases with hands-on coding exercises to give you an idea of the possibilities and concepts behind the thought process. All the code examples in the book are presented in Python 3.8. We choose Python because it is truly the best language for data science, with a plethora of packages and integrations with scikit-learn, deep learning libraries, etc.

We will start by introducing anomaly detection, and then we will look at legacy methods of detecting anomalies that have been used for decades. Then we will look at deep learning to get a taste of it. Then we will explore autoencoders and variational autoencoders, which are paving the way for the next generation of generative models. Following that, we will explore generative adversarial networks (GANs) as a way to detect anomalies, delving directly into generative AI.

Then we'll look at long short-term memory (LSTM) models to see how temporal data can be processed. We will cover temporal convolutional networks (TCNs), which are excellent for temporal data anomaly detection. We will also touch upon the transformer architecture, which has revolutionized the field of natural language processing as another means for temporal anomaly detection. Finally, we will look at several examples of anomaly detection in various business use cases.

In addition, all coding examples will be provided in TensorFlow 2/Keras, with accompanying PyTorch equivalents, on the GitHub repository for this book. You will combine all this extensive knowledge with hands-on coding using Jupyter notebook-based exercises to experience the knowledge firsthand and see where you can use these algorithms and frameworks. Best of luck, and welcome to the world of deep learning!

## CHAPTER 1

Introduction to Anomaly Detection

In this chapter, you will learn about anomalies in general, the categories of anomalies, and anomaly detection. You will also learn why anomaly detection is important, how anomalies can be detected, and the use case for such a mechanism.

In a nutshell, this chapter covers the following topics:

- What is an anomaly?

- Categories of different anomalies

- What is anomaly detection?

- Where is anomaly detection used?

What Is an Anomaly?

Before you get started with learning about anomaly detection, you must first understand what exactly you are targeting. Generally, an **anomaly** is an outcome or value that deviates from what is expected, but the exact criteria for what determines an anomaly can vary from situation to situation.

Anomalous Swans

To get a better understanding of what an anomaly is, let's take a look at some swans sitting by a lake (Figure 1-1).

Figure 1-1. *A couple swans by a lake*

Let's say that we want to observe these swans and make assumptions about the color of the swans at this particular lake. Our goal is to determine what the normal color of swans is and to see if there are any swans that are of a different color than this (Figure 1-2).

Figure 1-2. *More swans show up, all of which are white*

We continue to observe swans for a few years and all of them have been white. Given these observations, we can reasonably conclude that every swan at this lake should be white. The very next day, we are observing swans at the lake again. But wait! What's this? A black swan has just flown in (Figure 1-3).

Figure 1-3. *A black swan appears*

Considering our previous observations, we thought that we had seen enough swans to assume that the next swan would also be white. However, the black swan defies that assumption entirely, making it an *anomaly*. It's not really an outlier, which would be, for example, a really big white swan or a really small white swan; it's a swan that's entirely a different color, making it an anomaly. In our scenario, the overwhelming majority of swans are white, making the black swan extremely rare.

In other words, given a swan by the lake, the probability of it being black is very small. We can explain our reasoning for labeling the black swan as an anomaly with one of two approaches (though we aren't limited to only these two approaches).

First, given that a vast majority of swans observed at this particular lake are white, we can assume that, through a process similar to inductive reasoning, the normal color for a swan here is white. Naturally, we would label the black swan as an anomaly purely based on our prior assumptions that all swans are white, considering that we'd only seen white swans before the black swan arrived.

Another way to look at why the black swan is an anomaly is through probability. Now assume that there is a total of 1000 swans at this lake and only two are black swans; the probability of a swan being black is 2 / 1000, or 0.002. Depending on the *probability threshold*, meaning the lowest probability for an outcome or event that will be accepted as normal, the black swan could be labeled as anomalous or normal. In our case, we will consider it an anomaly because of its extreme rarity at this lake.

Anomalies as Data Points

We can extend this same concept to a real-world application. In the following example, we will take a look at a factory that produces screws and attempt to determine what an anomaly could be in this context and individual screws are sampled from each batch and are tested to ensure a certain level of quality is maintained. For each sampled screw, assume that the density and tensile strength (how resistant the screw is to breaking under stress) are measured.

Figure 1-4 is an example graph of various sampled screws with the dotted lines representing the range of densities and tensile strengths allowed. The solid lines form a bounding box where any value of tensile strength and density inside it is considered good.

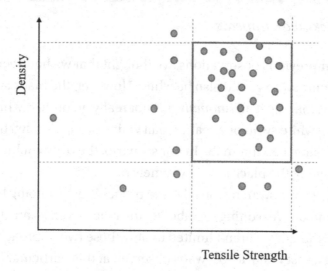

Figure 1-4. *Density and tensile strength in a batch of screw samples*

The intersections of the dotted lines have created several different regions containing data points. Of interest is the bounding box (solid lines) created from the intersection of both sets of dotted lines since it contains the data points for samples deemed acceptable (Figure 1-5). Any data point outside of that specific box will be considered anomalous.

4

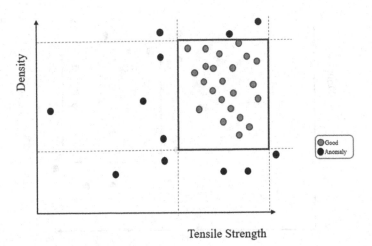

Figure 1-5. *Data points are identified as "good" or "anomaly" based on their location*

Now that we know which points are and aren't acceptable, let's pick out a sample from a new batch of screws and check its data to see where it falls on the graph (Figure 1-6).

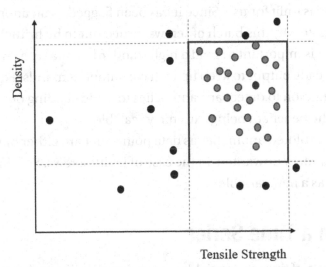

Figure 1-6. *A new data point representing the new sample screw is generated, with the data falling within the bounding box*

The data for this sample screw falls within the acceptable range. That means that this batch of screws is good to use since its density as well as tensile strength is appropriate for use by the consumer. Now let's look at a sample from the next batch of screws and check its data (Figure 1-7).

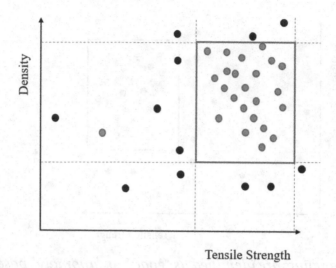

Figure 1-7. *A new data point is generated for another sample, but this falls outside the bounding box*

The data falls far outside the acceptable range. For its density, the screw has abysmal tensile strength and is unfit for use. Since it has been flagged as an anomaly, the factory can investigate why this batch of screws turned out to be brittle. For a factory of considerable size, it is important to hold a high standard of quality as well as maintain a high volume of steady output to keep up with consumer demands. For a monumental task like that, automation to detect any anomalies to avoid sending out faulty screws is essential and has the benefit of being extremely scalable.

So far, we have explored anomalies as data points that are either out of place, in the case of the black swan, or unwanted, in the case of faulty screws. So what happens when we introduce time as a new variable?

Anomalies in a Time Series

With the introduction of time as a variable, we are now dealing with a notion of *temporality* associated with the data sets. What this means is that certain patterns are dependent on time. For example, daily, monthly, or yearly occurrences are time-series patterns as they present on regular time intervals.

To better understand time series–based anomalies, let's look at a few examples.

Personal Spending Pattern

Figure 1-8 depicts a random person's spending habits over the course of a month.

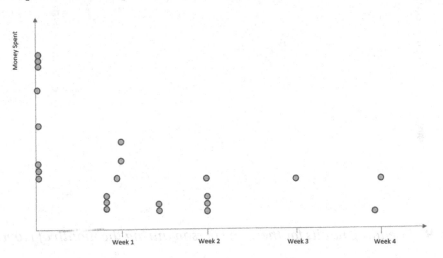

Figure 1-8. *Spending habits of a person over the course of a month*

Assume the initial spike in expenditures at the start of the month is due to the payment of bills such as rent and insurance. During the weekdays, our example person occasionally eats out, and on the weekends goes shopping for groceries, clothes, and various other items. Also assume that this month does not include any major holidays.

These expenditures can vary from month to month, especially in months with major holidays. Assume that our person lives in the United States, in which the holiday of Thanksgiving falls on the last Thursday of the month of November. Many U.S. employers also include the Friday after Thanksgiving as a day off for employees. U.S. retailers have leveraged that fact to entice people to begin their Christmas shopping by offering special deals on what has colloquially become known as "Black Friday." With that in mind, let's take a look at our person's spending pattern in November (Figure 1-9). As expected, a massive spike in purchases occurred on Black Friday, some of them quite expensive.

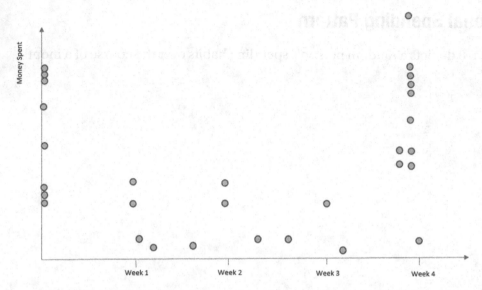

Figure 1-9. *Spending habits for the same person during the month of November*

Now assume that, unfortunately, our person has had their credit card information stolen, and the criminals responsible for it have decided to purchase various items of interest to them. Using the same month as in the first example (Figure 1-8; no major holidays), the graph in Figure 1-10 depicts what could happen.

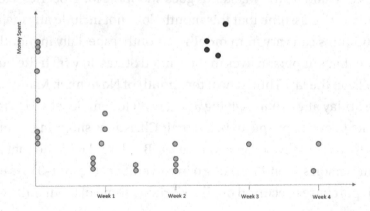

Figure 1-10. *Purchases in the person's name during the same month as in Figure 1-8*

Let's assume we have a record of purchases for this user going back many years. Thanks to this established prior history, this sudden influx in purchases would be flagged as anomalies. Such a cluster of purchases might be normal for Black Friday or in the weeks before Christmas, but in any other month without a major holiday, it looks out of place. In this case, our person might be contacted by the credit card company to confirm whether or not they made the purchases.

Some companies might even flag purchases that follow normal societal trends. What if that TV wasn't really bought by our person on Black Friday? In that case, the credit card company's software can ask the client directly through a phone app, for example, whether or not thcy actually bought the item in question, allowing for some additional protection against fraudulent purchases.

Taxi Cabs

As another example of anomalies in a time series, let's look at some sample data for taxi cab pickups and drop-offs over time for a random city and an arbitrary taxi company and see if we can detect any anomalies.

On an average day, the total number of pickups can look somewhat like the pattern shown in Figure 1-11.

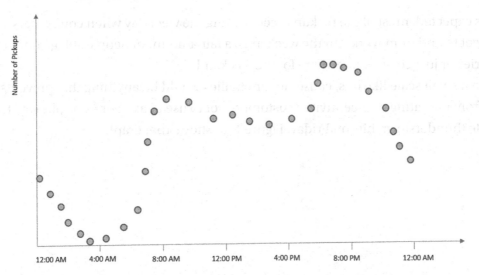

Figure 1-11. *Number of pickups for a taxi company throughout the day*

From the graph, we see that there's a bit of post-midnight activity that drops off to near zero during the late-night hours. However, customcr traffic picks up suddenly around morning rush hour and remains high until the early evening, when it peaks during evening rush hour. This is essentially what an average day looks like.

Let's expand the scope out a bit more to gain some perspective of passenger traffic throughout the week (Figure 1-12).

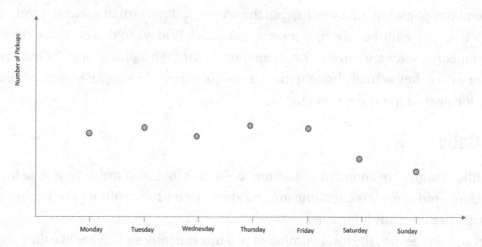

Figure 1-12. *Number of pickups for a taxi company throughout the week*

As expected, most of the pickups occur during the weekday when commuters must get to and from work. On the weekends, a fair amount of people still go out to get groceries or just go out somewhere for the weekend.

On a small scale like this, causes for anomalies would be anything that prevents taxis from operating or incentivizes customers not to use a taxi. For example, say that a terrible thunderstorm hits on Friday. Figure 1-13 shows that graph.

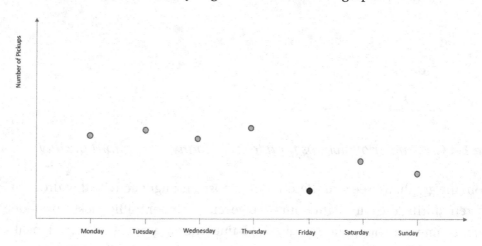

Figure 1-13. *Number of pickups for a taxi company throughout the week, with a heavy thunderstorm on Friday*

The thunderstorm likely influenced some people to stay indoors, resulting in a lower number of pickups than unusual for a weekday. However, these sorts of anomalies are usually too small-scale to have any noticeable effect on the overall pattern.

Let's take a look at the data over the entire year, as shown in Figure 1-14.

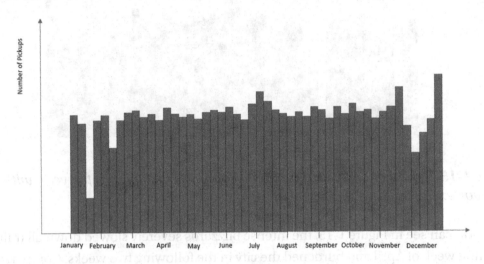

Figure 1-14. *Number of pickups for a taxi company throughout the year*

The largest dips occur during the winter months when snowstorms are expected. These are regular patterns that can be observed at similar times every year, so they are not an anomaly. But what happens to customer traffic levels when a relatively rare polar vortex descends on the city in early April and unleashes several intense blizzards? Figure 1-15 shows the graph.

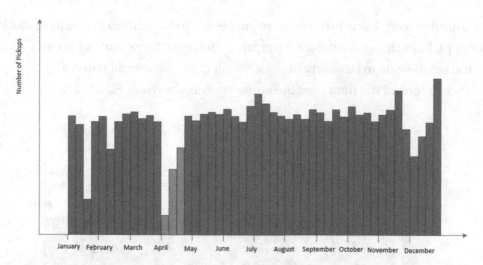

Figure 1-15. *Number of pickups for a taxi company throughout the year, with a polar vortex descending on the city in April*

As you can see in Figure 1-15, the intense blizzards severely slowed down all traffic in the first week of April and burdened the city in the following two weeks. Comparing this graph to the graph shown in Figure 1-14, there's a clearly defined anomaly in April caused by the polar vortex. Since this pattern is extremely rare for the month of April, it would be flagged as an anomaly.

Categories of Anomalies

Now that you have more perspective of what anomalies can be in various situations, you can see that they generally fall into these broad categories:

- Data point–based anomalies

- Context-based anomalies

- Pattern-based anomalies

Data Point–Based Anomalies

Data point–based anomalies may seem comparable to outliers in a set of data points. However, as previously mentioned, anomalies and outliers are not the same thing, though these terms are sometimes used interchangeably. **Outliers** are data points that are expected to be present in the data set and can be caused by unavoidable random errors or from systematic errors relating to how the data was sampled. **Anomalies** would be outliers or other values that one doesn't expect to exist. These data anomalies might be present wherever a data set of values exist.

As an example of a data set in which data point–based anomalies may exist, consider a data set of thyroid diagnostic values, where the majority of the data points are indicative of normal thyroid functionality. In this case, anomalous values represent sick thyroids. While they are not necessarily outliers, they have a low probability of existing when taking into account all the normal data.

We can also detect individual purchases totaling excessive amounts and label them as anomalies since, by definition, they are not expected to occur or have a very low probability of occurrence. In this case, they are labeled as potentially fraudulent transactions, and the card holder is contacted to ensure the validity of the purchase.

Basically, we can say this about the difference between anomalies and outliers: we should expect a data set to include outliers, but we should not expect it to include anomalies. Though the terms "anomaly" and "outlier" are sometimes interchanged, anomalies are not always outliers, and not all outliers are anomalies.

Context-Based Anomalies

Context-based anomalies consist of data points that might seem normal at first but are considered anomalies in their respective contexts. Returning to our earlier personal spending example, we might expect a sudden surge in purchases near certain holidays, but these purchases could seem unusual in the middle of August. The person's high volume of purchases on Black Friday was not flagged because it is typical spending behavior for people on Black Friday. However, if the purchases were made in a month where it is out of place given previous purchase history, it would be flagged as an anomaly. This might seem similar to the example presented for data point–based anomalies, but the distinction for context-based anomalies is that the individual purchase does not have to be expensive. If a person never buys gasoline because they own an electric car, sudden purchases of gasoline would be out of place given the context. Buying gasoline is normal behavior for many people, but in this context, it is an anomaly.

Pattern-Based Anomalies

Pattern-based anomalies are patterns and trends that deviate from their historical counterparts, and they often occur in time-series or other sequence-based data. In the earlier taxi cab company example, the customer pickup counts for the month of April were pretty consistent with the rest of the year. However, once the polar vortex hit, the numbers tanked visibly, resulting in a huge drop in the graph, labeled as an anomaly.

Similarly, when monitoring network traffic in the workplace, expected patterns of network traffic are formed from constant monitoring of data over several months or even years for some companies. If an employee attempts to download or upload large volumes of data, it generates a certain pattern in the overall network traffic flow that could be considered anomalous if it deviates from the employee's usual behavior.

As another example of pattern-based anomalies, if an external hacker decided to hit a company's website with a **distributed denial-of-service (DDoS) attack**—an attempt to overwhelm the server that handles network flow to a certain website in an attempt to bring the entire website down or stop its functionality—every single attempt would register as an unusual spike in network traffic. All of these spikes are clearly deviants from normal traffic patterns and would be considered anomalous.

Anomaly Detection

Now that you have a better understanding of the different types of anomalies, we can proceed to discuss approaches to creating models to detect anomalies. This section presents a few approaches we can take, but keep in mind we are not limited to just these methods.

Recall our reasoning for labeling the black swan as an anomaly. One reason was that since all the swans we have seen thus far were white, the single black swan is an obvious anomaly. A statistical way to explain this reasoning is that as of the most recent set of observations, we have one black swan and tens of thousands of white swans. Thus, the probability of occurrence of the black swan is one out of tens of thousands of all observed swans. Since this probability is so low, it would make the black swan an anomaly just because we do not expect to see it at all.

The anomaly detection models we will explore in this book follow these approaches either by training on unlabeled data, training on normal, nonanomalous data, or training on labeled data for normal and anomalous data. In the context of identifying swans, we would be told which swans are normal and which swans are anomalies.

So, what is anomaly detection? Quite simply, **anomaly detection** is the process in which an advanced algorithm identifies certain data or data patterns to be anomalous. Falling under anomaly detection are the tasks of outlier detection, noise removal, novelty detection, event detection, change point detection, and anomaly score calculation. In this book, we will explore all of these as they are all basically anomaly detection methods. The following tasks of anomaly detection are not exhaustive, but are some of the more common anomaly detection tasks today.

Outlier Detection

Outlier detection is a technique that aims to detect anomalous outliers within a given data set. As previously discussed, three methods that can be applied to this situation are to train a model only on normal data to identify anomalies (by a high reconstruction error, described next), to model a probability distribution in which anomalies would be labeled based on their association with really low probabilities, or to train a model to recognize anomalies by teaching it what an anomaly looks like and what a normal point looks like.

Regarding the high reconstruction error, think of it this way: the model trains on a set of normal data and learns the patterns corresponding to normal data. When exposed to an anomalous data point, the patterns do not line up with what the model learned to associate with normal data. The reconstruction error can be analogous to the deviance in learned patterns between the anomalous point and the normal points the model trained on. We will formally go over reconstruction error in Chapter 6. Going back to the example of the swans, the black swan is different based on the patterns that we learned by observing swans at the lake, and was thus anomalous because it did not follow the color pattern.

Noise Removal

Noise removal involves filtering out any constant background noise in the data set. Imagine that you are at a party and you are talking with a friend. There is a lot of background noise, but your brain focuses on your friend's voice and isolates it because that's what you want to hear. Similarly, a model learns to efficiently encode the original sound to represent only the essential information. For example, encoding the pitch of your friend's voice, the tempo, the vocal inflections, etc. Then, it is able to reconstruct the original sound without the anomalous interference noise.

This can also be a case where an image has been altered in some form, such as by having perturbations, loss of detail, fog, etc. The model learns an accurate representation of the original image and outputs a reconstruction without any of the anomalous elements in the image.

Novelty Detection

Novelty detection is very similar to outlier detection. In this case, a novelty is a data point outside of the training set the model was exposed to, that was shown to the model to determine if it is an anomaly or not. The key difference between novelty detection and outlier detection is that in outlier detection, the job of the model is to determine what is an anomaly within the training data set. In novelty detection, the model learns what is a normal data point and what isn't and tries to classify anomalies in a new data set that it has never seen before.

Examples can include quality assurance in factories to make sure new batches of created products are up to par, such as with the example of screws from earlier. Another case is network security. Incoming traffic data can be monitored to ensure there is no anomalous behavior going on. Both of these situations involve novelties (new data) to constantly predict on.

Event Detection

Event detection involves the detection of points in a time-series dataset that deviate anomalously from the norm. For example, in the taxi cab company example earlier in the chapter, the polar vortex reduced the customer pickup counts for April. These deviations from the norm for April were all associated with an anomalous event that occurred in the dataset, which an event detector algorithm would identify. Another example of event detection would be the tracking of sea-ice levels at the poles over time, forming a time series. An event detector algorithm could flag exactly when sea-ice levels deviate anomalously from the usual norm, such as occurred recently in July 2023, when the sea-ice levels were detected to be six standard deviations below the established average.

Change Point Detection

Change point detection involves the detection of points in the dataset where their statistical properties start to consistently deviate from the established norm given by the rest of the data. In other words, change point detection algorithms measure shifts in statistical trends. A good example is global average temperatures over time. A change point detection algorithm could identify periods of sustained warming where the statistical properties start to shift over time. A change point detection algorithm could identify accelerated warming periods that differ from the normal rate of warming.

Anomaly Score Calculation

Anomaly score calculation involves assigning a score of how anomalous a given data sample is. Rather than an outright determination of whether or not something is an anomaly, this is for when we want to assign some type of measure to how deviant from the normal a data sample may be. In some cases, anomaly score calculation is a direct prelude to anomaly detection, since you could assign a threshold that determines what are anomalies and what aren't by the anomaly score. In other cases, anomaly score calculation can help with understanding a data set more deeply and aid with data analysis by providing a more nuanced measure of the normalcy of a data point or a measure of how much a data sample deviates from the norm.

The Three Styles of Anomaly Detection

There are three overarching "styles" of anomaly detection:

- Supervised anomaly detection
- Semi-supervised anomaly detection
- Unsupervised anomaly detection

Supervised anomaly detection is a technique in which the training data has labels for both anomalous data points and normal data points. Basically, we tell the model during the training process if a data point is an anomaly or not. Unfortunately, this isn't the most practical method of training, especially because the entire data set needs to be processed and each data point needs to be labeled. Since supervised anomaly detection is basically a type of binary classification task, meaning the job of the model is to categorize data under one of two labels, any classification model can be used for the task, though not every model can attain a high level of performance. Chapter 9 provides an example in the context of a temporal convolutional network.

Semi-supervised anomaly detection involves partially labeling the training data set. Exact implementations and definitions for what "semi-supervised" entails may differ, but the gist of it is that you are working with partially labeled data. In the context of anomaly detection, this can be a case where only the normal data is labeled. Ideally, the model will learn what normal data points look like so that it can flag as anomalous data points that differ from normal data points. Examples of models that can use semi-supervised learning for anomaly detection include autoencoders, which you will learn about in Chapter 6.

Unsupervised anomaly detection, as the name implies, involves training the model on unlabeled data. After the training process, the model is expected to know what data points are normal and what points are anomalous within the data set. Isolation forest, a model we will explore in Chapter 4, is one such model that can be used for unsupervised anomaly detection.

Where Is Anomaly Detection Used?

Whether we realize it or not, anomaly detection is being utilized in nearly every facet of our lives today. Pretty much any task involving data collection of any sort could have anomaly detection applied to it. Let's look at some of the most prevalent fields and topics to which anomaly detection can be applied.

Data Breaches

In today's age of big data, where huge volumes of information are stored about users in various companies, information security is vital. Any information breaches must be reported and flagged immediately, but it is hard to do so manually at such a scale. Data leaks can range from simple accidents, such as an employee losing a USB stick that contains sensitive company information that someone picks up and accesses the data on, to intentional actions, such as an employee intentionally sending data to an outside party, or an attacker gaining access to a database via an intrusion attack. Several high-profile data leaks have been widely reported in news media, from Facebook / Meta, iCloud, and Google security breaches where millions of passwords or photos were leaked. All of those companies operate on an international scale, requiring automation to monitor everything in order to ensure the fastest response time to any data breach.

The data breaches might not even need network access. For example, an employee could email an outside party or another employee with connections to rival companies about travel plans to meet up and exchange confidential information. Of course, these emails would not be so obvious as to state such intentions directly. However, monitoring these emails could be helpful as a post-breach analysis to find out anyone suspicious from within the company, or as part of a real-time monitoring software to ensure data confidentiality compliance across teams for example. Anomaly detection models can sift through and process employee emails to flag any suspicious activity by employees. The software can pick up key words and process them to understand the context and decide whether or not to flag an employee's email for review.

The following are a few more examples of how anomaly detection software can detect an internal data breach:

- Employees may be assigned a specific connection to upload data to. For example, a mass storage drive that employees should not frequently access.

- An employee may regularly be accessing data as part of their work responsibilities. However, if they suddenly start downloading a lot of data that they shouldn't, then the anomaly detector may flag this.

- The detector would be looking at many variables, including who the user is, what data store is being accessed, what volume of data was transferred, how frequently did the user interact with this data store, etc. to find out if this recent interaction was a deviation from established patterns.

In this case, something won't add up, which the software will detect and then flag the employee. It could either turn out to be a one-off sanctioned event, which great, the model did its job but it was ok this time, or it could turn out that the employee somehow accessed data they shouldn't have, which would mean there was a data breach.

The key benefit to using anomaly detection in the workspace is how easy it is to scale up. These models can be used for small companies as well as large-scale international companies.

Identity Theft

Identity theft is another common problem in today's society. Thanks to development of online services allowing for ease of access when purchasing items, the volume of credit and debit card transactions that take place every day has grown immensely. However, this development also makes it easier to steal credit and debit card information or bank account information, allowing the criminals to purchase anything they want if the card isn't deactivated or if the account isn't secured again. Because of the huge volume of transactions, monitoring everything is difficult. However, this is where anomaly detection can step in and help, since it is highly scalable and can help detect fraud transactions the moment the request is sent.

As we discussed earlier, context matters. When a payment card transaction is made, the payment card company's anomaly detection software takes into account the card holder's previous history to determine if the new transaction should be flagged or not. Obviously, a series of high value purchases made suddenly would raise alarms immediately, but what if the criminals were smart enough to realize that and just make a series of purchases over time that won't put a noticeable hole in the card holder's account? Again, depending on the context, the software would pick up on these transactions and flag them again.

For example, let's say that someone's grandmother was recently introduced to Amazon and to the concept of buying things online. One day, unfortunately, she stumbles upon an Amazon lookalike website and enters her credit card information. On the other side, some criminal takes it and starts buying random things, but not all at once so as not to raise suspicion—or so he thought. The grandmother's identity theft insurance company starts noticing some recent purchases of batteries, hard drives, flash drives, and other electronic items. While these purchases might not be that expensive, they certainly stand out when all the prior purchases made by the grandmother consisted of groceries, pet food, and various decorative items. Based on this previous history, the detection software would flag the new purchases and the grandmother would be contacted to verify these purchases. These transactions might even be flagged as soon as an attempt to purchase is made. In this case, either the location of the purchaser or the nature of the transactions could raise alarms and stop the transaction from being successful.

Manufacturing

We have explored a use case of anomaly detection in manufacturing. Manufacturing plants usually have a certain level of quality that they must ensure their products meet before shipping them out. When factories are configured to produce massive quantities of output at a near constant rate, it becomes necessary to automate the process of checking the quality of various samples. Similar to the fictitious example of the screw manufacturer, manufacturing plants in real life might test and use anomaly detection software to ensure the quality of various metal parts, tools, engines, food, clothes, etc.

Networking

Perhaps one of the most important use cases for anomaly detection is in networking. The Internet is host to a vast array of various websites located on servers all around the world. Unfortunately, due to the ease of access to the Internet, many individuals access the Internet for nefarious purposes. Similar to the data leaks that were discussed earlier in the context of protecting company data, hackers can launch attacks on websites as well to leak their information.

One such example would be hackers attempting to leak government secrets through a network attack. With such sensitive information as well as the high volumes of expected attacks every day, automation is a necessary tool to help cybersecurity professionals deal with the attacks and preserve state secrets. On a smaller scale, hackers might attempt to breach a cloud network or a local area network and try to leak data. Even in smaller cases like this, anomaly detection can help detect network intrusion attacks as they happen and notify the proper officials. An example data set for network intrusion anomaly detection is the KDD Cup 1999 data set. This data set contains a large amount of entries that detail various types of network intrusion attacks as well as a detailed list of variables for each attack that can help a model identify each type of attack.

Medicine

Anomaly detection also has a massive role to play in the field of medicine. For example, models can detect subtle irregularities in a patient's heartbeat in order to classify diseases, or they can measure brainwave activity to help doctors diagnose certain conditions. Beyond that, they can help analyze raw diagnostic data for a patient's organ and process it in order to quickly diagnose any possible problems, similarly to the thyroid example discussed earlier.

Anomaly detection can even be used in medical imagery to determine whether a given image contains anomalous objects. For example, suppose an anomaly detection model was trained by exposing it only to MRI imagery of normal bones. When shown an image of a broken bone, it would flag the new image as an anomaly. Similarly, anomaly detection can even be extended to tumor detection, allowing for the model to analyze every image in a full-body MRI scan and look for the presence of abnormal growth or patterns.

Video Surveillance

Anomaly detection also has uses in video surveillance. Anomaly detection software can monitor video feeds and flag any videos that capture anomalous action. While this might seem dystopian, it can certainly help catch criminals and maintain public safety on busy streets and other transportation systems. For example, this type of software could potentially identify a mugging in a street at night as an anomalous event and alert the nearest police department. Additionally, this type of software can detect unusual events at crossroads, such as an accident or some unusual obstruction, and immediately call attention to the footage.

Environment

Anomaly detection can be used to monitor environmental conditions as well. For example, anomaly detection systems are used to monitor heavy-metal levels in rivers to pick up potential spills or leaks into the water supply. Another example is air quality monitoring, which can detect anything from seasonal pollen to wildfire smoke coming in from far away. Additionally, anomaly detection can be utilized to monitor soil health in agricultural or environmental survey cases to gauge the health of an ecosystem. Any dips in soil moisture or specific nutrient levels could indicate some kind of problem.

Summary

Generally, anomaly detection is utilized heavily in medicine, finance, cybersecurity, banking, networking, transportation, and manufacturing, but it is not limited to those fields. For nearly every case imaginable involving data collection, anomaly detection can be put to use to help users automate the process of detecting anomalies and possibly removing them. Many fields in science can benefit from anomaly detection because of the large volume of raw data collection that goes on. Anomalies that would interfere with the interpretation of results or otherwise introduce some sort of bias into the data could be detected and removed, provided that the anomalies are caused by systematic or random errors.

In this chapter, we discussed what anomalies are and why detecting anomalies can be very important to the data processing we have at our organizations.

Next, Chapter 2 introduces you to core data science concepts that you need to know to follow along for the rest of the book.

CHAPTER 2

Introduction to Data Science

This chapter introduces you to the basic principles of the data science workflow. These concepts will help you prepare your data as necessary to be fed into a machine learning model as well as understand its underlying structure through analysis.

You will learn how to use the libraries Pandas, Numpy, Scikit-Learn, and Matplotlib to load, manipulate, and analyze data. Furthermore, you will learn how to perform data I/O; manipulate, transform, and process the data to your liking; analyze and plot the data; and select/create relevant features for the machine learning modeling task.

In a nutshell, this chapter covers the following topics:

- Data I/O
- Data manipulation
- Data analysis
- Data visualization
- Data processing
- Feature engineering/selection

Note   Code examples are provided in Python 3.8. Package versions of all frameworks used are provided. You will need some type of program to run Python code, so make sure to set this up beforehand. The code repository for this book is available at `https://github.com/apress/beginning-anomaly-detection-python-deep-learning-2e/tree/master`.

The repository also includes a requirements.txt file to check your packages and their versions.

Code examples for this chapter are available at `https://github.com/apress/beginning-anomaly-detection-python-deep-learning-2e/blob/master/Chapter%20 2%20Introduction%20to%20Data%20Science/chapter2_datascience.ipynb`. Navigate to "Chapter 2 Introduction to Data Science" and then click chapter2_datascience. ipynb. The code is provided as a .py file as well, though it is the exported version of the notebook.

We will be using JupyterLab (`https://jupyter.org`) to present all of the code examples.

Data Science

"Data science" is quite the popular term and buzzword nowadays, so what exactly is it? In recent times, the term "data science" has come to represent a wide range of roles and responsibilities. Depending on the company, a data scientist can be expected to perform anything from data processing (often at scale, dipping into "big data" territory), to statistical analysis and visualization, to training and deploying machine learning models. In fact, data scientists often perform two or more of these roles at once.

This chapter focuses on concepts from all three roles, walking you through the process of preparing and analyzing the dataset before you explore the modeling aspect in Chapter 3.

Be advised that this will be a brief, high-level walkthrough over the most relevant functionality that these various data science packages offer. Each package is so comprehensive that a full book would be required to cover it in depth. Therefore, you are encouraged to explore each package's online documentation, as well as other guides and walkthroughs, to learn as much as you can.

Dataset

A popular introductory dataset for budding data scientists is the Titanic Dataset, available at Kaggle: `https://www.kaggle.com/c/titanic`. You can also find this dataset hosted on this book's repository at `https://github.com/apress/beginning-anomaly-detection-python-deep-learning-2e/blob/master/data/train.csv`.

Kaggle is a great place to find datasets. It also hosts machine learning modeling competitions, some of which even have prize money. Kaggle is an excellent resource for practicing your skills, and you are encouraged to do so! If you would like to practice the concepts that you learn in this book, search Kaggle for various anomaly detection datasets.

To download the Titanic dataset from the Kaggle web site, follow the instructions provided next. If you prefer, Kaggle offers an API that you can install through PIP, available at `https://github.com/Kaggle/kaggle-api`.

1. Go to `https://www.kaggle.com/c/titanic`, where you are greeted by something that looks like Figure 2-1.

Figure 2-1. *Overview page for the Titanic dataset on Kaggle (as it looks as of April, 2023)*

2. Click the Data tab. You should see a brief description of the dataset as well as a Data Explorer, as shown in Figure 2-2.

3. Click Download All. You will be prompted to sign in. Create an account or log in with an alternate option. After logging in, you will be returned to the Overview page.

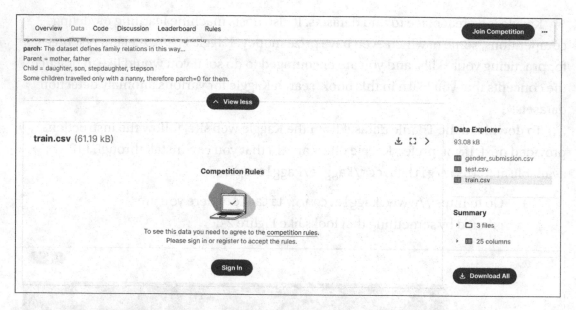

Figure 2-2. *Click Download All to download the data as a zip file*

4. Return to the Data tab, scroll down, and click Download All. A zip
    file will download.

5. Extract this zip file anywhere that you'd like and make a note of
    the directory path.

After you have extracted the zip file, you are almost ready to start processing this
dataset in Python. First, make sure you have an IDE to develop Python code with. To
easily follow the examples in this book, you are recommended to use Jupyter Notebook
or JupyterLab.

Next, make sure you have the following libraries and versions installed, though you
might also want to check the requirements.txt file available on the GitHub repository or
use it to prepare your environment:

- pandas==2.0.0

- numpy==1.22.2

- scikit-learn==1.2.2

- matplotlib==3.7.1

It is not necessary to have the exact same versions, but keep in mind that older versions may not contain features we explore later in the book. Newer versions should be fine unless they are significantly more recent, which might result in reworked syntax or features and thus introduce incompatibility.

You can easily check the version in Python. Figure 2-3 introduces code to import these packages and print their versions.

```
import pandas
import numpy
import sklearn
import matplotlib

print(f"Pandas version: ", pandas.__version__)
print(f"Numpy version: ", numpy.__version__)
print(f"Scikit-learn version: ", sklearn.__version__)
print(f"Matplotlib version: ", matplotlib.__version__)
```

Figure 2-3. *Code to import pandas, numpy, sklearn, and matplotlib and print their versions*

You should see output similar to that displayed in Figure 2-4.

```
Pandas version:   2.0.0
Numpy version:   1.22.2
Scikit-learn version:   1.2.2
Matplotlib version:   3.7.1
```

Figure 2-4. *The text output of running the code in Figure 2-3*

Pandas, Scikit-Learn, and Matplotlib

Pandas is a data science framework for Python that lets you freely manipulate and analyze data to your liking. It is an expansive, comprehensive framework offering a lot of functionality, and should suit your data processing needs. Until you start to scale into the gigabyte realm with your datasets, Pandas can be quick and efficient.

Scikit-Learn is a machine learning library that covers a wide range of functionality related to machine learning modeling, ranging from data processing functions to various machine learning methods

Matplotlib is a data visualization library that lets you build customized plots and graphs. It integrates well with pandas and scikit-learn, and even lets you display images.

NumPy is a numerical computation library with highly efficient implementations for various computational tasks. You can represent your data as numpy arrays and perform quick, vectorized computations, linear algebra operations, and more. NumPy integrates very well with many popular Python packages, including pandas, scikit-learn, and matplotlib.

With all of these packages, you should be able to perform comprehensive data analysis. However, keep in mind that there are other packages available for Python that you may want to check out to complement your data analysis, such as **statsmodels** (statistical modeling library), **seaborn** (another plotting library, which we will use in subsequent chapters), **plotly** (yet another plotting library but for interactive graphs), and **scipy** (scientific computing library, which lets you perform various statistical tests)

Data I/O

With our environment set up, let's jump straight into the content. Before we conduct any type of data analysis, we need to actually have data. There are a myriad of ways to load data in Pandas, but we will keep it simple and load from a csv file.

For the sake of convenience, let's reimport our libraries with aliases, as shown in Figure 2-5.

```
import pandas as pd
import numpy as np
import matplotlib.pyplot as plt
```

Figure 2-5. *Importing pandas, numpy, and matplotlib's pyplot as aliases, for convenience*

Once you have executed this code, let's move on to loading our dataset.

Data Loading

First, make sure that the path to your dataset is defined, like in Figure 2-6.

```
# Path to the training data
data_path = '../data/train.csv'
```

Figure 2-6. *Path to the Titanic training dataset defined*

You may optionally define this to be the full, absolute path, just to make sure pandas will find this file if you are having problems. Here, the data folder resides in a directory level above our notebook, as it contains data files common to every chapter.

To load the data, we will use `pd.read_csv(data_path)`, as shown in Figure 2-7. The method `read_csv()` reads a csv file and all the data contained with it. It also can read from other input formats, including JSON, SAS7BDAT, and Excel files. You can find more details in the Pandas documentation: `https://pandas.pydata.org/docs/reference/index.html`.

```
# pd.read_csv() returns a Pandas DataFrame object
df = pd.read_csv(data_path)
```

Figure 2-7. `pd.read_csv()` *takes in the path as a parameter and returns a pandas DataFrame*

If this runs without producing any errors, you now have a pandas dataframe loaded. An easy way to visualize this dataframe is to use `df.head(N)`, where *N* is an *optional* parameter to define how many rows you want to see. If you don't pass *N*, the default is five displayed rows. Simply run the line of code shown in Figure 2-8.

```
df.head(2)
```

Figure 2-8. *Calling* `.head(2)` *on df to display two rows of df*

You should see output that looks like Table 2-1.

Table 2-1. *Output of Executing the Code in Figure 2-8*

	PassengerId	Survived	Pclass	Name	Sex	Age	SibSp	Parch	Ticket	Fare	Cabin	Embarked
0	1	0	3	Braund, Mr. Owen Harris	male	22.0	1	0	A/5 21171	7.2500	NaN	S
1	2	1	1	Cumings, Mrs. John Bradley (Florence Briggs Th...	female	38.0	1	0	PC 17599	71.2833	C85	C

To get the table's dimensions, run the following:

```
df.shape
```

This returns a tuple (*M*, *N*) with the table's dimensions. It shows *M* rows and *N* columns. For this example, you should see the following printed output:

```
(891, 12)
```

Data Saving

To save your dataset, call `df.to_csv(save_path)`, where `save_path` is a variable that contains a string path to where you want to save the dataframe. (There are many other output formats available besides csv.) As an example, run the code shown in Figure 2-9.

```
df2 = df.head(2)
df2.to_csv('two_rows.csv', index=False)
```

Figure 2-9. `df.head(2)` *returns two rows of* `df`*, which is saved as* `df2`*. This is then saved as a csv to* `two_rows.csv`*, with the parameter* `index=False` *passed. This parameter tells pandas not to save the index as part of the csv, so an extra column is not introduced into the data where there was none before*

The parameter **index=False** tells pandas to not save the dataframe index to the csv. Pandas creates an index when you load in data, which you can override with a custom index if it is relevant. You can change this to `index=True` (which is the default) and see how that changes the csv output.

DataFrame Creation

Besides loading data from a specific source, you can create a dataframe from scratch given a list or a dictionary. This is very useful to do when you are conducting experiments in an iterative manner over several different variable settings and you want to save the data into a nice table.

The code shown in Figure 2-10 creates a dataframe of arbitrary metrics.

```
metric_rows = [ [0.9, 0.2, 0.3], [0.8, 0.3, 0.2] ]
metrics = pd.DataFrame(metric_rows, columns=['Model1', ↵
'Model2', 'Model3'])
metrics
```

Figure 2-10. *Creating a dataframe from a list of lists (two rows, three columns)*

Note   The ↵ symbol in Figure 2-10 (and subsequent code displays) indicates that the code has been truncated and that it's still the same line. So `'Model1'`, `'Model2'`, `'Model3'])` is the actual ending of this line.

The output should look like Table 2-2.

Table 2-2. *Output of Executing the Code in Figure 2-10*

	Model1	Model2	Model3
0	0.9	0.2	0.3
1	0.8	0.3	0.2

To create a dataframe from a dictionary, execute the code shown in Figure 2-11. In this format, the keys of the dictionary are the columns themselves, and the values are lists with the data corresponding to the keys.

```
metric_dict = {'Model1': [0.9, 0.8], 'Model2': [0.2, 0.3], ↵
'Model3':[0.3, 0.2]}
metrics = pd.DataFrame(metric_dict)
metrics
```

Figure 2-11. *Creating a dataframe from dictionaries*

After executing this code, you should once again see the dataframe in Table 2-2, as this code creates the same output as the code shown in Figure 2-10.

Now that you know the very basics of data I/O in pandas, let's move on to the many ways we can manipulate this data to our liking.

Data Manipulation

Pandas will let you do just about anything imaginable with your data. To get you started, this section covers the following data manipulation procedures that you can perform with pandas:

- **Selecting**: Selecting and slicing a dataframe by specific rows/columns

- **Filtering**: Filtering rows based on specific conditions

- **Sorting**: Sorting in ascending/descending order on one or more columns

- **Applying functions**: Applying a custom function over dataframe columns or rows, and running aggregate functions

- **Grouping**: Grouping dataframes, iterating through groups, and performing group aggregations

- **Combing dataframes**: Merging or concatenating dataframes

- **Manipulating columns**: Creating, renaming, and dropping columns

These are some of the most common data manipulation procedures used in any software pertaining to data, though pandas does let you have full control over how you want to iterate through the data as, in essence, a pandas dataframe is just a multidimensional array.

Select

Before we start selecting, let's find out what columns are even in the dataframe. We can do this by running Figure 2-12a.

```
df.columns
```

Figure 2-12a. *Outputting the columns of the dataframe df*

You should see text output like Figure 2-12b:

```
Index(['PassengerId', 'Survived', 'Pclass', 'Name', 'Sex', 'Age', 'SibSp',
       'Parch', 'Ticket', 'Fare', 'Cabin', 'Embarked'],
      dtype='object')
```

Figure 2-12b. *Text output of executing df.columns*

If you need to, you can typecast this to a list by executing the following:

`list(df.columns)`

Before we start selecting anything, don't forget that pandas is **case-sensitive**, meaning you must match the column name exactly as what is in the dataframe to pick that same column.

Let's pick the column 'Name' and select this column. The best convention for picking individual dataframes (since it accounts for columns with spaces) is to run `df[column_name]`, where `column_name` is a string containing some column name that is present in the dataframe. Run the code shown in Figure 2-13 to select the first two rows of the Name column.

```
df['Name'].head(2)
```

Figure 2-13. *Selecting a single column and the first two rows of the result*

You should see the output in Figure 2-14.

```
0                              Braund, Mr. Owen Harris
1    Cumings, Mrs. John Bradley (Florence Briggs Th...
Name: Name, dtype: object
```

Figure 2-14. *The output of executing the code in Figure 2-13*

This looks different than how it appeared in Table 2-1, doesn't it? This is because when you select individual columns, the returned result is a pandas **Series**, not a dataframe. A Series and a dataframe do contain a lot of shared functionality, but they fundamentally differ in some cases. A Series is better treated as a more explicitly indexed list of values, while a dataframe is a list of lists.

We can pick a single column and get a dataframe result by using the following convention:

```
df[list_of_columns]
```

list_of_columns is a list of strings corresponding to the names of various columns present in df. To pick just one column, just pass a list of one column name, like in Figure 2-15.

```
df[['Name']].head(2)
```

Figure 2-15. *Selecting a column as a list of strings to get a dataframe output*

You should now see the output shown in Table 2-3.

Table 2-3. *Output of Executing the Code in Figure 2-15*

	Name
0	Braund, Mr. Owen Harris
1	Cumings, Mrs. John Bradley (Florence Briggs Th...

Now this looks like the outputs in Tables 2-1 and 2-2 because this is a dataframe result. A dataframe can be thought of as a list of lists, although in this case, each list only has one value in it.

We just follow this same convention to select multiple columns, as shown in the code line in Figure 2-16.

```
df[['PassengerId', 'Name', 'Survived']].head(2)
```

Figure 2-16. *Selecting multiple columns*

The output should look like Table 2-4.

Table 2-4. *Output of Executing the Code in Figure 2-16*

	PassengerId	Name	Survived
0	1	Braund, Mr. Owen Harris	0
1	2	Cumings, Mrs. John Bradley (Florence Briggs Th...	1

We can also select rows using the following:

- iloc

- loc

iloc is integer-based indexing and operates the same as how you'd index lists. *Integer-based indexing* means that rows are selected by their integer position starting from the first row, which is zero-indexed. So to pick the first row, we have to select the index 0, just like a list.

iloc also allows us to select columns. The general format of iloc is

```
.iloc[row_start:row_end, column_start:column_end]
```

where the start and end indices are integers. This is called *slicing*, where we specify what kind of "slice" of the DataFrame we want with these indices. In Python lists, you cannot slice columns this way if you have a multidimensional list (list of lists, where each list has several entries), but in pandas (and numpy) you can slice like this.

loc is different, as it is more label-based indexing. Slicing is done explicitly according to the index values. This is where the index comes into play. You may have noticed the unnamed column on the leftmost side of these table outputs—this is the index, and pandas zero-indexes it by default. You can change this and even name the column, which will be demonstrated shortly.

Let's slice the first five rows using .iloc, as shown in Figure 2-17.

```
df.iloc[:5]
```

Figure 2-17. *Slicing the first five rows using iloc. If no start index is given for iloc, it's assumed to start at zero, just like in Python*

The output should look like Table 2-5.

Table 2-5. *Output of Executing the Code in Figure 2-17*

	PassengerId	Survived	Pclass	Name	Sex	Age	SibSp	Parch	Ticket	Fare	Cabin	Embarked
0	1	0	3	Braund, Mr. Owen Harris	male	22.0	1	0	A/5 21171	7.2500	NaN	S
1	2	1	1	Cumings, Mrs. John Bradley (Florence Briggs Th...	female	38.0	1	0	PC 17599	71.2833	C85	C
2	3	1	3	Heikkinen, Miss. Laina	female	26.0	0	0	STON/O2. 3101282	7.9250	NaN	S
3	4	1	1	Futrelle, Mrs. Jacques Heath (Lily May Peel)	female	35.0	1	0	113803	53.1000	C123	S
4	5	0	3	Allen, Mr. William Henry	male	35.0	0	0	373450	8.0500	NaN	S

We can slice the columns as well, like in Figure 2-18.

```
# First five rows, first 4 columns
df.iloc[:5, :4]
```

Figure 2-18. *Slicing the first five rows and first four columns*

The output should look like Table 2-6.

Table 2-6. *Output of Executing the Code in Figure 2-18*

	PassengerId	Survived	Pclass	Name
0	1	0	3	Braund, Mr. Owen Harris
1	2	1	1	Cumings, Mrs. John Bradley (Florence Briggs Th...
2	3	1	3	Heikkinen, Miss. Laina
3	4	1	1	Futrelle, Mrs. Jacques Heath (Lily May Peel)
4	5	0	3	Allen, Mr. William Henry

So how would **.loc** work in this situation? .loc is very similar, but operates on direct labels instead of indices. With loc, you have the following:

```
.loc[start_label:end_label, start_label:end_label]
```

.loc is inclusive, meaning it also includes the end_label.

Go ahead and try out the following line of code to see how it differs from .iloc:

```
df.loc[:5]
```

Let's construct a new index to demonstrate how loc can work with different labels. Figure 2-19 shows the code.

```
view = df.iloc[:5].copy()
view.index = ['a', 'b', 'c', 'd', 'e']
view
```

Figure 2-19. *Creating a copy of the first five rows using* `.copy()` *so that the original df has no chance of being changed. A new index is assigned to this copy, which is called* `view`, *composed of 'a,' 'b,' 'c,' 'd,' 'e'*

The output of the code in Figure 2-19 should look like Table 2-7. Notice that the index is different now.

Table 2-7. Output of Executing the Code in Figure 2-19

	PassengerId	Survived	Pclass	Name	Sex	Age	SibSp	Parch	Ticket	Fare	Cabin	Embarked
a	1	0	3	Braund, Mr. Owen Harris	male	22.0	1	0	A/5 21171	7.2500	NaN	S
b	2	1	1	Cumings, Mrs. John Bradley (Florence Briggs Th...	female	38.0	1	0	PC 17599	71.2833	C85	C
c	3	1	3	Heikkinen, Miss. Laina	female	26.0	0	0	STON/O2. 3101282	7.9250	NaN	S
d	4	1	1	Futrelle, Mrs. Jacques Heath (Lily May Peel)	female	35.0	1	0	113803	53.1000	C123	S
e	5	0	3	Allen, Mr. William Henry	male	35.0	0	0	373450	8.0500	NaN	S

Now run the code in Figure 2-20. Indexing columns requires direct labels too, but you can slice the columns based on the order in which they appear in `df.columns`.

```
view.loc['b':'d', 'Name':'Ticket']
```

Figure 2-20. *Including only rows b to d and columns Name to Ticket using loc*

The output of this code is shown by Table 2-8. Loc knows how to index between 'b' and 'd' because they appear in order. The same goes for how loc is able to correctly index between 'Name' and 'Ticket'. It's not that it alphanumerically enumerates between 'b' and 'd', so keep that in mind.

Table 2-8. *Output of Executing the Code in Figure 2-20*

	Name	Sex	Age	SibSp	Parch	Ticket
b	Cumings, Mrs. John Bradley (Florence Briggs Th...	female	38.0	1	0	PC 17599
c	Heikkinen, Miss. Laina	female	26.0	0	0	STON/O2. 3101282
d	Futrelle, Mrs. Jacques Heath (Lily May Peel)	female	35.0	1	0	113803

If you ever want to reset the index back into a zero-indexed form, just call `df.reset_index(drop=True)`, where `df` is a stand-in for any dataframe result.

You could use the following:

```
view.loc['b':'d', 'Name':'Ticket'].reset_index(drop=True)
```

This will show the same as Table 2-8 except that the indices are 0, 1, 2 on the left. The `drop=True` means you don't want the old index to be inserted as a new column into the dataframe result, which Pandas would do by default otherwise.

With that, you have the basics of selecting rows and columns in Pandas.

Filtering

Filtering allows you to pick rows that satisfy certain conditions. This is useful if you want to pick out rows pertaining to individuals above a certain minimum age, who are married, whose statuses are one of several possible categories, and so forth.

At its core, the way filtering works is

```
df[boolean_list]
```

This `boolean_list` is simply a list of true or false values with a total length matching the length of `df`. If this `boolean_list` is a pandas Series, then it also has the corresponding index for each Boolean value. If the Boolean value is `True`, pandas will keep the row in the resulting dataframe. If it is `False`, then this row is not included.

Go ahead and try the following line of code to develop an intuition for what is going on under the hood:

```
df.head()[[True, False, False, False, True]]
```

We are "filtering" the `df.head()` result manually by passing a hard-coded Boolean list.

With that being said, recall that something like `df['Name']` produces a pandas Series result. If you try something like `df['Age'] > 50`, you will see that the result is a Series of Boolean values. Using this same intuition, let's first start with a basic filter by selecting everyone over the age of 50, as shown in Figure 2-21.

```
df[df['Age'] > 50].head()
```

Figure 2-21. *Filtering the dataframe to have entries with age over 50 and displaying the first five rows*

You should see Table 2-9 as the output. This dataframe is a filtered version of `df` that contains only rows where the `Age` value is greater than 50.

Table 2-9. Output of Executing the Code in Figure 2-21

	PassengerId	Survived	Pclass	Name	Sex	Age	SibSp	Parch	Ticket	Fare	Cabin	Embarked
6	7	0	1	McCarthy, Mr. Timothy J	male	54.0	0	0	17463	51.8625	E46	S
11	12	1	1	Bonnell, Miss. Elizabeth	female	58.0	0	0	113783	26.5500	C103	S
15	16	1	2	Hewlett, Mrs. (Mary D Kingcome)	female	55.0	0	0	248706	16.0000	NaN	S
33	34	0	2	Wheadon, Mr. Edward H	male	66.0	0	0	C.A. 24579	10.5000	NaN	S
54	55	0	1	Ostby, Mr. Engelhart Cornelius	male	65.0	0	1	113509	61.9792	B30	C

We can also combine different Boolean conditions and create the exact filtering condition we want by using the following operators:

- &: The and operator

- |: The or operator

- ~: The negation operator

To negate a condition, simply use

~(condition)

To join two conditions, make sure each condition is surrounded by parentheses, and do either of the following:

(condition1) & (condition2) for and

(condition1) | (condition2) for or

This time, to find all women over the age of 50, let's run the code in Figure 2-22.

```
df[(df['Age'] > 50) & (df['Sex'] == 'female')].head()
```

Figure 2-22. *Chaining two different Booleans together with the & operator*

Note   This all resolves to a single Boolean list at the end, selecting only women who are over the age of 50.

The resulting table should look like Table 2-10.

Table 2-10. Result of Executing the Code in Figure 2-22

	PassengerId	Survived	Pclass	Name	Sex	Age	SibSp	Parch	Ticket	Fare	Cabin	Embarked
11	12	1	1	Bonnell, Miss. Elizabeth	female	58.0	0	0	113783	26.5500	C103	S
15	16	1	2	Hewlett, Mrs. (Mary D Kingcome)	female	55.0	0	0	248706	16.0000	NaN	S
195	196	1	1	Lurette, Miss. Elise	female	58.0	0	0	PC 17569	146.5208	B80	C
268	269	1	1	Graham, Mrs. William Thompson (Edith Junkins)	female	58.0	0	1	PC 17582	153.4625	C125	S
275	276	1	1	Andrews, Miss. Kornelia Theodosia	female	63.0	1	0	13502	77.9583	D7	S

Notice how the index values are jumbled up after the filtering. Don't forget that you can call `.reset_index(drop=True)` to reset this anytime. The `drop=True` means you don't want the old index to be inserted as a new column into the dataframe result.

If the column contained string values, you can just use the following:

```
df[column].str
```

which grants access to functions like these:

- `df[column].str.contains(pattern)`

- `df[column].str.startswith(pattern)`

- `df[column].str.endswith(pattern)`

You can find additional functions here: `https://pandas.pydata.org/docs/reference/api/pandas.Series.str.capitalize.html`.

You can even apply regex patterns.

Figure 2-23 shows a selection where we are picking everyone whose name contains "Mrs.".

```
df[df['Name'].str.contains("Mrs.")].head()
```

Figure 2-23. *Selecting all rows where the name contains "Mrs.", selecting every married woman*

You can also check to see if the values exist as one of the elements within a list. To do so, use `df[column_name].isin(iterable_of_values)`, where `iterable_of_values` is a list of values that you want to match.

Notice that the values in the Embarked column seem to be 'S' or 'C' quite commonly from the displayed tables. Let's use the negation operator to find any rows where the value in the 'Embarked' column is neither 'S' nor 'C'. Refer to Figure 2-24.

```
df[~df['Embarked'].isin(['S', 'C'])].head()
```

Figure 2-24. *Filtering so that the result has 'Embarked' values that are neither 'S' nor 'C'*

Notice that the result contains the value 'Q', meaning we successfully filtered out 'S' and 'C' values.

Finally, understand that we can also **filter and replace** values. Let's replace the values in the Sex column to be numeric equivalents of the existing labels of 'male' and 'female'. This is a commonly done step for categorical columns like 'Sex' so that the resulting values can be used in a machine learning model. Machine learning models cannot learn from text directly—the text needs to be converted into equivalent numeric representations.

A good convention is to use loc:

```
df.loc[boolean_list, column] = replacement
```

We are selecting only the rows where the Boolean is applicable, selecting the column whose values to replace, and then replace it with a new value. Refer to Figure 2-25 to see how to do this.

```
df2 = df.copy()
df2.loc[df2['Sex'] == 'male', 'Sex'] = 0
df2.loc[df2['Sex'] == 'female', 'Sex'] = 1
df2.head()
```

Figure 2-25. *Replacing 'male' with 0 and 'female' with 1*

Table 2-11 shows the output table. As you can see, 'male' and 'female' have been replaced with numeric equivalents.

Table 2-11. *Result of Executing the Code in Figure 2-25*

	PassengerId	Survived	Pclass	Name	Sex	Age	SibSp	Parch	Ticket	Fare	Cabin	Embarked
0	1	0	3	Braund, Mr. Owen Harris	0	22.0	1	0	A/5 21171	7.2500	NaN	S
1	2	1	1	Cumings, Mrs. John Bradley (Florence Briggs Th...	1	38.0	1	0	PC 17599	71.2833	C85	C
2	3	1	3	Heikkinen, Miss. Laina	1	26.0	0	0	STON/O2. 3101282	7.9250	NaN	S
3	4	1	1	Futrelle, Mrs. Jacques Heath (Lily May Peel)	1	35.0	1	0	113803	53.1000	C123	S
4	5	0	3	Allen, Mr. William Henry	0	35.0	0	0	373450	8.0500	NaN	S

With this, you now know how to filter rows to your liking, and how to replace the values that you have filtered.

Sorting

Sometimes, you will find that it is important to sort your data, whether you are matching dataframes with different indices by a specific sorted order of rows, or are sorting a time-Series dataframe by increasing time.

You can simply use the following convention:

`df.sort_values(column, ascending=True)`

If you pass nothing but `column`, pandas by default assumes `ascending=True`.

Figure 2-26 shows the code for sorting by the column 'Fare' in ascending order (low to high).

```
df.sort_values('Fare', ascending=True)
```

Figure 2-26. *Sorting values for the column 'Fare' by ascending order*

Table 2-12 shows the output for running the code in Figure 2-26. As you can see, Fare is sorted in ascending order. (Some rows were omitted for clarity.)

Table 2-12. *Result of Executing the Code in Figure 2-26*

	PassengerId	Survived	Pclass	Name	Sex	Age	SibSp	Parch	Ticket	Fare	Cabin	Embarked
271	272	1	3	Tornquist, Mr. William Henry	male	25.0	0	0	LINE	0.0000	NaN	S
597	598	0	3	Johnson, Mr. Alfred	male	49.0	0	0	LINE	0.0000	NaN	S
302	303	0	3	Johnson, Mr. William Cahoone Jr	male	19.0	0	0	LINE	0.0000	NaN	S
...	...	...	...	...	...	...	...	...	...	...	...	...
341	342	1	1	Fortune, Miss. Alice Elizabeth	female	24.0	3	2	19950	263.0000	C23 C25 C27	S
737	738	1	1	Lesurer, Mr. Gustave J	male	35.0	0	0	PC 17755	512.3292	B101	C
258	259	1	1	Ward, Miss. Anna	female	35.0	0	0	PC 17755	512.3292	NaN	C
679	680	1	1	Cardeza, Mr. Thomas Drake Martinez	male	36.0	0	1	PC 17755	512.3292	B51 B53 B55	C

We can also pass a list of columns to sort, like so:

```
df.sort_values(list_of_values, ascending=list_of_booleans)
```

The list of values and the list of Booleans correspond one to one. If `list_of_values` is `['a', 'b', 'c']`, and `list_of_booleans` is `[True, False, True]`, then we are saying that:

1. 'a' is in ascending order.

2. 'b' is in descending order.

3. 'c' is in ascending order.

With several columns, 'a' will be sorted first. Ties are then broken by sorting 'b', and ties with 'a' and 'b' are broken by 'c'. A tie in column 'a' is when two or more rows have the exact same value for 'a'.

Consider the column Survived. Many people either survived or didn't survive the sinking of the Titanic, so if we sort by 'Survived', we will have many rows that are in an ambiguously sorted order.

If we additionally sort by 'Name', then within the sorted order of 'Survived', the remaining rows are then sorted by 'Name' as well.

Figure 2-27 shows how to sort by columns `['Pclass', 'Age']` and ascending list `[True, False]`.

```
df.sort_values(['Pclass', 'Age'], ascending=[True, False])
```

Figure 2-27. *Sorting by 'Pclass' and 'Age' and keeping 'Pclass' ascending and 'Age' descending*

Table 2-13 shows the results of executing the code in Figure 2-27. Notice that Pclass is in ascending order, whereas Age is in descending order.

Table 2-13. *Result of Executing the Code in Figure 2-27*

	PassengerId	Survived	Pclass	Name	Sex	Age	SibSp	Parch	Ticket	Fare	Cabin	Embarked
630	631	1	1	Barkworth, Mr. Algernon Henry Wilson	male	80.0	0	0	27042	30.0000	A23	S
96	97	0	1	Goldschmidt, Mr. George B	male	71.0	0	0	PC 17754	34.6542	A5	C
493	494	0	1	Artagaveytia, Mr. Ramon	male	71.0	0	0	PC 17609	49.5042	NaN	C
745	746	0	1	Crosby, Capt. Edward Gifford	male	70.0	1	1	WE/P 5735	71.0000	B22	S
54	55	0	1	Ostby, Mr. Engelhart Cornelius	male	65.0	0	1	113509	61.9792	B30	C
...	...	...	...	...	...	...	...	...	...	...	...	...
859	860	0	3	Razi, Mr. Raihed	male	NaN	0	0	2629	7.2292	NaN	C
863	864	0	3	Sage, Miss. Dorothy Edi"h "Do"ly"	female	NaN	8	2	CA. 2343	69.5500	NaN	S
868	869	0	3	van Melkebeke, Mr. Philemon	male	NaN	0	0	345777	9.5000	NaN	S
878	879	0	3	Laleff, Mr. Kristo	male	NaN	0	0	349217	7.8958	NaN	S
888	889	0	3	Johnston, Miss. Catherine Hel"n "Car"ie"	female	NaN	1	2	W./C. 6607	23.4500	NaN	S

With that, you have learned the basics of sorting values.

Applying Functions

We can apply functions to each individual value within a column, or we can use aggregate functions, which will condense the entire column into some kind of metric. Examples of aggregate functions are sum(), .mean(), .std(), .min(), and .max().

The notation to apply a function is

```
series.apply(function)
```

Recall that we can obtain a Series by selecting only one column by a string.

We can apply any compatible function, including typecasting. We can typecast the column 'Fare' to an integer, for example, by executing the code shown in Figure 2-28.

```
# making a copy of df
df2 = df.copy()
df2['Fare'] = df2['Fare'].apply(int)
df2.head(2)
```

Figure 2-28. *Creating a copy of df so we preserve the original data without needing to reload it from disk*

The output of executing the code in Figure 2-28 is shown in Table 2-14. Notice that Fare is now an integer, whereas it was a float before.

Table 2-14. *Output of Executing the Code in Figure 2-28*

PassengerId	Survived	Pclass	Name	Sex	Age	SibSp	Parch	Ticket	Fare	Cabin	Embarked
0	1	0	3	Braund, Mr. Owen Harris	male	22.0	1	0	A/5 21171	7	NaN	S
1	2	1	1	Cumings, Mrs. John Bradley (Florence Briggs Th...	female	38.0	1	0	PC 17599	71	C85	C

You can also create custom functions and use them in `.apply()`. Figure 2-29 shows an example of an arbitrary operation to be performed on 'Fare'.

```
def custom_func(x):
    return 2*x + x

df2['Fare'] = df2['Fare'].apply(custom_func)
df2.head(2)
```

Figure 2-29. *Applying some mathematical operations to 'Fare' as defined by our own custom function*

The output is as shown in Table 2-15.

Table 2-15. *Output of Applying the Custom Function to the Column 'Fare'*

	PassengerId	Survived	Pclass	Name	Sex	Age	SibSp	Parch	Ticket	Fare	Cabin	Embarked
0	1	0	3	Braund, Mr. Owen Harris	0	22.0	1	0	A/5 21171	21	NaN	S
1	2	1	1	Cumings, Mrs. John Bradley (Florence Briggs Th...	1	38.0	1	0	PC 17599	213	C85	C

We can create new columns through aggregate functions. Figure 2-30 shows an example demonstrating the use of axis=1 to indicate that we want to apply the function over entire rows. Each row will be passed into the function, and we can access the columns individually.

```
df2['Fare_Survived'] = df2.apply(lambda x: x['Fare'] * ↵
x['Survived'], axis=1)
df2.head(2)
```

Figure 2-30. *Creating a new column using a custom function on the entire row with the* axis=1 *parameter*

This way, we can create a new column 'Fare_Survived' by multiplying two different column values together. The output of this code is captured in Table 2-16.

Table 2-16. *Output of Executing the Code in Figure 2-30, with New Column*

	PassengerId	Survived	Pclass	Name	Sex	Age	SibSp	Parch	Ticket	Fare	Cabin	Embarked	Fare_Survived
0	1	0	3	Braund, Mr. Owen Harris	male	22.0	1	0	A/5 21171	21	NaN	S	0
1	2	1	1	Cumings, Mrs. John Bradley (Florence Briggs Th...	female	38.0	1	0	PC 17599	213	C85	C	213

However, there is a more efficient way of performing the same computation. We will do what is called a *vectorized operation*, which refers to performing operations on an entire pandas Series at once. This is quicker and scales better than the `.apply()` method.

Figure 2-31 demonstrates how you can perform the equivalent operation in a vectorized manner.

```
df2['Fare_Survived_Vectorized'] = df2['Fare'] * ↵
df2['Survived']
df2.head(2)
```

Figure 2-31. *Notice how the multiplication between 'Fare' and 'Survived' now takes place between the entire column for both of them at once. This is how to perform a vectorized operation*

With that, let's move on to aggregate functions. Aggregate functions are functions that perform some type of computation on the entire column to reduce it to one number.

Simple aggregates are functions like `.sum()`, `.mean()`, `.std()`, `.min()`, and `.max()`, all of which can be applied to a pandas Series or a pandas dataframe. Figure 2-32 demonstrates the application of these aggregate functions to a Series.

```
df['Age'].mean(), df['Fare'].std(), df['Age'].max()
```

Figure 2-32. *Various aggregations applied to several Series*

The output is captured in Figure 2-33.

```
(29.69911764705882, 49.693428597180905, 80.0)
```

Figure 2-33. *The output of executing the code in Figure 2-32*

Figure 2-34 demonstrates the application of the `.sum()` function on a dataframe. Figure 2-35 captures the resulting output, which is a Series.

```
df[['Age', 'Fare', 'SibSp']].sum()
```

Figure 2-34. *Sum aggregate on a dataframe of three columns*

```
Age          21205.1700
Fare         28693.9493
SibSp          466.0000
dtype: float64
```

Figure 2-35. *Result of executing the code in Figure 2-34*

Just like .apply(), we can specify axis=1 to use the aggregate function across the entire row. Figure 2-36 demonstrates how you can sum the values for each row across three different columns to condense it down to one column. The result is a Series, and we can call .head() on a Series to show only the first five rows.

```
df[['Age', 'Fare', 'SibSp']].sum(axis=1).head(5)
```

Figure 2-36. *Summing across three columns, applying the sum() function across each row*

Figure 2-37 shows the output of running the code in Figure 2-36.

```
0      30.2500
1     110.2833
2      33.9250
3      89.1000
4      43.0500
dtype: float64
```

Figure 2-37. *The resulting Series after we apply the sum aggregation row-wise, not on an entire column. By doing it row-wise using axis=1, the sum aggregate condenses the three columns of each row into one single column by summing all the values together*

Grouping

We can group the dataframe by certain columns and iterate over each group. For example, if we have a dataframe of houses, and color is an attribute, we can group by house color and explore each individual group of color at a time. Other data processing software may not enable you to have this much control (or, at least, not control that is this straightforward), but in pandas, you can access each group directly and iterate over all of them.

The general notation is

```
grouped_obj = df.groupby(column)
```

grouped_obj is an iterable object. To iterate through it, we can execute the code shown in Figure 2-38. This enables us to access both what the grouped values are (name) and the grouped dataframe itself (group). As shown in Figure 2-39, the grouped dataframe, group, will contain only values where the grouped columns are equal to the values in name. If we were to group by color, and we had 'red', 'green', and 'blue' to pick from, then each iteration would produce a name being one of 'red', 'green', or 'blue' and a dataframe called 'group' that consists only of each of those respective colors.

```
grouped = df.groupby('Pclass')
for name, group in grouped:
    print('\nPclass: ', name)
    print(group[['Pclass', 'Name']].head(2))
```

Figure 2-38. *Code to iterate through a grouped dataframe and display the individual grouped dataframes*

```
Pclass:  1
   Pclass                                              Name
1       1  Cumings, Mrs. John Bradley (Florence Briggs Th...
3       1          Futrelle, Mrs. Jacques Heath (Lily May Peel)

Pclass:  2
   Pclass                                  Name
9       2  Nasser, Mrs. Nicholas (Adele Achem)
15      2      Hewlett, Mrs. (Mary D Kingcome)

Pclass:  3
   Pclass                      Name
0       3  Braund, Mr. Owen Harris
2       3  Heikkinen, Miss. Laina
```

Figure 2-39. *The output of executing the code in Figure 2-38*

We can group by multiple columns as well. Figure 2-40 shows how to do this, as well as how to print out the grouped values and the grouped dataframe. If you have multiple group columns, then you will iterate through all present combinations of unique values in the multiple group columns each time.

```
grouped = df.groupby(['Survived', 'Pclass'])
for name, group in grouped:
    print('\n[Survived, Pclass]: ', name)
    print(group[['Survived', 'Pclass', 'Name']].head(2))
```

Figure 2-40. *Multiple grouped columns*

The output of Figure 2-40 is shown in Figure 2-41. The full output has been truncated, but you should see something similar.

```
[Survived, Pclass]:  (0, 1)
    Survived  Pclass                                      Name
6          0       1            McCarthy, Mr. Timothy J
27         0       1  Fortune, Mr. Charles Alexander

...

[Survived, Pclass]:  (1, 3)
    Survived  Pclass                                                    Name
2          1       3                          Heikkinen, Miss. Laina
8          1       3  Johnson, Mrs. Oscar W (Elisabeth Vilhelmina Berg)
```

Figure 2-41. *Truncated output of the code in Figure 2-40, showing just the first group's and last group's output value*

It is also possible to group the dataframes and then perform aggregations. Doing so will compute and display the corresponding aggregate for each group. Recall from the previous section that we did aggregations like sum, mean, max, etc., but if we group and then perform the same aggregations, it will first group the dataframe into different chunks, then perform the aggregation on each chunk, and then put together the final result. Figure 2-42 shows a sum aggregation on several numeric columns grouped by whether or not the passenger survived.

```
df[['Fare', 'Survived', 'Age']].groupby('Survived').sum()
```

Figure 2-42. *Selecting a few numeric columns, grouping by one column, and then computing sum*

The output of Figure 2-42 is captured in Table 2-17.

Table 2-17. *Output of Executing the Code in Figure 2-42*

	Fare	Age
Survived		
0	12142.7199	12985.50
1	16551.2294	8219.67

We can perform grouped aggregations with multiple grouped columns. The code shown in Figure 2-43 modifies the code in Figure 2-42 to group by an additional column.

```
df[['Fare', 'Pclass', 'Survived', 'Age']].groupby(['Survived', ↵
'Pclass']).sum()
```

Figure 2-43. *Grouping by an additional column but using the code from Figure 2-42*

The output of Figure 2-43 is shown by Table 2-18.

Table 2-18. *Output of Executing the Code in Figure 2-43*

		Fare	Age
Survived	**Pclass**		
0	**1**	5174.7206	2796.50
	2	1882.9958	3019.00
	3	5085.0035	7170.00
1	**1**	13002.6919	4314.92
	2	1918.8459	2149.83
	3	1629.6916	1754.92

With that, we have covered the basics of grouping and grouped aggregate functions.

Combining DataFrames

There are several ways to combine dataframes. You might want to merge or join them (the functionality for merge and join is very similar—we will only cover merge) or concatenate (stack) dataframes together.

First, let's prepare two separate dataframes. They will both contain the column PassengerId, but their other columns will be different. The goal is to unify both tables into one dataframe so that for every passenger ID, the row contains columns from both dataframes together. In other words, we are zipping up the two dataframes only where passenger ID is the same for both tables. This is called *joining* or *merging* two tables together. Figure 2-44 shows the code to do this.

```
df1 = df[['PassengerId', 'Pclass', 'Fare']]


# We are reversing the order of this dataframe by using [::-1] to
demonstrate left vs right joins
df2 = df[['PassengerId', 'Name', 'Age', 'Survived']][::-1]


df1.head(2), df2.head(2)
```

Figure 2-44. *Creating two dataframes with one column in common but all other columns different*

The output of executing this code is captured in Figure 2-45.

```
(    PassengerId  Pclass     Fare
 0             1       3   7.2500
 1             2       1  71.2833,
      PassengerId                     Name   Age  Survived
 890           891    Dooley, Mr. Patrick  32.0         0
 889           890  Behr, Mr. Karl Howell  26.0         1)
```

Figure 2-45. *There are two distinct dataframes now*

Let's perform a default join without specifying how we want to join the two dataframes. By default, Pandas performs an inner join, which zips up the rows where we specify (using the on= parameter) and orders them by the left dataframe's order. When we refer to left and right dataframes, we are joining the dataframe we call "right" to the dataframe on the "left." For certain join methods like left join, it's important which dataframe is left and which is right. In a left join specifically, all rows from the left dataframe are preserved, and columns from the right dataframe are appended to these rows only if they are matching entries. Otherwise, a null value is inserted for columns being added on the right for nonmatching rows.. Refer to Figure 2-46 to see the inner join code on the column PassengerId.

```
# Default inner join. Inner join: combine the columns of rows where the
on='column' values are the same, order by df1 (left) dataframe
df1.merge(df2, on='PassengerId').head()
```

Figure 2-46. *When we want to specify a join, we add the parameter how="inner", for example. If we do not specify this, Pandas will perform an inner join by default. The merge itself returns a dataframe;* `.head()` *is used only for conciseness*

The output of Figure 2-46 is captured by Table 2-19. Both dataframes are zipped where PassengerId is equal, and the results are ordered by how the rows appear in df1 (left dataframe).

Table 2-19. *Output of Executing the Code in Figure 2-46*

	PassengerId	Pclass	Fare	Name	Age	Survived
0	1	3	7.2500	Braund, Mr. Owen Harris	22.0	0
1	2	1	71.2833	Cumings, Mrs. John Bradley (Florence Briggs Th...	38.0	1
2	3	3	7.9250	Heikkinen, Miss. Laina	26.0	1
3	4	1	53.1000	Futrelle, Mrs. Jacques Heath (Lily May Peel)	35.0	1
4	5	3	8.0500	Allen, Mr. William Henry	35.0	0

Figure 2-47 shows you how to perform a right join, which zips up the dataframes in a similar fashion but orders the rows by how they appear in the "right" dataframe (df2).

```
# Right join, meaning merge the dataframe as usual but order by the right
dataframe (df2)'s values
df1.merge(df2, on='PassengerId', how='right').head()
```

Figure 2-47. *Specifying a right join by passing how=*`'right'` *as a parameter*

The output of this is captured by Table 2-20. Notice that the ordering matches df2's ordering. This is because a right join orders the rows according to how they were presented in the "right" dataframe, or df2 in this case.

Table 2-20. *Output of Executing the Code in Figure 2-47*

	PassengerId	Pclass	Fare	Name	Age	Survived
0	891	3	7.75	Dooley, Mr. Patrick	32.0	0
1	890	1	30.00	Behr, Mr. Karl Howell	26.0	1
2	889	3	23.45	Johnston, Miss. Catherine Helen "Carrie"	NaN	0
3	888	1	30.00	Graham, Miss. Margaret Edith	19.0	1
4	887	2	13.00	Montvila, Rev. Juozas	27.0	0

We can also specify a left join by passing how=`'left'` as a parameter, but the result should look identical to Table 2-19.

There are several different ways to specify how you want to join two dataframes. You are encouraged to explore this further, since the joining methods go far beyond just simple 'inner', 'left', and 'right' joins. However, as of now in Pandas, 'left', 'right', 'outer', 'inner', and 'cross' joins are what are supported by the `.merge()` function.

Next, let's look at how to concatenate two dataframes. To do this, we will first split up the original dataframe into one dataframe for male passengers and another dataframe for female passengers. Refer to Figure 2-48 for this code.

```
df_male = df[df['Sex'] == 'male']
df_female = df[df['Sex'] == 'female']
```

Figure 2-48. *Creating two different dataframes, each grouped by either 'male' or 'female'*

To perform the concatenation, we simply need to follow this convention:

`pd.concat(list_of_dataframes).`

Figure 2-49 shows how to merge `df_male` and `df_female`.

```
pd.concat([df_male, df_female])
```

Figure 2-49. *Merging `df_male` and `df_female`, the result of which is another dataframe that contains the concatenated result*

Refer to Table 2-21 to see the output.

Table 2-21. *Concatenation Result from Executing the Code in Figure 2-49*

	PassengerId	Survived	Pclass	Name	Sex	Age	SibSp	Parch	Ticket	Fare	Cabin	Embarked
0	1	0	3	Braund, Mr. Owen Harris	male	22.0	1	0	A/5 21171	7.2500	NaN	S
4	5	0	3	Allen, Mr. William Henry	male	35.0	0	0	373450	8.0500	NaN	S
5	6	0	3	Moran, Mr. James	male	NaN	0	0	330877	8.4583	NaN	Q
6	7	0	1	McCarthy, Mr. Timothy J	male	54.0	0	0	17463	51.8625	E46	S
7	8	0	3	Palsson, Master. Gosta Leonard	male	2.0	3	1	349909	21.0750	NaN	S
...	...	...	...	...	...	...	...	...	...	...	...	...
880	881	1	2	Shelley, Mrs. William (Imanita Parrish Hall)	female	25.0	0	1	230433	26.0000	NaN	S
882	883	0	3	Dahlberg, Miss. Gerda Ulrika	female	22.0	0	0	7552	10.5167	NaN	S
885	886	0	3	Rice, Mrs. William (Margaret Norton)	female	39.0	0	5	382652	29.1250	NaN	Q
887	888	1	1	Graham, Miss. Margaret Edith	female	19.0	0	0	112053	30.0000	B42	S
888	889	0	3	Johnston, Miss. Catherine Helen "Carrie"	female	NaN	1	2	W./C. 6607	23.4500	NaN	S

As you can see in Table 2-21, the dataframes are stacked together (note how the indices are not continuous and skip a few numbers). If one of your dataframes has an extra column that the other dataframes don't have, the column will be null in the rows belonging to the other dataframes.

Creating, Renaming, and Dropping Columns

Finally, let's go over how to create a new column (formally), rename columns, and drop them from the dataframe entirely.

Recall from the earlier discussion of .apply() that we create a new column like so:

```
df['new_col'] = df['some_col'].apply(some_func)
```

We can use the function .insert() to customize where we want to insert the column.

First, let's copy our original dataframe, as shown in Figure 2-50. This ensures that anything we do to df2 will never affect df.

```
df2 = df.copy()
```

Figure 2-50. *Copying the original dataframe, df, into a new variable, df2*

Let's now apply some transformation to the 'Age' column and store the Series result in a variable. Refer to Figure 2-51.

```
age_class = df2['Age'].apply(lambda x: 'child' if x < 18 else 'adult')
age_class.head(2)
```

Figure 2-51. *Applying a label transformation to the column 'Age' and storing the result in age_class*

You should see something like Figure 2-52.

```
0    adult
1    adult
Name: Age, dtype: object
```

Figure 2-52. *Result of executing the code in Figure 2-51*

Let's now insert this Series as a column in df2. Refer to Figure 2-53. The parameters are as follows:

df2.insert(position, column_name, column_values)

```
# Add to the end
df2.insert(df2.shape[-1], 'Age_Classification', age_class)
df2.head(2)
```

Figure 2-53. *Inserting the column to the end of df2*

position is anywhere from 0 to len(df2.columns), allowing you to specify where to insert the new column anywhere in the dataframe. The variable column_name lets you specify a custom name, and column_values is the Series that you are inserting.

You should see something like Table 2-22, in which the new column is inserted successfully to the end of df2.

Table 2-22. *Result of Executing the Code in Figure 2-53*

	PassengerId	Survived	Pclass	Name	Sex	Age	SibSp	Parch	Ticket	Fare	Cabin	Embarked	Age_Classification
0	1	0	3	Braund, Mr. Owen Harris	male	22.0	1	0	A/5 21171	7.2500	NaN	S	adult
1	2	1	1	Cumings, Mrs. John Bradley (Florence Briggs Th...	female	38.0	1	0	PC 17599	71.2833	C85	C	adult

Suppose now that we want to rename the column Age_Classification to Age_Status. To do this, we can create a dictionary of $k{:}v$ mappings where k is some column that currently exists in the dataframe, and v is what we want to rename it to.

Refer to Figure 2-54 for an example of how to rename a column. The general convention is

```
df.rename(columns={'existing_column': 'renamed_column'})
```

```
df2 = df2.rename(columns={'Age_Classification': 'Age_Status'})
df2.head(2)
```

Figure 2-54. *Renaming Age_Classification to Age_Status*

which results in a dataframe with the new name applied. You can rename multiple columns this way by having more entries in the dictionary you're using to rename the columns.

The output of Figure 2-54 is captured by Table 2-23. Notice that the renaming was successful.

Table 2-23. *Result of Executing the Code in Figure 2-54*

	PassengerId	Survived	Pclass	Name	Sex	Age	SibSp	Parch	Ticket	Fare	Cabin	Embarked	Age_Status
0	1	0	3	Braund, Mr. Owen Harris	male	22.0	1	0	A/5 21171	7.2500	NaN	S	adult
1	2	1	1	Cumings, Mrs. John Bradley (Florence Briggs Th...	female	38.0	1	0	PC 17599	71.2833	C85	C	adult

Finally, let's drop this column altogether. The general convention is

`df.drop(column_name, axis=1)`

Refer to Figure 2-55 for how to drop 'Age_Status' from the dataframe.

```
df2 = df2.drop('Age_Status', axis=1)
df2.head(2)
```

Figure 2-55. *Dropping the column 'Age_Status' from df2 entirely*

The output of Figure 2-55 is captured by Table 2-24. Notice that 'Age_Status' was successfully dropped.

Table 2-24. *Result of Executing the Code in Figure 2-55*

	PassengerId	Survived	Pclass	Name	Sex	Age	SibSp	Parch	Ticket	Fare	Cabin	Embarked
0	1	0	3	Braund, Mr. Owen Harris	male	22.0	1	0	A/5 21171	7.2500	NaN	S
1	2	1	1	Cumings, Mrs. John Bradley (Florence Briggs Th...	female	38.0	1	0	PC 17599	71.2833	C85	C

With this, you should be familiar enough with the Pandas API and what it can allow you to do to keep up with the rest of the code in the book. Should anything new or advanced be presented, it will be explained.

Now that you've learned how to manipulate data to your liking, you are ready to explore how to perform data analysis.

Data Analysis

Pandas offers a variety of functionality to analyze your data (and examine relationships, underlying patterns, etc.). It also easily integrates with other Python modules dedicated to statistical functionality, such as scipy and statsmodels, to let you perform more detailed statistical procedures to understand your data better.

Value Counts

Let's start with the function .value_counts(), which displays all unique values and their frequency counts. The code shown in Figure 2-56 is used to display such a count for the column 'Embarked.'

```
df['Embarked'].value_counts(dropna=False)
```

Figure 2-56. *Displaying the frequency counts for the column 'Embarked' with the parameter* dropna=False *specifying not to filter out frequency counts for any null values*

Refer to Figure 2-57 to see the corresponding output of Figure 2-56.

```
Embarked
S      644
C      168
Q       77
NaN      2
Name: count, dtype: int64
```

Figure 2-57. *The output of executing the code in Figure 2-56. The function value_ counts() gives you frequency counts*

By calling .unique(), we can obtain all unique values present in a column directly. Refer to Figure 2-58.

```
df['Embarked'].unique()
```

Figure 2-58. *Calling .unique() on a Pandas Series returns a numpy array of unique values present in this Series*

The output of executing the code in Figure 2-58 is shown in Figure 2-59.

```
array(['S', 'C', 'Q', nan], dtype=object)
```

Figure 2-59. *An array of unique values is returned*

Pandas .describe() Method

To get simple summary statistics of each column, we can call .describe() to give us the following:

- count
- mean
- std
- min
- max
- 25th percentile
- 50th percentile (median)
- 75th percentile

Figure 2-60 shows the command to call .describe() on a dataframe of several columns. This will result in a summary for each column.

```
df[['Fare', 'Age']].describe()
```

Figure 2-60. *Calling .describe() on a dataframe of only two columns*

The output table is captured by Table 2-25.

Table 2-25. *Statistical Summary Table Output of Calling* `.describe()`

	Fare	Age
count	891.000000	714.000000
mean	32.204208	29.699118
std	49.693429	14.526497
min	0.000000	0.420000
25%	7.910400	20.125000
50%	14.454200	28.000000
75%	31.000000	38.000000
max	512.329200	80.000000

Pandas Correlation Matrix

The Pandas correlation matrix tells us the correlation coefficient between each combination of columns. A **correlation coefficient** measures the strength of a linear relationship between two variables. The correlation coefficient ranges from –1 to 1, where:

- A coefficient of 1 indicates that two variables x and y share a perfectly positive linear relationship. As x increases by a fixed amount, y also increases by a fixed amount. Think of the line y = 2x.

- A coefficient of –1 indicates that two variables x and y share a perfectly negative linear relationship. As x increases by a fixed amount, y decreases by a fixed amount. Think of the line y = –2x.

- A coefficient of 0 indicates that there is no measurable linear relationship between x and y. It does not mean there isn't a nonlinear relationship, just that there was no linear relationship.

The closer the coefficient is to 1 or –1, the more linearly correlated the two variables will be. If you have a correlation coefficient of 0.8, then as x increases, y tends to increase as well, just not as reliably or as consistently as if the coefficient were 1 instead.

Typically, in a machine learning task, you are using some "feature" variables to make a prediction based on some "target" variable. For example, taking cost of living, some stock price, interest rate, ZIP code, etc. (feature variables) are used to predict housing prices (target variable). More formally, you might see feature variables referred to as "explanatory variables" or "predictor variables" and see the target variable referred to as the "dependent variable."

We can use the correlation matrix to determine what predictor variables have the highest linear correlation coefficients with the dependent variable. If you have some variables with nearly linear correlation with your dependent variable, then your prediction task is likely to perform quite well.

Furthermore, if multiple explanatory variables have high correlation with each other, you're better off dropping some of them to avoid redundant information. For example, if you have someone's pre-tax salary and post-tax salary as two separate variables, you're not really gaining much information from having both of them at once. All you'd be doing is introducing more "noisiness" into the learning task and making it harder for the model to learn. Therefore, it can be beneficial to drop predictor variables that are highly correlated with one another.

With that out of the way, let's explore how to generate and view a correlation matrix in Pandas. You want to use the method .corr(). Feel free to refer to the Pandas documentation to look at different ways of generating correlation coefficients, which you can specify as a parameter.

Figure 2-61 shows how to do this. First, pick a few numeric columns. (Correlation can be computed only between numeric columns. If you want to include string columns, you must numerically encode them in some form. We will cover this shortly.)

```
df2 = df[['Survived', 'Pclass', 'Age', 'SibSp', 'Parch', 'Fare']]
df2.corr()
```

Figure 2-61. *Selecting numeric columns and creating a correlation matrix using* `.corr()`*. The result is a pandas dataframe*

Refer to Table 2-26 to see the output correlation matrix (returned by Pandas as a dataframe).

Table 2-26. *Output Correlation Matrix Resulting from Execution of the Code in Figure 2-61*

	Survived	Pclass	Age	SibSp	Parch	Fare
Survived	1.000000	-0.338481	-0.077221	-0.035322	0.081629	0.257307
Pclass	-0.338481	1.000000	-0.369226	0.083081	0.018443	-0.549500
Age	-0.077221	-0.369226	1.000000	-0.308247	-0.189119	0.096067
SibSp	-0.035322	0.083081	-0.308247	1.000000	0.414838	0.159651
Parch	0.081629	0.018443	-0.189119	0.414838	1.000000	0.216225
Fare	0.257307	-0.549500	0.096067	0.159651	0.216225	1.000000

Ignore the diagonal columns—of course, a column is perfectly linearly correlated with itself. Of particular interest might be the column 'Survived' and the correlation it has with the other columns. Pclass and Fare seem to be the most correlated with Survived, though the magnitude of this correlation is not that high. We can call this a *weak correlation*, whereas a *strong correlation* is some coefficient with a magnitude greater than 0.7, for example.

Visualization

To visualize your data, you can create plots using matplotlib, a library for Python that allows you to create customized charts to your liking. Libraries like seaborn add on to the functionality of matplotlib in terms of features. We will use seaborn later on in Chapter 5 and 6 for its heatmap, a useful function to let us visualize various machine learning performance metrics.

The following sections cover a few basic things you should know about creating a chart.

Line Chart

A **line chart** is useful for graphing time Series data or any type of data that progresses along some type of axis—for example, plotting mathematical functions. Figure 2-62 shows an example of how to generate some white noise by sampling a normal distribution and plotting a line chart.

```
# Sample random noise and plot
noise = np.random.normal(size=100)
plt.plot(noise)
```

Figure 2-62. *Creating some random noise and plotting it*

`plt.plot(sequence)` creates a basic line chart, where `sequence` is some kind of iterable like a list or numpy array that holds the data to be plotted. You can also pass `plt.plot(seq_x, seq_y)`, but if you pass only one sequence, matplotlib implicitly creates a range from zero to the length of the sequence to be used along the x axis.

The chart resulting from executing the code in Figure 2-62 should look like Figure 2-63. Keep in mind that it won't exactly match, because we are taking a random sample of noise. So as long as your graph looks similar in style to Figure 2-63, then it works.

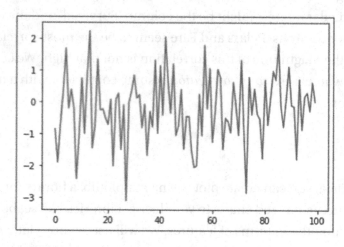

Figure 2-63. *A simple line plot of random normal noise*

Chart Customization

Now let's look at how to customize our plots as well as plot a second line within the same graph. Refer to the code listing in Figure 2-64.

```
noise1 = np.random.normal(size=100)
noise2 = np.random.normal(size=100)


# plt.figure(figsize=(a, b)) lets you configure how large you want the ↵
plot to be
plt.figure(figsize=(10,5))


# plt.title(x) lets you add a title to the plot
plt.title("Plot of White Noise")


# plt.xlabel(x) adds a label to the x dimension on the plot
plt.xlabel("Sample Number")


# plt.ylabel(y) adds a label to the y dimension on the plot
plt.ylabel("Value")


# You can plot multiple series at once, they will all try and show up on
the same plot
plt.plot(noise1)
plt.plot(noise2)


# A legend allows us to label the plots for convenience and readability. ↵
Pass a list of names corresponding with the order of the plotting code, so↵
here Plot1 corresponds with the series noise1
```

Figure 2-64. *Code to plot two different random noise Series. Additionally, we can configure plot size, title, x and y axis labels, and add a legend*

The output of executing the code in Figure 2-64 is shown in Figure 2-65.

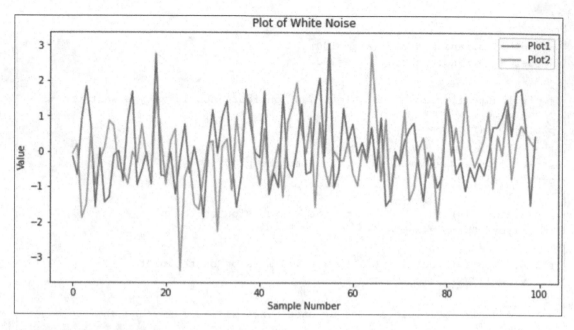

Figure 2-65. *The output of executing the code in Figure 2-64*

Note With this basic amount of configuration, we can freely plot anything we'd like to in later sections. Examples include plotting the loss curves of multiple different models (visually evaluating model performances against one another) or just evaluating two time Series plots side by side. An example of this would be plotting your time Series predictions vs. the actual time Series data for a certain month.

The following list describes the code in Figure 2-64:

- `plt.figure(figsize=(a,b))` lets you configure the size of the plot. The parameters `a, b` refer to the width and height in inches of the plot.

- `plt.title(your_title)` lets you add a title to the plot.

- `plt.xlabel(xlabel_string)` and `plt.ylabel(ylabel_string)`, respectively, let you add a label to the x and y axes.

- `plt.legend(list_of_names)` lets you add a legend. The list of string names must correspond to the chronological order of plots.

Scatter Plot

You can also graph other types of plots, such as a scatter plot. Figure 2-66 shows you how to create a scatter plot. Scatter plots are useful when you want to just get a sense of how the data is distributed against each other on various columns. For example, you can look at housing prices vs. ZIP codes to get a sense of the distribution of data across each ZIP code.

```
x = np.random.random(size=100)
y = np.random.random(size=100)

plt.scatter(x, y)
```

Figure 2-66. *Generating random noise on the x and y axes and creating a scatter plot*

The output of executing the code in Figure 2-66 is shown in Figure 2-67.

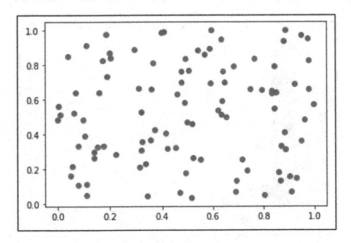

Figure 2-67. *For a scatter plot, you need to explicitly provide both the x and y data*

Remember that the previous customization functions will apply here, so you can title your plot, label your axes, adjust figure size, and much more. Matplotlib is very customizable, so take some time to explore it further on your own.

Histogram

You can also plot a histogram. This is useful when you want to understand the frequency counts across your dataset to find out the distribution of your data. The way it works is that the overall data range is binned into several subintervals. If a data point falls in this interval, the interval's tally is increased by one. The rest of the data is counted in this manner, resulting in a frequency graph. Figure 2-68 shows the code to plot a histogram using the 'Fare' column in the dataframe.

```
plt.hist(df['Fare'])
```

Figure 2-68. *Creating a histogram of fares to find the overall distribution of passenger fares paid*

The result should look like Figure 2-69. Keep in mind that this histogram is highly customizable. You can set your own number of bins to more finely display the frequencies so that too many values aren't lumped under the same bin.

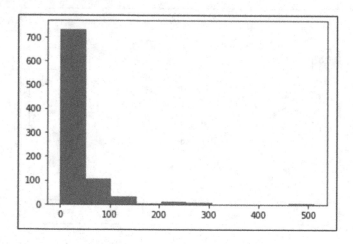

Figure 2-69. *A histogram of frequencies as the fare data falls into each respective bin*

Bar Graph

A bar graph is very useful for displaying the data over multiple categories. For example, you can display how many anomalies you have and how many normal data points you have in a dataset, if you know that information. The code in Figure 2-70 displays the male to female data frequencies using a bar chart.

```
plt.bar(df['Sex'].unique(), df['Sex'].value_counts())
```

Figure 2-70. *Creating a bar graph*

The first parameter tells matplotlib what unique values there are, and the second parameter gives the frequency counts for male and female data, respectively.

Figure 2-71 shows the resulting bar graph.

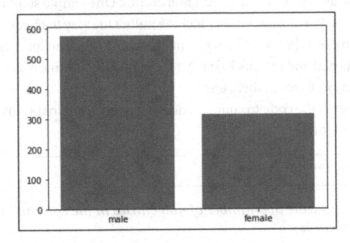

Figure 2-71. *The resulting bar graph created by the code in Figure 2-70*

Note   Keep in mind that these charts are all highly customizable. This section has provided only the bare minimum code to show you some of what is possible to do in matplotlib.

We'll encounter some of these basic graphs and plot types later in the book. Don't forget that matplotlib offers so many more plot types that can be quite useful in your data analysis or visualization.

Data Processing

Data processing is a very crucial element of data science in general. Real-world data is almost never perfect, especially if you're working with large swathes of stored data in a big database or if you're working with a live stream of data. At some point, you're going to have to deal with missing values, outliers, erroneous entries, or the need to transform your data into a better, more model-friendly format.

One of the most common data imperfections you're likely to encounter is the null, which occurs when no entry exists where there should be one. In these cases, you'll need to address data with such missing elements because a machine learning model won't know how to directly interpret it.

Nulls

There are many strategies for dealing with null entries. One simple solution is to either drop the column itself or drop rows that include nulls. One way to check how many nulls a column has is to use df[your_column].isna().sum(). The function .isna() returns True if the value is null and returns False if it is not null. By summing it, you can get a count of how many null values there are.

Figure 2-72 shows the code to compare how many nulls are in the column 'Age' with the total number of rows.

```
df['Age'].isna().sum(), len(df['Age'])
```

Figure 2-72. *Comparing the number of null entries in the column 'Age' with the total number of records*

You should see an output like (177, 891). So now we know that the 'Age' column has 177 nulls, which is quite a lot considering this means almost 20% of the data is null for 'Age'.

Again, one simple method is to drop all the rows that are null. We can do this by executing the code in Figure 2-73.

```
df.dropna(subset=['Age'])
```

Figure 2-73. *Dropping rows and specifying that we want to only drop rows with nulls belonging to a column in the list passed to subset*

We are explicitly including 'Age' in the subset of columns from which to drop rows if a null is encountered. If we did not pass any parameters to .dropna(), it would drop rows where a single null is detected anywhere. This would reduce our total data count drastically.

Sometimes, you won't want to drop nulls because it would reduce the data count by too much. Perhaps just one column is missing values and the other columns are relatively fine. You could also *interpolate* values—fit various functions along the columns to approximate what should go in between or fill them in by some algorithmic method like the average of the closest non-null values—which is a separate topic by itself with plenty of depth.

We can simply fill in the null values (**impute**) by calculating the average age by first computing the average of the 'Age' column and then using the `.fillna()` function, as shown in Figure 2-74.

```
avg_age = df['Age'].mean()

df2 = df.copy()
df2['Age'] = df2['Age'].fillna(avg_age)

# Checking that the ages were filled in correctly by Sex
df2[(df['Age'].isna())].head()
```

Figure 2-74. *Calculating the average age, imputing null values with this calculated average, and then displaying the imputed null values using the index of the original null values for the 'Age' column*

The output of executing the code in Figure 2-74 is displayed in Table 2-27. If you looked at these same rows in df, you'd notice that 'Age' would be null for these specific rows. However, in df2, we imputed the nulls with the average age. Table 2-27 shows that we have successfully replaced the nulls.

Table 2-27. Output of Executing the Code in Figure 2-74

	PassengerId	Survived	Pclass	Name	Sex	Age	SibSp	Parch	Ticket	Fare	Cabin	Embarked
5	6	0	3	Moran, Mr. James	male	29.699118	0	0	330877	8.4583	NaN	Q
17	18	1	2	Williams, Mr. Charles Eugene	male	29.699118	0	0	244373	13.0000	NaN	S
19	20	1	3	Masselmani, Mrs. Fatima	female	29.699118	0	0	2649	7.2250	NaN	C
26	27	0	3	Emir, Mr. Farred Chehab	male	29.699118	0	0	2631	7.2250	NaN	C
28	29	1	3	O'Dwyer, Miss. Ellen "Nellie"	female	29.699118	0	0	330959	7.8792	NaN	Q

Categorical Encoding

In your data, you might have many string values that you'd want to model. For example, sex was logged as 'male' or 'female', but machine learning models can't directly interpret text. They operate on numbers, so we will need to transform these strings into numerical representations. There are two ways to do this:

- Map the labels to a one-to-one numerical equivalent

- Craft one-hot vectors and add them as columns to encode the categories

For the first method, we can use scikit-learn to help us create a label encoder to automatically map our categorical labels into numeric equivalents. We can also switch these back to the original categorical values using this same label encoder object. To get started with the label encoder, refer to Figure 2-75.

```
from sklearn.preprocessing import LabelEncoder

df2 = df.copy()
df2['Embarked'].unique()
```

Figure 2-75. *Importing the label encoder and finding out the unique values present in 'Embarked' to see what we must encode. 'Embarked' contains a null as well, which will also be mapped by the label encoder*

You should see something like this:

```
array(['S', 'C', 'Q', nan], dtype=object)
```

To actually fit the label encoder on our data, we execute the code shown in Figure 2-76.

```
embarked_encoder = LabelEncoder()
embarked_encoder.fit(df2['Embarked'])

embarked_encoder.classes_
```

Figure 2-76. *Instantiating a label encoder object and fitting it to our column of values. Then we print out the classes*

You should see something like this:

```
array(['C', 'Q', 'S', nan], dtype=object)
```

Now we can go about transforming the values in our dataframe, as shown in Figure 2-77.

```
df2['Embarked'] = embarked_encoder.transform(df2['Embarked'])
df2.head()
```

Figure 2-77. *Transforming the existing values in 'Embarked' into numerical equivalents using the embarked_encoder object*

The output should look like Table 2-28.

Table 2-28. *The 'Embarked' Column with Values Transformed into Numeric Equivalents*

	PassengerId	Survived	Pclass	Name	Sex	Age	SibSp	Parch	Ticket	Fare	Cabin	Embarked
0	1	0	3	Braund, Mr. Owen Harris	male	22.0	1	0	A/5 21171	7.2500	NaN	2
1	2	1	1	Cumings, Mrs. John Bradley (Florence Briggs Th...	female	38.0	1	0	PC 17599	71.2833	C85	0
2	3	1	3	Heikkinen, Miss. Laina	female	26.0	0	0	STON/O2. 3101282	7.9250	NaN	2
3	4	1	1	Futrelle, Mrs. Jacques Heath (Lily May Peel)	female	35.0	1	0	113803	53.1000	C123	2
4	5	0	3	Allen, Mr. William Henry	male	35.0	0	0	373450	8.0500	NaN	2

If you ever want to reverse the transformation, use the `.inverse_transform()` function, as shown in Figure 2-78.

```
df2['Embarked'] = embarked_encoder.inverse_transform(df2['Embarked'])
df2.head()
```

Figure 2-78. *Reversing the transformation so that the 'Embarked' values go back to what they originally were*

You should see that the values are back to normal.

We can also perform one-hot encoding to encode the categorical values in a column. One-hot encoding produces a vector of values where each entry corresponds exactly to a specific categorical class. The vector will have 0s everywhere except for a specific index that corresponds to what class it is, where the value is 1.

To illustrate this, pretend we have only three categorical values: 'cat', 'dog', and 'mouse.' A one-hot vector for a data point belonging to 'dog' would be [0, 1, 0]. A one-hot vector for the label 'mouse' would be [0, 0, 1], and the one-hot vector for 'cat' would be [1, 0, 0]. There should be only one 1 across the entire vector, belonging to a specific class. If the label is Null, then the vector can be all 0s, with no 1 present at all.

When we do one-hot encoding as a transformation of the column itself, we are now creating a column for every categorical label where the value is only ever 1 (if the entry belongs to this class) or 0 everywhere else. This is a more scalable approach to categorical encoding, especially if you have many categories. Furthermore, the sparse data (**sparse** means there are many 0 values and few real values) format can also make the learning task potentially easier because it may more easily be able to discriminate two different data points belonging to different classes. Differentiating between 1 and 0 is easier than differentiating between 1, 2, 3, 4, etc.

To perform categorical encoding in this one-hot manner, we can use the function `pd.get_dummies()`, which creates separate columns for every unique category encountered where the values are only ever `True` or `False`. Each row has only one `True` value across the entire row because it's supposed to be one-hot.

The code in Figure 2-79 shows how to use `.get_dummies()` to create one-hot columns for the 'Embarked' column. The `prefix` argument adds that value as a prefix to the created columns. Otherwise, each created column would be named after the categorical value it belongs to. It is best to provide a prefix, which can simply be the column name itself for convenience and clarity.

```
# Prefix column lets us prepend a fixed name to the columns, otherwise ↵
the column name is just going to be whatever value corresponds with it. ↵
Go ahead and try it without the prefix argument to see the result
pd.get_dummies(df2['Embarked'], prefix='Embarked').head(2)
```

Figure 2-79. *Creating one-hot columns for the 'Embarked' column*

The output should look like Table 2-29. Notice that null is missing as a column, but in this case, if there isn't a single True value in the entire row, then the row represents a null value. So if the entire row is False (or 0), then 'Embarked' was null for that row.

Table 2-29. *One-Hot Columns Created for 'Embarked'*

	Embarked_C	Embarked_Q	Embarked_S
0	False	False	True
1	True	False	False

You'll notice in Table 2-29 that the values are not 1 or 0 but rather Booleans. If needed, you can convert these to numbers explicitly by typecasting as an int or float.

So how do we combine this with our original dataframe? This is where joining comes in handy. All we need to do is to join this one-hot dataframe with the original dataframe. The index should be the same. Refer to Figure 2-80.

```
df2.join(pd.get_dummies(df2['Embarked'], prefix='Embarked'))
```

Figure 2-80. *Joining the original dataframe with the derived one-hot dataframe. Implicitly, Pandas joins on the index where two rows have the same index*

The output should look like Table 2-30. Here, we are only displaying .head(2).

Table 2-30. *Joining the Original Dataframe with the One-Hot Dataframe to Add the One-Hot Columns Back In*

	PassengerId	Survived	Pclass	...	Embarked	Embarked_C	Embarked_Q	Embarked_S
0	1	0	3	...	S	False	False	True
1	2	1	1	...	C	True	False	False

Don't forget to drop 'Embarked' before you do any modeling; the whole purpose was to replace it with one-hot columns.

Scaling and Normalizing

Scaling/normalizing your data into a common range can be very helpful to facilitating the modeling task. Suppose you have two columns, one with a range up to 10,000 and the other with a range only up to 1. The model might consider the values for the former column much more heavily than the values for the latter column. This is purely because adjustments made to the model will reflect much more heavily upon the columns with large-magnitude numbers.

If you think of a simple y=mx+b model, where the domain of x includes some massive numbers up to the range of 1 million, then incremental updates to the slope m even around 0.1 will lead to massive changes in the predicted y-value. Now imagine if the domain of x was scaled down to something like (–1, 1) while the distribution was kept roughly the same. Incremental updates would no longer lead to large fluctuations in the predictions. When multiple columns are all in similar ranges, the learning task is much better because the model can make more efficient learning steps to fit the data properly.

More formally, the perks of scaling and normalization include the following:

- **Bias reduction**: The model will not be so swayed by outliers or columns with large-magnitude values over the rest of the data, which will enable the model to more accurately model the data.

- **Stability**: Sometimes, it's possible for the model's learning process to become unstable and totally diverge (arrive at a bad solution). By normalizing the scale of the data, the model's learning task will be much more straightforward, and it might even converge to a better solution.

So how do we scale/normalize data? There are quite a few methods, but two in particular are standard scaling (computing z-scores across the column) and min-max scaling (scaling the values to a smaller range by using the minimum and maximum values found in the column). We will use scikit-learn to help with this.

The following is the formula for Z-Score as used by StandardScaler to rescale the values. The variable x is the input, u is the mean of the column, and s is the standard deviation of the column.

$$z = (x - u) / s$$

Let's get started with StandardScaler. Check out the code in Figure 2-81.

```
from sklearn.preprocessing import StandardScaler

df2 = df.copy()
age_scaler = StandardScaler()


# Note, when fitting StandardScaler, it expects an array of shape ↵
(M, 1). So either do df2['Age'].to_numpy().reshape(-1,1)
# age_scaler.fit(df2['Age'].to_numpy().reshape(-1,1))
age_scaler.fit(df2[['Age']])
# df2['Age'] = age_scaler.transform(df2['Age'].to_numpy().reshape(-1,1))
df2['Age'] = age_scaler.transform(df2[['Age']])
df2.head()
```

Figure 2-81. *StandardScaler has very similar syntax to the label encoder. Using the StandardScaler, we normalize the 'Age' column*

You should see output like Table 2-31, which shows that the 'Age' column has had standard scaling applied to it so that the data follows a normal distribution.

Table 2-31. *Standard Scaling Applied to the 'Age' Column*

	PassengerId	Survived	Pclass	Name	Sex	Age	Cabin	...	Embarked
0	1	0	3	Braund, Mr. Owen Harris	male	-0.530377	NaN	...	S
1	2	1	1	Cumings, Mrs. John Bradley (Florence Briggs Th...	female	0.571831	C85	...	C
2	3	1	3	Heikkinen, Miss. Laina	female	-0.254825	NaN	...	S
3	4	1	1	Futrelle, Mrs. Jacques Heath (Lily May Peel)	female	0.365167	C123	...	S
4	5	0	3	Allen, Mr. William Henry	male	0.365167	NaN	...	S

Standard scaling does not necessarily reduce your range to (–1,1). It normalizes the data, so it's possible for the data to have values up to four standard deviations from the norm. The transformation gets the data to have a mean of 0 and standard deviation of 1.

Scikit-learn offers many other standardization techniques, each with its own effect on the data. It's worth exploring and trying these out, as there is no catch-all scaling method and the nature of your data can require various techniques.

Feature Engineering and Selection

Feature engineering is about creating meaningful features from data. It can involve data processing of raw data, cleaning up nulls, and also the creation of new columns of data either by using a combination of existing columns/data or incorporating extraneous data from other sources. Sometimes, domain expertise can be helpful as well in creating new features.

Feature selection is more about selecting only the features that are relevant and useful to the modeling task. Not every column or variable will be useful, as we touched upon in the Pandas Correlation Matrix section.

Let's create some new features and see if we can get any useful information out of them. In particular, we will create the following columns:

- **Title**: Extract Miss., Mrs., Mr., or any other titles.

- **Age_Group**: Grouping by age such as child, adult, young adult, etc.

- **Family_Count**: Parch is the number of parents and children of the passenger on the ship. SibSp is the count of siblings of this passenger. Together, we can combine these into a new metric.

Let's define some functions to help us extract these features. Refer to Figure 2-82.

```
df2 = df.copy()


def get_title(x):
    # Every name is separated into last, first format
    last, first = x.split(',')

    # Remove any space after comma
    first = first.strip()

    # Always a dot after title
    title = first.split('.')[0].strip() + '.'

    return title

df2['Title'] = df2['Name'].apply(get_title)
```

Figure 2-82. *Creating df2, a copy of df, and applying* get_title() *to create a column just for titles*

Next, Figure 2-83 continues feature creation by creating the 'Age_Group' column and 'Family_Count' column.

```
def age_group(x):
    if x < 13:
        return 'child'
    elif x < 18:
        return 'teen'
    elif x < 25:
        return 'young_adult'
    elif x < 50:
        return 'adult'
    else:
        return 'old'

df2['Age_Group'] = df2['Age'].apply(age_group)
df2['Family_Count'] = df2['SibSp'] + df2['Parch']
```

Figure 2-83. *Mapping the ages to create different categories of age classification and creating 'Family_Count' by summing 'SibSp' and 'Parch'*

By calling `df2.head()`, you should see something like Table 2-32.

Table 2-32. *Observing the Newly Created Features*

	PassengerId	Survived	Pclass	Name	...	Title	Age_Group	Family_Count
0	1	0	3	Braund, Mr. Owen Harris	...	Mr.	young_adult	1
1	2	1	1	Cumings, Mrs. John Bradley (Florence Briggs Th...	...	Mrs.	adult	1
2	3	1	3	Heikkinen, Miss. Laina	...	Miss.	adult	0
3	4	1	1	Futrelle, Mrs. Jacques Heath (Lily May Peel)	...	Mrs.	adult	1
4	5	0	3	Allen, Mr. William Henry	...	Mr.	adult	0

Let's now apply the label encoder to these columns where appropriate. Refer to Figure 2-84.

```
sex_label_encoder = LabelEncoder()
df2['Sex'] = sex_label_encoder.fit_transform(df2['Sex'])

embarked_label_encoder = LabelEncoder()
df2['Embarked'] = embarked_label_encoder.fit_transform(df2['Embarked'])

title_label_encoder = LabelEncoder()
df2['Title'] = title_label_encoder.fit_transform(df2['Title'])

agegroup_label_encoder = LabelEncoder()
df2['Age_Group'] = agegroup_label_encoder.fit_transform(df2['Age_Group'])

df2 head(2)
```

Figure 2-84. *Applying a label encoder to the columns 'Sex,' 'Embarked,' 'Title,' and 'Age_Group'*

Your output should look like Table 2-33.

Table 2-33. *Truncated Table Output with the Relevant Columns Having Their Values Mapped to Numerical Equivalents*

	PassengerId	Name	...	Sex	Title	Age_Group	Family_Count
0	1	Braund, Mr. Owen Harris	...	1	11	4	1
1	2	Cumings, Mrs. John Bradley (Florence Briggs Th...	...	0	12	0	1

Now we can finally look at the correlation matrix to see what columns are most highly correlated with 'Survived'. Refer to Figure 2-85.

```
select_list = ['Survived', 'Pclass', 'Sex', 'Age', 'SibSp', 'Parch',
'Embarked', 'Title', 'Age_Group', 'Family_Count']
df2[select_list].corr()
```

Figure 2-85. *Selecting specific columns to display in the correlation matrix. Categorical columns can't directly be in the matrix*

Your correlation matrix should look something like Table 2-34.

Table 2-34. *Correlation Matrix (Truncated)*

	Survived	Pclass	...	Age_Group	Family_Count
Survived	1.000000	-0.338481	...	-0.052325	0.016639
Pclass	-0.338481	1.000000	...	0.124159	0.065997
Sex	-0.543351	0.131900	...	-0.029088	-0.200988
Age	-0.077221	-0.369226	...	-0.298501	-0.301914
SibSp	-0.035322	0.083081	...	-0.013418	0.890712
Parch	0.081629	0.018443	...	-0.054641	0.783111
Embarked	-0.163517	0.157112	...	-0.039660	0.064701
Title	-0.193635	0.029099	...	-0.080130	-0.199883
Age_Group	-0.052325	0.124159	...	1.000000	-0.036469
Family_Count	0.016639	0.065997	...	-0.036469	1.000000

If we look only at the column 'Survived', we find that 'Sex' and 'Pclass' are the two most strongly correlated values due to those columns having the highest magnitude of correlation.

We also want to make sure that features are not too intercorrelated with one another. If this is the case, then we are introducing redundancy, because it's likely more of the same type of information if the correlation is high. If we take a look at 'Family_Count', 'SibSp', and 'Parch', the correlation between the three is really high. This makes sense considering we directly derived 'Family_Count' from the latter two columns, but this illustrates how 'Family_Count' is a redundant variable. We can drop 'Family_Count' and keep 'SibSp' and 'Parch' or we can drop the latter two and just keep 'Family_Count'. There are many approaches to take.

Finally, let's try one-hot encoding our columns and see if expanding out the categorical features can help us understand the data relation better. Refer to Figure 2-86.

```
df2 = df2.join(pd.get_dummies(df2['Sex'], prefix='Sex'))
df2 = df2.join(pd.get_dummies(df2['Embarked'], prefix='Embarked'))
df2 = df2.join(pd.get_dummies(df2['Title'], prefix='Title'))
df2 = df2.join(pd.get_dummies(df2['Age_Group'], prefix='Age_Group'))
df2.head(2)
```

Figure 2-86. *Creating one-hot encoded columns for 'Sex', 'Embarked', 'Title', and 'Age_Group'*

Your output will look something like Table 2-35. There are now 43 columns as a result of the expansion. Keep in mind that your column count may explode if you have a lot of categories. You can use the correlation matrix to help you trim and drop the column count by eliminating columns that are very poorly correlated. But be careful, because a lack of linear correlation does not imply there is no nonlinear correlation, and a high correlation may not imply a causation either.

Table 2-35. Expanded Dataframe

	PassengerId	Survived	...	Title_15	Title_16	Age_Group_0	Age_Group_1	Age_Group_2	Age_Group_3	Age_Group_4
0	1	0	...	False	False	False	False	False	False	True
1	2	1	...	False	False	True	False	False	False	False

We've created a lot of new columns, so the updated correlation matrix will be massive. Let's observe the updated correlation matrix by picking out the column 'Survived' and checking those correlation values. The code is shown in Figure 2-87. You can examine the full correlation matrix in more detail if you'd like. Don't forget that the correlation matrix result is a dataframe itself, so you can select columns, filter values, sort the table, etc.

```
df2 = df2.drop(['PassengerId', 'Name', 'Ticket', 'Cabin'], axis=1)
df2.corr()[['Survived']]
```

Figure 2-87. *Dropping unneeded columns and only picking the 'Survived' column in the resulting correlation matrix*

Your output from executing the code in Figure 2-87 should look somewhat like Table 2-36. Again, the full table will be quite long.

Table 2-36. *Updated Correlation Matix Truncated to Survived Column*

	Survived
Survived	1.000000
Pclass	-0.338481
Sex	-0.543351
...	...
Age_Group_3	0.043789
Age_Group_4	-0.037627

It looks like by doing the one-hot encoding, we have discovered some features that are more correlated with 'Survived' than others. Examples include Title_11, Title_12, and Title_8. If we reverse transform these with the label encoder as shown in Figure 2-88, we will immediately understand why.

```
title_label_encoder.inverse_transform([8, 11, 12])
```

Figure 2-88. *Inverse transforming the three title values of relevance*

The output given is the following:

```
array(['Miss.', 'Mr.', 'Mrs.'], dtype=object)
```

So now we can see why these titles were more highly correlated than the other titles. You can directly extract the sex of the passenger from these titles alone, which makes sense why Title_11 (corresponding to 'Mr.') has such a higher correlation with 'Survived' than the rest, because the passengers who survived are biased towards being female.

With that, you have now completed basic feature engineering and selection. There is far more to data science in general, but for the purposes and scope of this book, you have made it past the basics that you will need to know to follow along with the rest of the book.

Summary

You should now be more familiar with how to use Pandas, scikit-learn, and matplotlib to load data, transform it to your liking, prepare it for machine modeling, and conduct statistical and visual analysis. With this knowledge, you will be able to follow along with the rest of the book and have a basic foundation to expand upon for future projects.

In Chapter 3, we will go over what machine learning is at a high level as well as cover important concepts that will apply to the rest of the book, to deep learning, and beyond.

CHAPTER 3

Introduction to Machine Learning

This chapter introduces you to the basic principles of machine learning, including the machine learning workflow. These concepts will apply to nearly every modeling task you may come across, and they extend into deep learning modeling as well. This is a high-level theoretical introduction to machine learning, since the practical material and implementation of these machine learning principles will be covered in the subsequent chapters.

By the end of this chapter, you will understand the basic machine learning modeling workflow, what machine learning modeling is, how to evaluate and interpret model performances, and how to deal with getting the model to properly learn the task and perform just as well on brand-new data.

In a nutshell, this chapter covers the following topics:

- Introduction to machine learning
- Data splitting
- Modeling and evaluation
- Overfitting and bias-variance tradeoff
- Hyperparameter tuning
- Validation

Note   Code examples are provided in Python 3.8. The code repository for this book is available at `https://github.com/apress/beginning-anomaly-detection-python-deep-learning-2e/tree/master`.

© Suman Kalyan Adari, Sridhar Alla 2024

There is also a requirements.txt file available in the repository to check your packages and their versions.

Coding examples for this chapter are available at `https://github.com/apress/beginning-anomaly-detection-python-deep-learning-2e/blob/master/Chapter%20 3%20Introduction%20to%20Machine%20Learning/chapter3_machinelearning.ipynb`.

Navigate to "Chapter 3 Introduction to Machine Learning" and then click chapter3_ machinelearning.ipynb. The code is provided as a .py file as well, though it is the exported version of the notebook.

We will be using JupyterLab to present all of the code examples.

Machine Learning

Artificial intelligence (AI) and machine learning (ML) have become popular topics of discussion in business, academia, and the media in recent years. This is not without reason. Thanks to computing power and algorithmic developments, machine learning has really taken off and has permanently transformed society. You may not know it, but machine learning powers nearly every part of your day, including online content/product recommendations, voice assistants, Google Search, and even resource consumption forecasting (done by energy companies and cable companies, for example).

Most of these advancements come primarily from deep learning, a subfield of machine learning that focuses on the concept of the artificial neuron. We will cover deep learning in more detail in Chapter 5. However, there are fundamental aspects of machine learning that apply both to traditional ML algorithms (that don't rely on neural networks) and to deep learning models, which we will cover in this chapter. In this chapter, we will go over what machine learning is, what the machine learning workflow looks like, how you can evaluate machine learning models, and how you can tune them to perform more optimally.

Introduction to Machine Learning

In its simplest sense, **machine learning** is the process of teaching a model how to perform a task by providing it with data. Instead of explicitly coding the model to do the task, we program the mechanisms by which this model can learn, improve, and repeat the task so that by the end of training, the model is able to perform whatever task we want it to.

You can think of it as providing a child (who serves as a blank slate and as the model, knowing nothing about the topic) with a proper learning environment and materials and letting the child study only that topic for a few days. By the end, the child will hopefully have picked up how to perform whatever task was taught by the learning material.

In the context of machine learning, a **model** is an advanced function that takes the input data and maps it to a specific output. When a model is **learning**, it is adjusting its internal mechanisms to better perform the task on the data it is given. When we say a model performs well or not, we define some metrics to evaluate the output predictions of these machine learning models to understand how well they're performing. For example, accuracy is a simple metric by which we can measure predictive performance.

The concept of machine learning is often confused with the concepts of artificial intelligence and deep learning, so let's define the other two terms.

Artificial intelligence is a vast field of computer science that is focused on solving high-level, complex tasks that usually only humans can perform. Artificial intelligence programs are typically systems composed of many components that all work together to help complete the various tasks, which can range from being a voice assistant like Siri, Alexa and Cortana, to a powerful general purpose AI like ChatGPT which can write code, create picture and more, to self-driving cars. Each of these systems relies on various machine learning model components to work. Self-driving cars require a vision model to detect objects from video feed and classify them as pedestrians, cars, road signs, etc., and require another machine learning model to be in charge of making decisions such as "How much should I steer by at this given moment?"

Deep learning is a subfield of machine learning. It inherits all the typical modeling workflow processes, but deep learning is strictly concerned with the concept of the artificial neuron, which is a generous abstraction of a biological neuron. Despite that, it has proven immensely successful as a learning model, and paired with the explosion in computational power in the past decade, deep learning models have become incredibly powerful and capable. Perhaps the pinnacle of deep learning as of this writing is OpenAI's ChatGPT.

Figure 3-1 depicts how the three fields are related to one another.

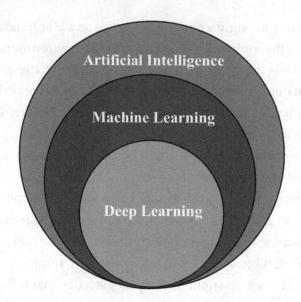

Figure 3-1. *Relationship between artificial intelligence, machine learning, and deep learning*

Note that all three of these fields overlap with data science, since modeling can be a part of data science as well. Note also that AI itself is partially overlapped by other, related fields, such as robotics, reinforcement learning (which itself is overlapped with robotics, ML, and DL), statistics, and so on. AI is truly an interdisciplinary field.

Back to the abstracted definition of machine learning, we know that we have a machine and that it is trying to learn something given data. So, what does this workflow typically look like? Whether you're dealing with machine learning models or deep learning models, a modeling workflow involves the following:

- **Problem statement**: What is the task you are trying to solve?

- **Data collection**: You first need the relevant data before you can teach the model anything.

- **Data processing**: Processing the data and manually annotating (if necessary) to show the model what correct performance on the task looks like. This is where you clean the data, analyze it, and perform feature engineering.

- **Data splitting**: Partitioning the dataset into training, testing, and validation components. The idea is to train on the training split, evaluate on the testing split, and make improvements using the validation split. We will go over this in more detail in the next section.

- **Model training**: Training the model on the training dataset. There are many intricacies involved in model training, especially if you use an iterative method and/or deep learning models.

- **Model evaluation**: Evaluating your model's performance on testing data and examining any faults with the process so far. The blind performance (testing data is unseen by the model) can indicate a lot about the effectiveness of the model.

- **Model tuning**: Optimizing the performance by tuning certain high-level parameters that can affect the learning ability of the model.

This is more or less the typical workflow of a machine learning model, whether you're using a traditional machine learning model or a deep learning model.

Speaking of deep learning and machine learning, if deep learning is so popular nowadays, why even bother with machine learning? That's a fair question. To look at how machine learning performs against deep learning with respect to the amount of data, refer to Figure 3-2.

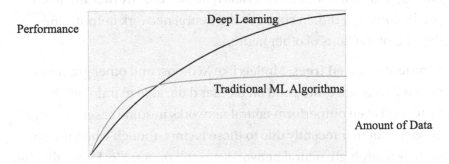

Figure 3-2. *The performance of deep learning algorithms and traditional machine learning algorithms as the amount of data increases*

Generally, with low data scales (up to the hundreds of thousands and to the very low millions), machine learning models might outperform deep learning models at the same tasks. However, it's not a hard-and-fast rule, just a general possible trend. And as you can see, deep learning far eclipses machine learning models with large amounts of data.

For a great reference as to what deep learning models on the upper end of that curve can look like, ChatGPT is by far the most advanced model (again, as of this writing), and it is a very large deep learning model with about 175 billion parameters that trained on dozens of terabytes of data.

There are some caveats you must consider when deciding the most appropriate model for your task. The following factors may make traditional ML models more suitable for usage than deep learning models:

- **Small dataset**: Datasets ranging from thousands of records to up to the low millions can be worth trying traditional machine learning models on.

- **Computational power**: Deep learning models can be trained and operated on the CPU but are far quicker on specialized hardware like GPUs (graphics processing units) and TPUs (tensor processing units). These resources can be expensive to run, since deep learning models train for many iterations and take time as the data and model complexity scale up.

- **Interpretability**: Neural networks are known to be "black-box" models whose inner workings are nearly impossible to decipher. On the other hand, traditional ML algorithms such as decision trees and regression algorithms can clearly show you why they made the predictions they did. In contrast, the neural network output can depend on millions of other neurons.

- **Gradient-boosted trees**: Models like XGBoost and other gradient-boosted trees are quite good on tabular data, able to train much faster and even outperform neural networks in some cases. They've become popular recently due to these factors, though when the data scale is really high, neural networks are still preferable. Depending on the machine, even up to the low millions of record might be able to be modeled by these algorithms.

Again, there isn't a hard-and-fast rule for when to use ML over DL; you'll just need to try out various solutions and find the ones that work best. What is more feasible over the other will depend on what compute power you have, how much data you have, how complex the task is, and so on. We will go over the basic framework of how you can do this later in the "Validation" section.

Data Splitting

The data processing step of the modeling workflow is covered in Chapter 2. Typically, you have some kind of collection of data relevant to your task. After the processing, analyzing, and feature engineering steps, it is ready to be used for training. However, before you immediately pass the data in, you want to split the overall dataset into **train**, **test**, and **validation** splits.

In practice, you might encounter the train and testing splits only, but the validation split is quite useful in a deep learning context. First, we will go over what each of these splits means:

- **Training split**: This should be the largest proportion of the overall dataset, and it is reserved purely for the model to train on.

- **Testing split**: This data should be completely separate from the training dataset, and the model must never train on this.

- **Validation split**: Acts similar to the test dataset, except it can be used in the training process to help perform a variety of optimization tasks. It should be sourced from the training split.

Figure 3-3 depicts how these splits are derived. The validation set is generated from the training split, to keep it completely separate from the testing split since the validation set is used in the training phase.

Figure 3-3. *Two possible ways to partition your data: a train-testing split, or a train-val-testing split. When deriving the validation split, you should first compute the train-test partitions, then take the train partition, and split it again into train-val splits. This way, the validation split is derived from the training split and is separate from the test data. This ensures absolutely no leakage of information when developing your models*

Your **training** set will typically comprise of 60–80% of your entire data split. Exactly what percentage you allocate usually depends on whether you're using a deep learning model or a traditional machine learning model. With deep learning models, you may benefit from higher training splits because deep learning models scale better with more data. This also helps to avoid overfitting, discussed later in this chapter.

The **testing** set will comprise anywhere from 10–20% of your data split. If your data count numbers are high, such as in the millions, you can get away with a lesser ratio reserved for test data, because the number of data samples in the test set will be high enough to adequately evaluate whether your model is performing the task about as well on brand-new data that it hasn't seen.

The **validation** set is optional, but it is also around 10–20% of your data split. The validation set is a small subset (derived from an original training split) that is useful for quick experimentation and prototyping, and in the case of models with iterative training methods (like deep learning), they can indicate whether the model is actively overfitting or not. We will cover the validation set in more detail in the final section of this chapter.

As for how these sets are used overall, the validation set is extremely useful for quickly determining which models are even appropriate for the data task. This way, you can create a list of candidate models for the task, use a validation training strategy like K-Fold cross validation (which will be covered soon), and gather the evidence to determine which model you should go with. Be careful, though, because some models, like deep learning models, might just need more data to out-scale their competitor models in performance.

Once you have a core set of models you want to train on the training split, it is important to ensure that they never train on the validation set nor the test set while they are actively being trained on the training set. You want to keep the evaluation data separate so that you can get a fair assessment of the model's performance on unseen data, which is an important goal of machine learning model training. This is what is known as **generalization**, which refers to the model's ability to learn the task in such a way that it achieves high performance conducting the task on data it has never seen.

Modeling and Evaluation

When it comes to machine learning modeling, there are several paradigms that dictate how you can train your model:

- **Supervised learning**: A model is given input data and corresponding correct outputs. Its goal is to learn as correct a mapping as possible from input to output. Models will try and match their predictions as closely as possible to the true outputs. Two overarching learning tasks fall under supervised learning:

- **Classification**: Labeling data as a specific category. For example, predicting if an image is of a cat or a dog. Many other tasks fall under classification, such as anomaly detection (normal vs. anomaly), multiple label classification (an image of a dog can be classified as a mammal and as a dog simultaneously), and more.

- **Regression**: Any real valued output prediction task, like housing price prediction, stock price forecasting, etc.

- **Unsupervised learning**: A model is given the input data and must learn structures and patterns within the data by itself. The following learning tasks fall under unsupervised learning:

  - **Clustering**: Grouping data points together by some shared commonalities.

  - **Representation learning/dimensionality reduction**: Learning to represent the same data in a smaller dimension while minimizing loss of information.

- **Semi-supervised learning**: Falls somewhere between supervised and unsupervised learning and can encompass many learning approaches. Typically it involves both a small bit of supervised learning and a small bit of unsupervised learning. Examples:

  - **Teacher-student model**: Hand-label a small portion of data and train a teacher model in a supervised manner. Then, have the teacher model create a larger training set by making predictions on many new, unseen examples. These input-output pairs can be algorithmically refined to eliminate uncertain predictions, and this new dataset is used to train a student model.

  - **Training on one class**: In the context of anomaly detection, where you may have many normal data points and few anomalous data points, you could train a model to recognize only normal data. When making predictions on new data (which can include anomalies), the model may have trouble making confident predictions on the anomalous data points. Because of this uncertainty, you can label those points as anomalies. We will explore this approach later with the One-Class SVM model.

What approach you use depends on your problem statement and your data distribution. If you have lots of unlabeled examples and can only annotate a few, then a semi-supervised approach may be best. If you have plenty of data that is easily labeled, then a supervised approach usually works best. If you do not want to annotate the data and want to let the model itself try and create categories (via clustering), unsupervised approaches would be best.

Regardless of the paradigm you follow, you need a way to measure error. **Error** is a measure of how wrong the model's predictions are. Error is also commonly referred to as the **loss** of a model. For **classification** tasks, a simple loss function would be the **cross-entropy loss/log loss** function, defined as such by Equation 3-1.

The value y_i is the prediction of the model for data input x_i. $P(y_i)$ represents the confidence probability with which the model predicts y_i. You can plug in $y_i = 1$ and then $y_i = 0$ to see how the same equation works for both predictions of 1 and 0. For a logistic regression model output, $p(y_i)$ will be a probability score from 0 to 1, where anything above 0.5 counts as a classification of 1 and otherwise it is 0. When you're plugging in values to compute the loss, $y_i = 1$ and $p(y_i)$ can be 0.65, or yi=0 and p(yi) can be 0.34.

$$Log\ Loss = -\frac{1}{N}\sum_{i=1}^{N} y_i * \log\big(p(y_i)\big) + (1 - y_i) * \log\big(1 - p(y_i)\big)$$

Equation 3-1. *The log loss function*

This is a loss function that is applicable if you're classifying only two labels, represented by a 0 or a 1.

For **regression** tasks, a simple loss function is the **mean-squared error (MSE)**, defined in Equation 3-2. It is simply the average of all the deviations between predictions and actual values squared.

$$MSE = \frac{1}{N}\sum_{i=1}^{N}\left(y_i - \widehat{y}_i\right)^2$$

Equation 3-2. *The mean-squared error (MSE) loss*

The objective for any learning task is to **minimize the loss**. In other words, the measured difference between the prediction and the true output must be as low as possible.

Now that we have covered the gist of loss functions, let's move on to performance metrics. Performance metrics can be as simple as classification accuracy, where you count how many predictions matched and divide it by the total number of predictions. For regression tasks, mean-squared error is itself a metric. However, there are a multitude of ways to measure your model's performance that go far beyond simple accuracy, especially for classification.

Classification Metrics

What's wrong with accuracy? Well, in the case of imbalanced data, you can have 99 normal points and 1 anomaly. If the anomaly detection model predicted all the normal points correctly but missed the anomaly, your accuracy is 99%. However, you failed to detect the anomaly, which is the entire purpose of an anomaly detection model, so in this sense, accuracy is extremely misleading. How can we do better?

This is where the **confusion matrix** comes into play, depicted in Figure 3-4.

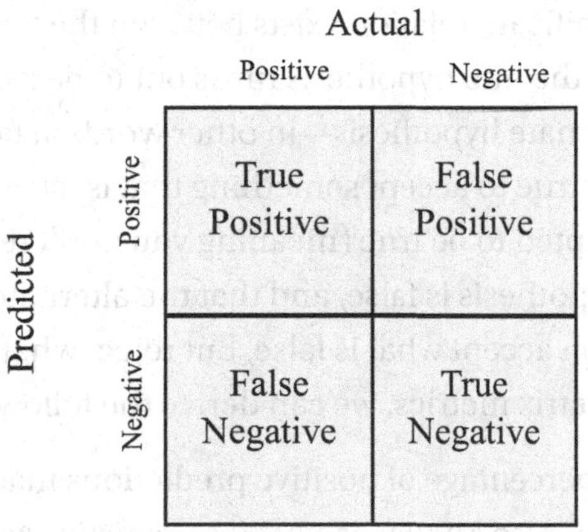

Figure 3-4. *The confusion matrix. The vertical axis corresponds to the predictions of the model. The horizontal axis corresponds to the true labels. The matrix will be populated by frequency counts of each type of prediction*

The four quadrants of the confusion matrix are defined as follows:

- **True positive (TP)**: Predicted positive, and the true label is positive

- **False positive (FP)**: Predicted positive, but the true label is negative

- **True negative (TN)**: Predicted negative, and the true label is negative

- **False negative (FN)**: Predicted negative, but the true label is positive

Let's set the context to be whether or not an animal in a zoo is diseased based on several markers in its blood sample. The model will take these markers as input features and make a prediction of either positive, meaning the animal has a disease, or negative, meaning the animal does not have a disease. With this context, let's put the confusion matrix metrics into perspective.

A **true positive** is whenever the model's prediction turns out to be correct. Here, if the model predicted the animal is sick, and it really is sick, it's a true positive prediction. A **true negative** is when the model predicts the animal is not sick, and it really is not sick. As for a **false positive**, it's when the model predicts the animal is sick, but it is not actually sick. Similarly, a **false negative** is when the model predicts the animal is not sick, yet it is actually sick.

In statistics, there are similar terms to false positive and false negative: **type I error** and **type II error**. These errors are used in hypothesis testing where there is a null hypothesis (which usually states that no relation exists between two observed phenomena) and an alternate hypothesis (which aims to disprove the null hypothesis, meaning a statistically significant relation exists between the two observations).

A **type I error** is when the null hypothesis turns out to be true, but you mistakingly reject it in favor of the alternate hypothesis—in other words, a false positive, since you reject what turns out to be true to accept something that is false. A **type II error** is when the null hypothesis is accepted to be true (meaning you don't reject the null hypothesis), but it turns out the null hypothesis is false, and that the alternate hypothesis is true. This is a false negative, since you accept what is false, but reject what is true.

From the confusion matrix metrics, we can derive the following additional metrics:

- **Precision**: The percentage of positive predictions made by the model that were correct. Also known as *positive predictive value*.

- **Recall**: Of all the truly positive instances in the dataset, this is the percentage the model got correct. Also known as *sensitivity* or *true positive rate*.

- **F1-measure**: The harmonic mean of precision and recall. The higher the precision and recall both are, the higher the F1-measure will be.

- **True negative rate**: Of all truly negative instances evaluated by the model, this is the percentage the model got correct. Also known as *specificity*.

- **Accuracy**: The number of correct predictions over all predictions.

Figure 3-5 shows the confusion matrix with the formulas for calculating **precision**, **recall**, and **accuracy**.

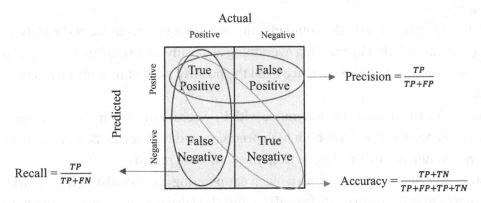

Figure 3-5. Confusion matrix with calculation formulas for precision, recall, and accuracy displayed

As for how to calculate the **F1-measure**, refer to Equation 3-3. To achieve higher F1-measure scores means the model's precision and recall must both be high, implying that the model makes correct positive predictions as well as correctly captures a high proportion of the truly positive data samples.

$$F1\ Measure = \frac{2 * Precision * Recall}{Precision + Recall}$$

Equation 3-3. F1-measure equation

For how to calculate the **true negative rate (TNR)**, refer to Equation 3-4. The TNR tells us how accurately the model predicts negative. It is similar to recall, except that of all the truly negative instances, it calculates the percentage of these negative instances correctly predicted to be negative.

$$TNR = \frac{TN}{TN + FP}$$

Equation 3-4. True negative rate equation

It is important to understand the tradeoff between precision and recall. Having one or the other can lead to some disastrous results. For example, it is possible for the model to have perfect precision, having perfect positive predictions. However, of all the positive instances, it maybe only got 15% of them right, giving it a recall of 0.15. This is terrible—the model is confident in its predictions, but barely identified anything from the positive data population.

On the other hand, a model could identify everything correctly from the truly positive data population by marking everything as positive. Its recall would be perfect, but its precision would be awful, since of all the positive predictions, only a fraction are correct.

This is why F1-Measure (synonymous with F1-Score) is a good metric: to have a high F1-score, it is necessary to have both high precision and high recall. Beyond the F1-score, we can derive further insights from the confusion matrix.

Using scikit-learn, we can graph a receiver operating characteristic (ROC) curve, which plots the true positive rate (recall) against the false positive rate (1 – true negative rate). Figure 3-6 shows an example of an ROC curve.

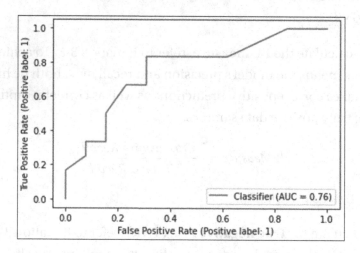

Figure 3-6. *Example of an ROC curve for a slightly weak classifier. AUC = 1 indicates a perfect classifier, but AUC = 0.76 indicates a weaker one with both increased false positives and not as many true positives predicted*

We can find the area under the ROC curve to derive a metric known as the area under curve (AUC) score. The higher the AUC score, the better the model can distinguish between the two classes. Intuitively, this makes sense, since the graph itself is plotting the model's positive prediction correctness against its false positive rate. The most ideal graph would have a TPR of 1.0 at an FPR of 0.0 and carry that 1.0 all the way across, looking like Figure 3-7.

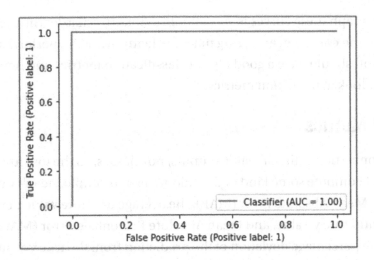

Figure 3-7. *An ROC curve with perfect AUC, meaning all predictions were correct. This means that for every FPR the TPR remained at 1.0*

This AUC score is known as the ROC-AUC score, which is the more commonly referenced AUC score. A good AUC score to target is around 0.95 or better.

There is another curve known as the precision-recall (PR) curve. We can similarly find the area under this curve, which scikit-learn calls the average precision (AP) score. Similar to the ROC-AUC score, an AP score of 1 indicates a perfect classifier that is able to perfectly distinguish between positive and negative classes.

The PR curve plots precision against recall, making it more friendly toward imbalanced datasets as it only concerns itself with the positive class. In anomaly detection, it is possible to have data heavily skewed toward the normal data, meaning the majority of your predictions would be true negatives. The PR curve deals with precision (correctness of model's positive predictions) and recall (proportion of all positive instances correctly captured by the model), so it is easy to see why it is more informative when we are concerned with pure performance on anomaly detection.

However, the ROC curve, which is concerned with recall/TPR (proportion of all positive instances correctly captured by the model) and the false positive rate (proportion of all negative data incorrectly predicted as positive), is also useful, as it can give us a picture of the model's performance on the overall data.

For our use cases, it is best to use both curves to get a solid picture of the model's performance on both anomaly detection and overall predictive performance on normal and anomaly data both.

The code to graph these curves is very simple thanks to scikit-learn, as covered in later chapters in the context of analyzing machine learning model performances.

With that, you should have a good idea of classification metrics and how to evaluate them. Let's now look at regression metrics.

Regression Metrics

Regression is concerned with real-value outputs, not classes, so the loss functions we will deal with all compute some kind of deviation. Beyond simple mean-squared error (MSE), we have Mean Absolute Error (MAE), the average absolute value deviation from the prediction and true y-value, and Mean Absolute Percentage Error (MAPE), the average absolute percentage deviation of the prediction from the true y-value.

The question of which metric to use will depend on your modeling task as well as how the data distribution looks. On a high level:

- **Mean-squared error** penalizes larger deviations more than smaller ones because of the squared element. This is good for penalizing outlier predictions. The formula is shown in Equation 3-5.

$$MSE = \frac{1}{N}\sum_{i=1}^{N}\left(y_i - \widehat{y}_i\right)^2$$

Equation 3-5. *The MSE equation, measuring the average squared deviance of prediction and true value*

- **Mean absolute error** penalizes all the values much more equally. Larger deviations won't incur as high an error as they would with MSE. This is more useful when prioritizing the bulk of the predictions over outliers. The formula is displayed in Equation 3-6.

$$MAE = \frac{1}{N}\sum_{i=1}^{N}\left|y_i - \widehat{y}_i\right|$$

Equation 3-6. *The MAE equation, measuring the average magnitude of deviance of prediction and true value*

- **Mean absolute percentage error** measures percentage deviations of the prediction from the true value. This is a potentially more useful metric if the data being modeled fluctuates a lot (like stock prices) or varies in scale a lot. The errors are also on the percentage scale and aren't measuring direct deviances, so they will not blow up the loss function if a deviation is large in magnitude. This is why it is potentially more stable if the data scale varies quite a bit. The formula is displayed in Equation 3-7.

$$MAPE = \frac{1}{N} \sum_{i=1}^{N} \left| \frac{y_i - \widehat{y_i}}{y_i} \right|$$

Equation 3-7. *The MAPE equation, measuring the magnitude of percentage error from the true value of the predicted value*

Now that you know more about proper metrics for the various tasks you will be performing later in this book, let's talk about the difference in performance of models on the training set and on blind, new data.

Overfitting and Bias-Variance Tradeoff

With the aforementioned performance metrics, we can now measure training, validation, and test loss. Ideally, your training and testing performance will be in the same neighborhood or match. If they don't differ, and your performance is good, then the model has learned the task properly and is generalizing well. However, you might often see that the test error will be markedly worse than the training error. To better formulate this situation, let's go over what the **bias-variance tradeoff** is.

First, what is bias? In simple terms, **bias** is a measure of how much the model's average prediction differs from the true value. If the model has not properly learned the task on a training set, it will have high loss, and thus high bias. If the model has learned the task very well, then its bias will be low.

In statistical terms, bias is the measure of the difference between the expected value of an estimator and the parameter it is trying to estimate. An estimator is some kind of function that seeks to approximate a specific parameter. Back to our machine learning sense, a model can fill the role of an estimator.

As for the formula of bias, refer to Equation 3-8. It measures the expected value of an estimator and its difference from the population parameter being estimated. If the bias amounts to zero, then the estimator $\hat{\theta}$ is an unbiased estimator of θ.

$$Bias\left[\hat{\theta}\right] = E\left[\hat{\theta}\right] - \theta$$

Equation 3-8. *The formula for bias*

Take, for example, a Gaussian/normal distribution of student heights at a university. The population parameters of this distribution are given by mu and sigma. However, say we want to estimate μ using $\hat{x}$. If we define $\hat{x}$ to be the mean formula, then it turns out that $\hat{x}$ is an unbiased estimator of μ, because if we plug in the values respectively into the bias equation, the ending bias becomes zero. When we say an estimator is **unbiased**, its bias is zero.

Similarly, a **biased** estimator has a nonzero bias. In practice, your machine learning models will likely all be biased, because a perfect model does not exist in a practical setting.

Variance is simply the variance formula in statistics but applied to the model's predictions. As a refresher, variance is defined as in Equation 3-9.

$$Var\left[\hat{\theta}\right] = E\left[\left(\hat{\theta} - E\left[\hat{\theta}\right]\right)^2\right]$$

Equation 3-9. *The variance formula for an estimator* $\hat{\theta}$

If the bias is zero, then the result of Equation 3-9 should exactly equal the mean-squared error. So, what is variance measuring here? The variance measure the average spread of the predictions from its mean. If the data is relatively noise-free (the data points do not fluctuate very far from the average prediction), then the variance itself will be lower. But if the individual predictions deviate a bit from the mean of the predictions, then the variance will be higher. Higher variance is not good because the added noisiness of the predictions can lead to higher loss.

To view high bias vs. low and high variance, refer to Figure 3-8. The context is archery and target practice.

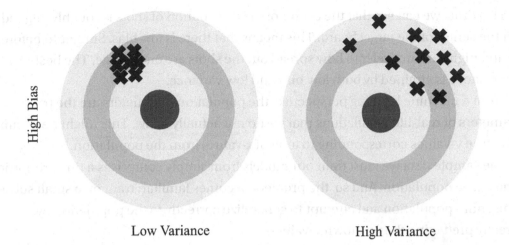

Figure 3-8. *High bias, low variance, and high variance are visually represented by target practice, a good example to illustrate how bias and variance relate to each other*

Referring to Figure 3-8, high bias is apparent in how all the shots landed away from the center. With low variance, the shots landed very close to each other. In a mathematical sense, the shots did not deviate much from the center of all the shots. In contrast, with high variance, the shots are much more spread out. It's still high bias because the center of those shots is quite far from the center of the target board. It's now high variance because the average deviation from the center of the shots is much higher.

Figure 3-9 shows what low bias looks like in both a low and high variance setting.

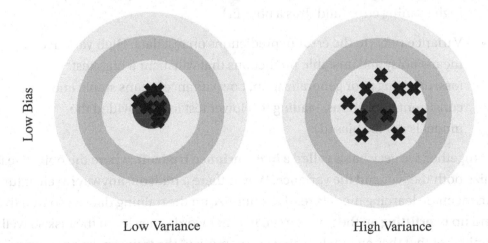

Figure 3-9. *Low bias, low variance, and high variance are visually represented by target practice*

This time, we can see that the center of the distribution of shots is roughly aligned with the center of the target board. This means that there is low bias. Similar to before, low and high variance refer to how spread out the shots are on average. The best performance is attained by both low bias and low variance.

From a machine learning perspective, the population parameters are the true parameters of real-life populations that we do not actually know. This might be as simple as the true y-values corresponding to a set of x-values from the population.

The sample data we must train our models from simply comprises a tiny proportion of the entire population. And so, the problem becomes familiar: train on a small subset of the entire population and attempt to generalize correctly to the population by correctly predicting the unknown y-values.

Since we do not know the entire population of y-values, it helps to evaluate on a test set, which is itself another sample of the overall population. However, because the distribution likely differs from the training set, it offers us a glimpse into how the model may perform on blind data. The larger our test set, the better we can evaluate the blind performance and generalization of the model.

With the terms bias and variance covered, let's now relate them to the training and testing sets more directly:

- Bias refers to the error of the machine learning model on the training set. It is a measure of how well, on average, the machine learning model correctly predicts the y-values in the training set. Low bias implies low training error (and thus good fit), while high bias implies high training error (and thus a poor fit).

- Variance refers to the error in predictions on test data. High variance means noisy and unstable predictions that will incur a high test loss, implying poor generalization. Low variance means stable and consistent predictions, leading to a lower test loss (provided the model is not too biased).

Put together, we get what is called a **bias-variance tradeoff**, where the objective is to minimize both the bias and the variance. Why is there a tradeoff, anyway? Well, it turns out that machine learning models tend to optimize on the training dataset so well that they end up **overfitting** on the training set. In other words, they learn the task so well on the training set that they end up learning to "memorize" the training data to maximize its performance. When evaluated on test data, the performance will differ quite a bit from what is seen on training data.

A similar example would be a student cramming and memorizing the problems from the homework, only to find out that the exam contains brand-new problems. The only way to have done well on the exam is to have a general understanding of the problems and how to solve them. So now, despite the student having 100% on the homework, they end up failing the exam because they did not properly learn the task.

Overfitting is influenced by a few parameters:

- **Model complexity**: If the model is very complex, it tends to learn a very complicated function mapping of input-output data that may work well on the training data, but it harms its test data performance.

- **Training time**: If a model is allowed to keep training forever, it eventually adjusts its internal parameters to just suit the training data. This is more apparent in models that learn iteratively, like neural networks.

- **Insufficient training data**: If there are too few data samples to learn from, the model might not understand how to generalize properly. And so, even if it does well on the training data, its testing performance suffers.

- **Poor data processing**: If one of the features is large in magnitude or is too noisy, the model might tend to just fit to the noise or features that overwhelm others. (If one column has large-magnitude data, the loss will be really high. The quickest path to reduce loss is to cater only to this column, which is not the behavior the model should be learning.) And so, once the test data (with different noise distributions) is evaluated on, the model's performance suffers.

Fortunately, there are many ways to avoid overfitting, including:

- **Loss penalty**: We can add penalties to the loss function to force the weights to be smaller, thus forcing the learning of more efficient mappings that are more robust to overfitting. This is called **regularization**.

- **Reduce complexity**: We can reduce the model complexity by having fewer learnable parameters in the model.

- **Data processing**: This will ensure that the data is clean and that the model won't learn poorly or be influenced by other factors in the data that will differ across training and test sets.

- **Hyperparameter tuning**: Tuning high-level parameters that dictate the model's learning process can lead to good gains in its generalization capabilities.

- **Adding training data**: Sometimes, simply adding more training data can help a model learn how to generalize better.

In contrast, **underfitting** occurs if a model has high bias on the training data. The following are factors for why a model may underfit:

- **Overly simple**: The model may not be complex enough to understand how to learn the task.

- **Not enough training**: The model may not have been trained long enough to have learned the task.

- **Poor data**: The data features may not have good predictive qualities. For example, what does the stock price have to do with the weather outlook? Additionally, the data may be far too noisy, which would prevent the model from learning because any signals from the data that may lead to learning the task are drowned out by the noise.

- **Harsh regularization**: If the regularization/loss penalty is too high, the model is penalized too harshly and cannot learn properly.

Regarding model complexity, we want to make sure the model has a sufficient quantity of parameters to be able to properly fit the data. Consider an example where we are trying to learn the true function $y = 0.12x^3 - 0.2x^2$. We have sampled the x space and injected artificial noise to test how different complexities of the model will fit on this noisy data and to see if they can learn to approximate the true underlying function. To recreate this code, run the code shown in Figure 3-10.

```
import numpy as np
import matplotlib.pyplot as plt

# Creating some x data and plotting a function 0.12x^3 - 0.2x^2
# the y training data to learn from has had random noise injected into it
to provide something to overfit to for high model complexity.
x = np.linspace(-2, 2, 500)
y = 0.12*x**3 -0.2*x**2 + 0.25*np.random.normal(size=len(x))
y_clean = 0.12*x**3 -0.2*x**2

x_test = np.linspace(-2, 2, 50)

def plot_poly_regression(x_test, degree=2):
    # Fitting the polynomial regression model based on the degree (number
of parameters)
    m = np.poly1d(np.polyfit(x, y, degree))
    y_test = m(x_test)
    y_line = m(x)

    # Plotting results
    plt.scatter(x, y, c='lightblue')
    plt.plot(x, y_line, c='black', label='Predicted Function')
    plt.plot(x, y_clean, c='blue', label='True Function')
    plt.scatter(x_test, y_test, c='red')
    plt.title(f"Polynomial Regression Degree = {degree}")
    plt.legend()
    plt.show()

# Underfitting
plot_poly_regression(x_test, 2)

# Good fit
plot_poly_regression(x_test, 8)

# Overfitting
plot_poly_regression(x_test, 64)
```

Figure 3-10. *Code to fit a polynomial regression model and plot it*

Figure 3-11 illustrates the model of degree 2, which is underfitting.

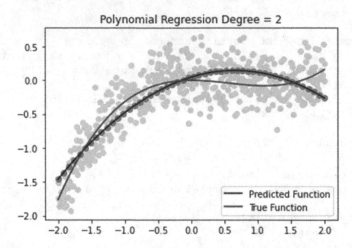

Figure 3-11. *Predictions of a model that is underfitting. It does not have enough parameters to adequately approximate the true function*

Figure 3-12 illustrates the model with degree 8, which has a good fit.

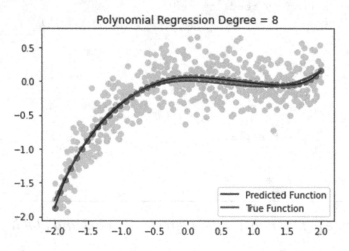

Figure 3-12. *The model has a much better fit due to a sufficient number of parameters. It is able to estimate the true function a lot better than the previous model*

Figure 3-13 illustrates a model with degree 64 that is overfitting. Notice how the predicted function is trying so hard to hit every sampled data point (highlighted as red dots) that it fails to best approximate the true function. This is because it is so influenced by the added noise, it attempts to include as much of the noisy data as possible into its model.

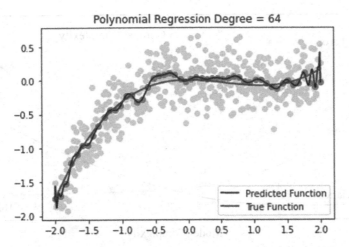

Figure 3-13. *A model with far too many parameters that is severely overfitting. The predicted function curves so much because it is trying to accommodate all the noisy outliers and fails to best approximate the true function*

We have previously referred to generalization a few times. Once again, **generalization** of a model is when we see similar performance and errors across the training and testing datasets. So, how can we understand the generalization capability of the model by viewing its performance in terms of bias and variance?

Let's look at the following scenarios and what they mean:

1. **Low bias, low variance**: The ideal setting. The model performs well on training and testing data.

2. **Low bias, high variance**: Overfitting. The model performs well on training data but not on testing data. Consider implementing strategies to reduce overfitting.

3. **High bias, low variance**: Underfitting. The model performs poorly on the training data, and somehow has low variance in its test data predictions. That means the model is really bad, like for example a model that predicts the same value no matter what the input is.

4. **High bias, high variance**: Underfitting. The model performs poorly on the training and testing data. A worst-case scenario of this is a model that predicts all random values.

Visually put together, observe the bias-variance tradeoff in Figure 3-14.

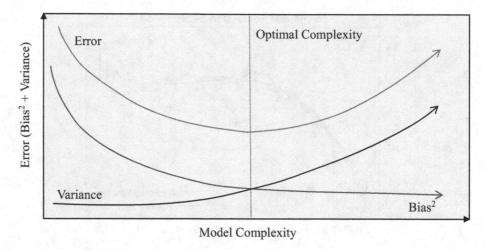

Figure 3-14. A graph depicting the Bias² , the variance, and the total error as model complexity increases. A similar curve exists for deep learning to find the optimal point to maximize generalization capability. The optimal complexity minimizes bias and variance both

As we saw earlier, the model complexity can vastly affect how well the model fits the data (bias), which can subsequently affect how it generalizes to new data (variance). Overall, this bias-variance tradeoff is likely the biggest component of machine learning model life cycles that you will struggle with, and it is this tradeoff that is critical to balance to produce good models.

Hyperparameter Tuning

Once you have a good-fitting model, you may seek to improve its performance. You can do this by adjusting the model's **hyperparameters**, which are high-level parameters that control the modeling behavior of the model as well as its other internal mechanisms. Examples for deep learning may include adjusting the number of layers, adjusting the learning rate of the model (concepts we will cover in Chapter 5), and more. For machine learning models like linear regression or polynomial regression, they can include the model degree, as we saw earlier.

To find the optimal hyperparameters, you will have to just search through possible configurations of the hyperparameter and find what fits best. This can be done manually, which is very tedious, or through a script, which can take a lot of time and computational resources but can find good hyperparameter settings for you.

A process known as **grid search** can help you determine what hyperparameters you should pick. Grid search is an algorithmic framework that lets you iterate through a custom range of possible hyperparameter values, finish modeling and evaluating with that hyperparameter setting, and store the relevant metrics. By the end of the overall search, you can easily pull up the stored metrics per hyperparameter and find out where you should search next.

A good strategy is to first cast a wide net encompassing a larger range of hyperparameters, find the region of values that performs better, and conduct a narrower search. Figure 3-15 illustrates this wide-to-narrow approach on the polynomial regression example from earlier, assuming that degree = 8 is the most optimal fit.

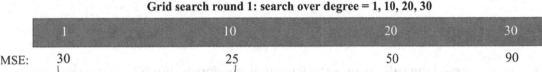

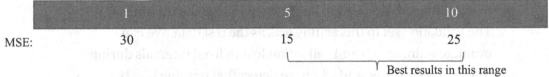

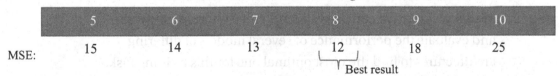

Figure 3-15. *Showing three rounds of grid search. The numbers in the blue box are specific hyperparameter settings that we searched over. The MSE values below the box are the corresponding obtained MSE loss values for using the parameter setting above it. As you can imagine, this is a computationally intensive process, which is why we need to be efficient about how we conduct grid search*

With more complex models and many more combinations of hyperparameters (if you search multiple hyperparameter settings for different hyperparameters, you will have a cartesian product) to search over, this can blow up in complexity and computational intensity very easily. This is why it is important to be efficient about your search ranges.

Furthermore, you do not want to make the mistake of training on the entire training dataset again and again. Past certain ranges, you will be waiting forever to make progress. This is why the **validation set** exists—it allows for quicker hyperparameter tuning.

We will now go over some validation set strategies.

Validation

We have discussed the **validation set** a few times previously, but it is time to go over what you might use validation sets for:

1. **Avoid overfitting**: In deep learning, we train neural networks on the training dataset many times over. This can easily lead to overfitting, so we need to know when to stop the training. The validation set in this setting acts as the test data. We can evaluate training loss and validation loss in fixed intervals during the training process, and when we detect that training loss is decreasing while validation loss is increasing (low bias, high variance), we can stop the training immediately.

2. **Model picking**: We can use validation sets to thoroughly test and evaluate the performance of several models of differing architectures to find the most optimal one for this training task.

3. **Hyperparameter tuning**: We can use the validation set during grid search to find optimal hyperparameter settings.

So how can we go about doing any of this? There are a couple popular strategies we can employ:

1. **Hold-out validation**: A validation set is derived from the training set and used for whatever purposes we need.

2. **K-fold cross-validation**: A more robust strategy where the entire
    dataset is partitioned into K separate bins. K rounds of training
    and evaluation are performed, and in each round of training, a
    unique bin serves as the evaluation dataset.

Hold-out validation is a quick and simple strategy, but it is subject to randomness
in the sense that a particular random split may result in validation data that is much
easier to predict on than usual. It might not have the typical noise profile of the training
data, for example. Most of the time, your situation may not be this drastic, but it is not a
thoroughly experimentally sound strategy.

K-fold cross-validation is a very thorough and robust strategy because it takes your
entire dataset and randomly partitions it into K different "folds" or bins. K rounds of
training are performed, where each unique bin of the K total bins serves as the testing/
evaluation split, while the rest of the bins are combined to form the training data. After
one round, the next bin (which previously has not served as testing data) is chosen, and
the process repeats.

All-in-all, you will have repeated the training-testing process K times. As you can
imagine, this is a computationally intensive strategy, but it is experimentally sound and
will ensure that your validation results are accurate and not subject to influences by
randomness. However, consider that if you combine this approach with grid search, you
not only will wait a long time for the results to finish, but also will be spending a lot of
time actively training the models. On large models that require specialized hardware to
train on, you want to be conservative about how long your models are training.

That being said, it is possible to combine the two strategies. You can apply K-fold
cross-validation on a specific random split of the full dataset for far more rapid
prototyping. This random split will not encompass the entire dataset you are given, but
you should make it sufficiently large.

Scikit-learn also provides easy functionality to implement K-fold cross-validation.
Combined with the data science principles you learned in Chapter 2, you can conduct
various experiments, interpret the results, and make adjustments to your approach as
needed, whether it is because of hyperparameter tuning, model selection, or anything
else that requires comparisons between two different models.

Summary

The fundamentals of machine learning that you have learned in this chapter should aid you in understanding the intuition behind models in the rest of this book and beyond. They apply to traditional machine learning modeling approaches as well as to deep learning modeling approaches. In Chapter 4, we will explore how to implement machine learning algorithms on datasets and perform anomaly detection.

CHAPTER 4

Traditional Machine Learning Algorithms

This chapter introduces you to the Isolation Forest and the One-Class Support Vector Machine algorithms and walks you through how to use them for anomaly detection. In the process, you will also practice incorporating the fundamental machine learning workflow and incorporating hyperparameter tuning using the validation set.

In a nutshell, this chapter covers the following topics:

- Isolation Forest

- One-Class Support Vector Machine

Note Code examples are provided in Python 3.8. The code repository for this book is available at `https://github.com/apress/beginning-anomaly-detection-python-deep-learning-2e/tree/master`.

The repository also includes a requirements.txt file to check your packages and their versions.

Code examples for this chapter are available in two locations. For the Isolation Forest algorithm, go to `https://github.com/apress/beginning-anomaly-detection-python-deep-learning-2e/blob/master/Chapter%204%20Traditional%20Machine%20Learning%20Algorithms/chapter4_isolationforest.ipynb`.

For the One-Class Support Vector Machine, go to `https://github.com/apress/beginning-anomaly-detection-python-deep-learning-2e/blob/master/Chapter%204%20Traditional%20Machine%20Learning%20Algorithms/chapter4_ocsvm.ipynb`.

© Suman Kalyan Adari, Sridhar Alla 2024

Navigate to "Chapter 4 Traditional Machine Learning Algorithms" and then click chapter4_isolationforest.ipynb and chapter4_ocsvm.ipynb. The code is provided as .py files as well, though it is the exported version of the notebook.

We will be using JupyterLab to present all of the code examples.

Traditional Machine Learning Algorithms

"Traditional" machine learning algorithms are non-neural network architectures like k-nearest neighbors, support vector machines, random forests, decision trees, and so on. These models are still quite popular, especially in forms such as XGBoost (gradient boosted trees), which remain quite competitive with neural networks on structured, tabular data. Other traditional popular algorithms that remain in use include clustering algorithms and trees. Though deep learning models may receive the spotlight, traditional machine learning algorithms have not fallen out of favor and still have good uses, especially in the cases we discussed in Chapter 3.

In particular, we are going to cover the Isolation Forest, One-Class Support Vector Machine (OC-SVM) and how we can perform anomaly detection with them.

Isolation Forest

Isolation Forest is an unsupervised learning algorithm that attempts to discover the underlying structure of the dataset and isolate outliers. It is a collection of individual tree structures that recursively partition the dataset according to its attributes. In each iteration of the process, a random feature is selected, and the data is split based on a randomly chosen value between the minimum and maximum of the chosen feature.

This process is repeated until the entire dataset is partitioned, forming an individual tree in the forest. Anomalies generally form much shorter paths from the root than normal data points since they are much more easily isolated. We can find the anomaly score by using a function of the data point involving the average path length, which is something scikit-learn provides with its implementation of Isolation Forest.

Example of an Isolation Forest

To understand better what an Isolation Forest does, let's look at an imaginary scenario. At a large lake, an irresponsible fish breeder has bred and released a mutated version of a native fish species. The following aspects differentiate the mutant species from the native species:

- Longer average length

- Larger average circumference

- Longer tail fin in proportion to its body length

This gives us three measurements we can take for each fish to distinguish the invasive, mutant species from the native species. This is akin to having three columns per data point in the X dataset.

For a visual representation of how the native and mutant species differ in this scenario, Figure 4-1 depicts what an average specimen of the native fish looks like and Figure 4-2 depicts what an average specimen of the mutant fish looks like.

Figure 4-1. *Example of average native fish species*

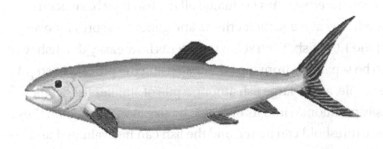

Figure 4-2. *Example of average invasive fish species*

As you can see, on average, the invasive species has a larger circumference, a longer body, and a longer tailfin. However, all of these measurements fall on a distribution, and quite a bit of overlap exists between the two species: some native fish are larger than average for the native species, and some invasive fish are smaller than average for the mutant species. This overlap makes it hard to immediately tell if a fish is invasive or native.

Ecologists working for a local government agency have become concerned that the infestation of the mutant species threatens the survival of native species in the lake. The ecologists have decided to conduct a survey of the fish populations to investigate the extent of the infiltration, and have hired a group of fishermen to assist them. Inspired by statistical methodology, the ecologists have asked the fishermen to catch 1,000 fish in the lake. The ecologists have also trained the fishermen to classify the fish as a native species (normal) or an invasive species (anomaly) based on the measurement of girth, overall length, and tail fin length. Based on the results of their findings, the ecologists can create a rough estimate of the distribution of invasive species to normal species in the lake overall.

On to the evaluations, the fishermen have agreed to the following process and will repeat this for a random split of several hundred fish out of the total 1,000:

1. Each fisherman selects a random subset of features, and the data is partitioned using randomly chosen values between the minimum and maximum of each selected feature. Some may emphasize the length and circumference, while others look at length and tailfin length, etc.

2. Each fisherman uses their intuition to separate fish that immediately seem anomalous. Some fish are harder to classify, requiring more careful scrutiny.

3. Once each fisherman has evaluated all the fish they have been allocated, they average their criteria and generate scores for every one of the 1,000 fish. Each score represents how easily that fish was able to be separated from the other fish using the process in step 1. For example, a very large fish among a pile of smaller fish would be easily partitioned into its own side by size alone: Based on these scores, a threshold can be set, and the fish can be evaluated as a normal or anomalous fish depending on the threshold.

In this example, each individual fisherman represents one isolation tree in the Isolation Forest. Given a random subset of the data, each tree attempts to randomly partition this by some randomly chosen subset of features. This is akin to each fisherman looking at one or two aspects of each fish in a small subset of the total 1,000 and trying to separate them out from each other. For example, a really large fish would clearly be put in a pile away from the rest, while the rest may need to be more carefully measured and sorted by size. The harder the fish are to separate from one another in their sample, the deeper the tree (and the longer the path length).

Once the trees have been formed, they come together to form a forest (ensemble). When the fishermen come together to assign each fish an anomaly score, each fisherman submits their own score based on their own criteria that they learned given their random batch. These anomaly scores are then averaged together to present a final anomaly score for each fish. In an Isolation Forest, each tree submits an anomaly score, which is derived as a function of the path length.

The easier the sample is to separate from the other data, the shorter the path length. The individual scores are averaged together to present an anomaly score that is effectively the average path length. If all of the trees found it easy to isolate a data point, its average path length and anomaly score will be low, thus making it likely to be an anomaly.

And so, for the fishermen, they individually create their own separation criteria, and as a group submit their final anomaly scores for each fish they caught. This is how the Isolation Forest operates.

However, this is not a perfect system; as previously discussed, due to the overlap of smaller than average mutant fish and larger than average native fish, some invasive fish will be misidentified as native fish, and some native fish will be misidentified as invasive fish. These cases, respectively, represent false negatives and false positives.

Anomaly Detection with an Isolation Forest

Let's dive right in and apply an Isolation Forest to a dataset. Once again, the Jupyter notebook for this section is available at the GitHub URL provided at the beginning of this chapter.

We will use the KDDCUP 1999 dataset, which contains an extensive amount of data representing a wide range of intrusion attacks. To simplify things, we will narrow the scope to only data related to HTTP attacks.

You can find the dataset at `https://github.com/apress/beginning-anomaly-detection-python-deep-learning-2e/blob/master/data/kddcup.data.gz`.

Download the kddcup.data.gz file and extract it. You should see a file named kddcup.data.corrected.

Let's get started with the imports. Execute the code shown in Figure 4-3. See Chapter 2 for a description of the imported packages.

```
import numpy as np
import pandas as pd
import matplotlib.pyplot as plt
from sklearn.ensemble import IsolationForest
from sklearn.linear_model import LogisticRegression
from sklearn.model_selection import train_test_split
from sklearn.preprocessing import LabelEncoder
```

***Figure 4-3.** Import statements required*

Once we have the imports, we are ready to load the data, process it, and begin our modeling.

Data Preparation

There are 42 columns that we must define. Execute the code shown in Figure 4-4 to load the dataframe and define the columns for the KDDCUP dataset.

```
columns = ["duration", "protocol_type", "service", "flag",
"src_bytes", "dst_bytes", "land", "wrong_fragment", "urgent",
        "hot", "num_failed_logins", "logged_in",
"num_compromised", "root_shell", "su_attempted", "num_root",
        "num_file_creations", "num_shells",
"num_access_files", "num_outbound_cmds", "is_host_login",
        "is_guest_login", "count", "srv_count", "serror_rate",
"srv_serror_rate", "rerror_rate", "srv_rerror_rate",
        "same_srv_rate", "diff_srv_rate",
"srv_diff_host_rate", "dst_host_count", "dst_host_srv_count",
        "dst_host_same_srv_rate", "dst_host_diff_srv_rate",
"dst_host_same_src_port_rate", "dst_host_srv_diff_host_rate",
        "dst_host_serror_rate", "dst_host_srv_serror_rate",
"dst_host_rerror_rate", "dst_host_srv_rerror_rate", "label"]

df = pd.read_csv("../data/kddcup.data.corrected", sep=",",
names=columns, index_col=None)
```

***Figure 4-4.** Code required to load the dataframe*

It is best to visit the Jupyter notebook on GitHub to get the column list necessary. Once you have the dataframe loaded, feel free to explore its shape and what the various data items in the columns look like. This is a very large dataset, comprised of 4,898,431 rows, but we will be operating only on the subset that involves HTTP attacks.

Figure 4-5 shows you how to filter the dataframe to obtain 623,091 rows.

```
# Filter to only 'http' attacks
df = df[df["service"] == "http"]
df = df.drop("service", axis=1)
```

Figure 4-5. *Code to filter the dataframe to only include HTTP attacks. The row count should be 623,091 rows with 41 columns*

Next, execute the code shown in Figure 4-6.

```
df["label"].value_counts()
```

Figure 4-6. *Getting the value counts of the 'label' column*

You should notice, as shown in Figure 4-7, that there are a lot of different types of anomalous attacks.

```
label
normal.         619046
back.             2203
neptune.          1801
portsweep.          16
ipsweep.            13
satan.               7
phf.                 4
nmap.                1
Name: count, dtype: int64
```

Figure 4-7. *Frequency counts of the different labels. The dataset is very imbalanced and is skewed heavily in favor of "normal"*

The first thing to note is that the dataset is very imbalanced and that there are very few anomalies by proportion. The second thing to note is that these are categorical labels. We need to encode them as a numeric format. To do this, we can simply assign 0 to a normal data point and 1 to an anomalous data point, since our only goal is to differentiate between an anomaly and a normal point. Figure 4-8 shows the code to do this.

```
# Label of 'normal.' becomes 0, and anything else becomes 1
and is treated as an anomaly.
df['label'] = df['label'].apply(lambda x: 0 if x=='normal.'
else 1)
df['label'].value_counts()
```

Figure 4-8. *Applying the numerical encoding and checking values counts*

You should see output like that shown in Figure 4-9.

```
label
0    619046
1      4045
Name: count, dtype: int64
```

Figure 4-9. *The numeric labels and their frequency counts. It is now easy to see by how much the normal data points outnumber the anomalies*

We also need to encode any other categorical columns. To do this, we can iterate through the columns, determine the datatype, and encode it using the label encoder from scikit-learn as necessary. A nice way to map the columns and data types together in a dictionary is captured in Figure 4-10.

```
datatypes = dict(zip(df.dtypes.index, df.dtypes))
```

Figure 4-10. *Creating a dictionary of df columns and their respective datatypes*

The attribute .dtypes returns a pandas Series of datatypes with the columns on the index. If we zip this up as shown in Figure 4-10, we get a dictionary of columns and their respective datatypes.

Run the code in Figure 4-11 next.

```
encoder_map = {}
for col, datatype in datatypes.items():
    if datatype == 'object':
        encoder = LabelEncoder()
        df[col] = encoder.fit_transform(df[col])
        encoder_map[col] = encoder
```

Figure 4-11. *Iterating through the columns of df and, if it is a string object, encoding it and saving the encoder object in case we need it later*

Now that we have dealt with all of the categorical variables and converted everything to numerical equivalents, we can check the correlation matrix. Run the code shown in Figure 4-12. You can explore the rest of the correlation matrix on your own if you would like.

```
# Check the variables with highest correlation with 'label'
df2 = df.copy()
label_corr = df2.corr()['label']
```

Figure 4-12. *Creating a copy of df (so that we have a restore point if needed) and saving the label column of the correlation matrix as its own variable*

We want to keep only the columns that have at least a 0.2 magnitude correlation with label. In other words, if it is at least weakly linearly correlated, we want to include those variables only. The following list provides a general rule of thumb to follow in terms of magnitude:

- **0.0 correlation**: No linear correlation

- **0.2 to 0.4**: Weak linear correlation

- **0.4 to 0.6**: Moderate linear correlation

- **0.6 to 0.8**: Strong linear correlation

- **0.8 to 1.0**: Very strong linear correlation

- **1.0 correlation**: Perfect linear correlation, like the relation $y = mx$, where m is a positive coefficient and all the points lie on a straight line

Correlation can also be negative. Positive correlation means as y increases, x also increases. Negative correlation means as y increases, x decreases. An example of –1.0 correlation is the relation $y = mx$, where m is a negative coefficient. The points all lie on a straight line.

We picked 0.2 as the threshold since practically no linear correlation exists below that point. Ultimately, this is another hyperparameter to tune and experiment with to see how this affects the end performance.

Run the code in Figure 4-13. Feel free to explore label_corr in its entirety.

```
# Filter out anything that has null entry or is not weakly correlated
train_cols = label_corr[(~label_corr.isna()) & (np.abs(label_corr) > 0.2)]
train_cols = list(train_cols[:-1].index)
train_cols
```

Figure 4-13. *Filtering out columns with null entries or no linear correlation. We also index* train_cols[:-1] *to remove label as its own correlation*

As a result of the correlation pruning, we have gone from 42 total columns to just 13, reducing the computational complexity of training our models as there's much less data to ingest for the model. You should see output like that shown in Figure 4-14.

```
['src_bytes',
 'hot',
 'num_compromised',
 'count',
 'serror_rate',
 'srv_serror_rate',
 'same_srv_rate',
 'diff_srv_rate',
 'dst_host_srv_count',
 'dst_host_same_srv_rate',
 'dst_host_diff_srv_rate',
 'dst_host_serror_rate',
 'dst_host_srv_serror_rate']
```

Figure 4-14. *Output of running the code in Figure 4-13. These are the columns with at least a weak linear correlation with 'label'*

You can check the correlations for yourself by doing label_corr[train_cols].

Let's now split our data up into train, test, and validation sets. Run the code in Figure 4-15.

```
labels = df2['label']
# Conduct a train-test split
x_train, x_test, y_train, y_test = train_test_split(df2[train_cols].values,
labels.values, test_size = 0.15, random_state = 42)
```

Figure 4-15. *Splitting the dataset into train and testing splits*

After that, execute the code in Figure 4-16 to create our validation set.

```
# Additional split of training dataset to create validation split
x_train, x_val, y_train, y_val = train_test_split(x_train, y_train,
test_size=0.2, random_state=42)
```

Figure 4-16. *Creating a validation set from the original training split*

Execute the code in Figure 4-17 to print out the shapes of our different sets.

```
print("Shapes")
print(f"x_train:{x_train.shape}\ny_train:{y_train.shape}")
print(f"\nx_val:{x_val.shape}\ny_val:{y_val.shape}")
print(f"\nx_test:{x_test.shape}\ny_test:{y_test.shape}")
```

Figure 4-17. *Code to print out the shapes of all the data splits we have created*

If all went well, it should match the output displayed in Figure 4-18.

```
Shapes
x_train:(423701, 13)
y_train:(423701,)

x_val:(105926, 13)
y_val:(105926,)

x_test:(93464, 13)
y_test:(93464,)
```

Figure 4-18. *The output shapes of the train, val, and test sets*

Training

Finally, we can get started with the Isolation Forest modeling task. Run Figure 4-19 to instantiate our model.

```
# Let's try out isolation forest with stock parameters. It may take a bit
of time to train.
isolation_forest = IsolationForest(random_state=42)
```

Figure 4-19. *With default parameters, instantiating an Isolation Forest model*

The code in Figure 4-20 trains on the training data and generates anomaly scores from the training data.

```
isolation_forest.fit(x_train)
anomaly_scores = isolation_forest.decision_function(x_train)
```

Figure 4-20. *Training the model and predicting anomaly scores*

We can visualize the anomaly scores and get an idea of the distribution of average path lengths. Remember, the Isolation *Forest* contains many individual trees that each have a specific path length, which is why we are averaging them all. We should see a left skew, meaning the tail is long to the left of the graph, while most of the data is distributed toward the right. Figure 4-21 shows how to plot this.

```
plt.figure(figsize=(10, 5))
plt.hist(anomaly_scores, bins=100)
plt.xlabel('Average Path Lengths')
plt.ylabel('Number of Data Points')
plt.show()
```

Figure 4-21. *Code to plot a histogram of the anomaly scores, derived from average path lengths of each sample in the data*

The output should look like the graph shown in Figure 4-22.

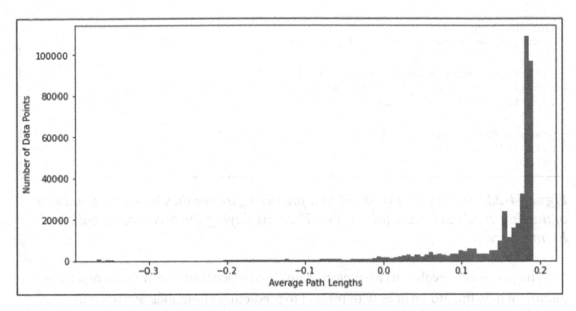

Figure 4-22. *A histogram showing the average path lengths (anomaly scores). Since most of the data is normal, there should be a left skew as we see in the graph. The little clump of outliers far to the left are obvious anomalies*

Given what we see in the graph, let's manually set a threshold. The idea is that any anomaly score below this threshold is counted as an anomaly, while any anomaly score above this threshold is counted as a normal point.

Run the code in Figure 4-23 to do this, as well as check the precision, recall, and F1-score of the resulting predictions. The negative path lengths are a result of the specific scoring algorithm used by Scikit-Learn's implementation of the Isolation Forest. Negative scores represent anomalies, and positive scores represent normal points. As we can clearly see a small clump of outliers to the left of –0.3, we will set a threshold of –0.3 based on this observation. Will this be enough to capture our anomalies? This is a tunable hyperparameter. Feel free to adjust it and notice the resulting change in performance.

```
threshold = -0.3
anomalies = anomaly_scores < threshold

precision = precision_score(y_train, anomalies)
recall = recall_score(y_train, anomalies)
f1 = f1_score(y_train, anomalies)

print(f"Precision: {precision}")
print(f"Recall: {recall}")
print(f"F1-Measure: {f1}")
```

Figure 4-23. *Setting the threshold and predicting anomalies based on what side of the threshold each score falls under. Then displaying the precision, recall, and F1-measure*

The precision, recall, and F1-measure are chosen because they help paint a clear picture of how the model does with respect to predicting anomalies. We want to maximize both precision (the model's anomaly prediction success rate) and the recall (the model's true anomaly capture rate). Only when both are maximized do we have a proper anomaly detector. The output of executing the code in Figure 4-23 is shown in Figure 4-24.

```
Precision: 1.0
Recall: 0.4050179211469534
F1-Measure: 0.576530612244898
```

Figure 4-24. *The model's predictive performance is poor, as indicated by the recall and F1-measure being so low. This is a bad anomaly detector*

Looking at Figure 4-24, the precision is 1.0. This means that we certainly did isolate the anomalies by picking our threshold, except it is clearly not enough, as we only captured ~40% of the anomalies.

This does not mean that the Isolation Forest model is bad, but rather that we now have to embark upon the hyperparameter tuning process to better fit the isolation model to our specific dataset and use case.

Hyperparameter Tuning

Isolation Forest depends quite heavily on the hyperparameter settings, which must be tailored to your specific dataset. It's not like a deep neural network that mostly learns by itself; most of the work with these traditional machine learning algorithms is getting the hyperparameter tuning just right.

With the Isolation Forest, the following are the hyperparameters of interest to tune:

- `max_samples`: How many samples to use to train each tree in the forest

- `max_features`: How many features per data point to use to train each tree in the forest

- `n_estimators`: How many trees should comprise the forest

There is also a feature, **contamination**, that tells the Isolation Forest what proportion of the data is contaminated by anomalies. But for this, we can calculate the proportion of anomalies from our training set and pass it in directly.

As for the method of hyperparameter tuning, we will use holdout validation combined with grid search. Holdout validation, given a sufficient data size, is good for getting ballpark estimates of what direction to move in with respect to hyperparameter adjustments. Holdout validation does a good job of massively boosting predictive capability. However, if you want to be experimentally thoroughly sound and robust, K-fold cross-validation is the way to go if you can afford the computational expenses or if your validation data size is not sufficient enough.

With that being said, let's calculate the `contamination` parameter first, as shown in Figure 4-25.

```
# estimate proportion of anomalies. It's about 0.0065, 0.0066
(y_train==1).sum() / len(y_train)
```

Figure 4-25. *Calculating the proportion of anomalies in the training dataset*

The output should be around 0.00658, which we can round up to 0.0066. Now, let's define one single experimental run as a function to make our code more modular. Refer to Figure 4-26.

```
# Given an isolation_forest instance, x, and y data, train and evaluate and
return results
def experiment(isolation_forest, x, y):
    isolation_forest.fit(x)

    anomaly_scores = isolation_forest.decision_function(x)

    # Using a stock Logistic Regression model to predict labels
    lr = LogisticRegression()
    lr.fit(anomaly_scores.reshape(-1, 1), y)

    preds = lr.predict(anomaly_scores.reshape(-1,1))

    pre = precision_score(y, preds)
    rec = recall_score(y, preds)
    f1 = f1_score(y, preds)
```

Figure 4-26. *Code to perform one experimental trial given an Isolation Forest model and the x and y data. The precision, recall, and F1-measure metrics are calculated and returned as a dictionary*

Notice that we are using a logistic regression model to predict the anomaly scores. This is to avoid manual tuning of the threshold and to get a model to find this threshold by itself. The logistic regression model takes in the input features and essentially serves as a threshold function, where if the threshold is passed, the output is 1, otherwise it is 0. The intention here is to have the model learn the threshold automatically and use that learned threshold to predict anomaly labels for future data.

Now let's start by tuning the max_samples parameter. We don't really know where to start, so let's begin with powers of two covering 256 and going up to 4096. Make sure to read the documentation to find out the ballpark you should start in when conducting your grid search. You don't want to be so far off that movement in any direction doesn't give you much gain.

Execute the code in Figure 4-27. It may take a bit of time.

```
# Perform experimental search for max_samples
validation_results = {}
max_samples = [2**f for f in [8, 9, 10, 11, 12]]
for max_sample in max_samples:

    # We are fixing the n_estimators to 50 to be quicker. n_jobs = -1 lets
us train on all cores
    isolation_forest = IsolationForest(n_estimators=50,
                                       max_samples = max_sample,
                                       n_jobs=-1,
                                       contamination = 0.0066,
                                       random_state=42)

    res = experiment(isolation_forest, x_val, y_val)
    validation_results[max_sample] = res
```

Figure 4-27. *Grid search over* max_sample *sizes from 256 to 4,096 to find out which setting provides the best result on the validation split*

Note that we have switched to using our validation set. This lets us perform faster prototyping than training on the entire training set every time.

If you run the code in Figure 4-28, you should see the results of this grid search displayed as the cell output.

```
# Printing out the results of the validation. The optimal setting is
between 512 and 4096
[(f, validation_results[f]['f1']) for f in validation_results.keys()]
```

Figure 4-28. *Code to navigate through the results dictionary and display the hyperparameter value and the corresponding F1-measure. Since F1-measure reflects both precision and recall, maximizing F1-measure will also maximize both precision and recall*

The output of Figure 4-28 is shown in Figure 4-29.

```
[(256, 0.5956175298804781),
 (512, 0.7676595744680851),
 (1024, 0.9541547277936963),
 (2048, 0.9316596931659693),
 (4096, 0.7769897557131601)]
```

Figure 4-29. *Just looking at* max_samples *= 1024, it looks like our results are almost good to go already, as the F1-measure is 0.954*

As you can see in Figure 4-28, just tuning one hyperparameter has made a world of difference. Let's see if we can keep tuning `max_samples` and arrive at a more precise value that might perform even better.

We will narrow our search to be in between 500 and 2000, since that seems to be the range where the F1-scores peak and start to reduce again.

Run the code in Figure 4-30.

```
# Repeat validation with a narrower range
validation_results = {}
max_samples = range(500, 2200, 200)
for max_sample in max_samples:

    # We are fixing the n_estimators to 50 to be quicker. n_jobs = -1 lets
us train on all cores
    isolation_forest = IsolationForest(n_estimators=50,
                                       max_samples = max_sample,
                                       n_jobs=-1,
                                       contamination = 0.0066,
                                       random_state=42)

    res = experiment(isolation_forest, x_val, y_val)
    validation_results[max_sample] = res
```

Figure 4-30. *Repeating hyperparameter grid search with a narrower range of*
max_samples candidate values

Now rerun the code in Figure 4-28 by either rerunning that cell or pasting the same code in a new cell. You should see output like that shown in Figure 4-31.

```
[(500, 0.714031971580817),
 (700, 0.7365771812080537),
 (900, 0.7617477328936522),
 (1100, 0.9549678341672624),
 (1300, 0.9338028169014084),
 (1500, 0.9550321199143469),
 (1700, 0.9323098394975576),
 (1900, 0.8123076923076923),
 (2100, 0.8586156111929308)]
```

Figure 4-31. *Output of running the code in Figure 4-28 after running the code in*
Figure 4-30

From Figure 4-31, we can see that the optimal range is somewhere within the range of 900–1100, and again from 1300–1700.

Let's repeat the tuning process one more time with a narrower range and see how much higher we can get with the F1-score. Execute the code in Figure 4-32.

```
# Repeat validation with a narrower range
validation_results = {}
max_samples = list(range(950, 1110, 10)) + list(range(1300, 1800, 100))
for max_sample in max_samples:

    # We are fixing the n_estimators to 50 to be quicker. n_jobs = -1 lets
us train on all cores
    isolation_forest = IsolationForest(n_estimators=50,
                                       max_samples = max_sample,
                                       n_jobs=-1,
                                       contamination = 0.0066,
                                       random_state=42)

    res = experiment(isolation_forest, x_val, y_val)
    validation_results[max_sample] = res
```

Figure 4-32. *Repeating the hyperparameter search with a more focused range of candidate values*

Once done, rerun the code in Figure 4-28. You should notice that the parameter setting of 950 had the best F1-score by far at 0.963. We can keep going with the hyperparameter tuning or accept the values we are seeing. After some point, we will see diminishing gains, so keep that in mind in the future. From our results, it seems that the setting of 950 actually performed the best somehow, so we will use this setting as we go forward with tuning the other hyperparameters.

Next, we will tune max_features. Execute the code in Figure 4-33.

```
# Tuning max_features
validation_results = {}
max_features = range(5, 8, 1)
for max_feature in max_features:
    max_feature = max_feature / 10.0

    # We are fixing the n_estimators to 50 to be quicker. n_jobs = -1 lets
us train on all cores
    isolation_forest = IsolationForest(n_estimators=50,
                                       max_samples = 950,
                                       max_features = max_feature,
                                       n_jobs=-1,
                                       contamination = 0.0066,
                                       random_state=42)


    res = experiment(isolation_forest, x_val, y_val)
    validation_results[max_feature] = res
```

Figure 4-33. *The parameter* max_features *is a fraction, which is why we divide by ten. We are searching from 0.5 to 0.7 in a 0.1 interval*

Once done, rerun the code in Figure 4-28. You should see that the max_features setting of 0.6 has given us an F1-score of 0.972.

We can try another round of hyperparameter tuning and see if a narrower range nets us more gains. Let's search from 0.55 to 0.70 with a 0.01 interval. Run the code in Figure 4-34.

```
# Tuning max_features with a narrower range
validation_results = {}
max_features = range(55, 71, 1)
for max_feature in max_features:
    max_feature = max_feature / 100.0

    # We are fixing the n_estimators to 50 to be quicker. n_jobs = -1 lets
us train on all cores
    isolation_forest = IsolationForest(n_estimators=50,
                                       max_samples = 950,
                                       max_features = max_feature,
                                       n_jobs=-1,
                                       contamination = 0.0066,
                                       random_state=42)


    res = experiment(isolation_forest, x_val, y_val)
    validation_results[max_feature] = res
```

Figure 4-34. *Trying a much narrower range of 0.55 to 0.70 with 0.01 intervals for* max_features

Once done, execute the code in Figure 4-28 again. You should see that 0.55 through 0.61 are the same values of approximately 0.9722, after which the scores drop. Let's just pick max_features = 0.55 and tune the final hyperparameter, n_estimators. The reason we chose max_features = 0.55 is just to pick the smallest hyperparameter setting that still netted us the most performance. There is a principle called Occam's razor which, when applied to machine learning, is where performance being equal, picking the simpler machine learning model is likely to be better.

This choice is consistent with the Occam's razor principle, which, when applied to machine learning models with relatively equal performance, means that the simplest machine learning model is likely the best choice. Picking a smaller ratio of random features to isolate by and still getting the same performance means it might be better to keep things simpler and pick the lowest value of max_features given the best performance. It's not a hard-and-fast rule, though.

Now execute the code in Figure 4-35 to conduct a broad search of n_estimators candidate values.

```
# Tuning max_features with a narrower range
validation_results = {}
n_estimators = range(10, 110, 10)
for n_estimator in n_estimators:

    # We are fixing the n_estimators to 50 to be quicker. n_jobs = -1 lets
us train on all cores
    isolation_forest = IsolationForest(n_estimators=n_estimator,
                                       max_samples = 950,
                                       max_features = 0.55,
                                       n_jobs=-1,
                                       contamination = 0.0066,
                                       random_state=42)


    res = experiment(isolation_forest, x_val, y_val)
    validation_results[n_estimator] = res
```

Figure 4-35. *Doing a broad search of n_estimators candidate values*

Once done, rerun Figure 4-28 to observe the results. It seems that n_estimators = 10 actually gives us the best F1-score, 0.9799, which is the strongest performance we have seen thus far. And so, with as little as ten estimators in the forest, the Isolation Forest is able to predict anomalies quite well on the validation set. Let's keep all these hyperparameter settings, train on the training data, and evaluate on the test data and see what performance we get.

Evaluation and Summary

With the discovered optimal hyperparameter settings, execute the code in Figure 4-36.

```
# Set our hyperparameters that we discovered during validation and train
isolation_forest = IsolationForest(n_estimators=10,
                                   max_samples = 950,
                                   max_features = 0.55,
                                   n_jobs=-1,
                                   contamination = 0.0066,
                                   random_state=42)

isolation_forest.fit(x_train)
anomaly_scores = isolation_forest.decision_function(x_train)


lr = LogisticRegression()
lr.fit(anomaly_scores.reshape(-1, 1), y_train)

preds = lr.predict(anomaly_scores.reshape(-1,1))

precision = precision_score(y_train, preds)
recall = recall_score(y_train, preds)
f1 = f1_score(y_train, preds)

print(f"Precision: {precision}")
print(f"Recall: {recall}")
```

Figure 4-36. *With all the optimal hyperparameter settings we have discovered, train on the training set and evaluate on the training set again*

You should see output like that shown in Figure 4-37.

```
Precision: 0.974349839686498
Recall: 0.9802867383512545
F1-Measure: 0.9773092728247273
```

Figure 4-37. *That is a pretty good performance on the training data. Let's see if this same performance holds for the test data*

This is good performance, but the predictions are made on the data it trained on. All we know at this point is that the model can at least fit to the training data. But will it generalize to new samples? Figure 4-38 contains the code to predict on the test data.

```
anomaly_scores = isolation_forest.decision_function(x_test)

preds = lr.predict(anomaly_scores.reshape(-1,1))

precision = precision_score(y_test, preds)
recall = recall_score(y_test, preds)
f1 = f1_score(y_test, preds)

print(f"Precision: {precision}")
print(f"Recall: {recall}")
print(f"F1-Measure: {f1}")
```

Figure 4-38. *Using the anomaly scores generated on the test data, and the logistic regression model trained to recognize anomalies given the training data, we make predictions and calculate metrics*

You should see output like that shown in Figure 4-39.

```
Precision: 0.9722703639514731
Recall: 0.9859402460456942
F1-Measure: 0.9790575916230367
```

Figure 4-39. *This is a very strong performance, as we got an F1-score of 0.979*

This is great performance. While we are at it, let's also check the roc_auc_score by running the following:

```
roc_auc_score(y_test, preds)
```

The output result should be a score of 0.9928. With an AUC score that high, and an F1-measure of 0.979, we have a strong anomaly detector. We can also check the confusion matrix. Execute the code in Figure 4-40.

```
conf_mat = confusion_matrix(y_test, preds)
ConfusionMatrixDisplay(conf_mat).plot()
plt.show()
```

Figure 4-40. *Creating a confusion matrix and using scikit-learn's built-in functionality to display it*

You should see something like Figure 4-41.

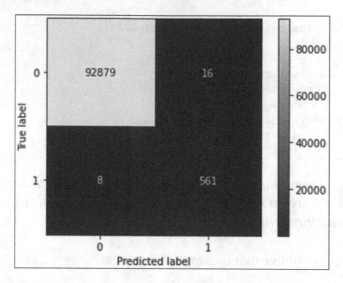

Figure 4-41. *The results are quite good, with 8 false negatives out of 569 anomalies, and 16 false positives out of almost 93,000 normal points*

Hopefully, you now understand how to implement your own Isolation Forest algorithm using scikit-learn, tune it with a holdout-validation and grid search strategy, and use the discovered hyperparameter settings to create a solid anomaly detector.

The Isolation Forest model does have some pros and cons that you should consider before you use it in practice.

Pros:

- **High-dimensional data**: The Isolation Forest can efficiently sample and handle high-dimensional data. The typical "curse of dimensionality" (higher dimension data requiring more data samples for models to learn the task) does not apply as strongly here and only lets the Isolation Forest isolate samples more easily.

- **Scalability**: The Isolation Forest can scale up to much higher data sizes due to its efficient implementation.

- **Anomaly scoring**: The Isolation Forest clearly gives you anomaly scores per sample, allowing you to interpret why a data point might be an anomaly vs. a normal data point as well as how strong of an anomaly a data point may be.

Cons:

- **Sensitive to hyperparameters**: As you saw earlier, the model is very sensitive to what kind of hyperparameters you set. That being said, once tuned, the model gets up to par.

- **Dense points**: When points are really close to each other, it may require more splits (and thus higher path lengths) to part the data. This higher path length may incorrectly imply that an anomaly that was difficult to separate is actually a normal point, when it is not.

- **Imbalanced dataset**: The parameter `contamination` must be set to help the Isolation Forest understand how many anomalies to expect.

- **Interpretability**: Though the anomaly scores are provided, it is hard to understand how a point is decided to be an anomaly or a normal point because of how many trees there are in the forest, how many partitions they've had to make, and so on.

All that being said, the Isolation Forest is a fantastic algorithm if you have large data sizes, high-dimensional data, and are willing to sit through the hyperparameter tuning process and carefully guide parameter selection.

One-Class Support Vector Machine

The **One Class Support Vector Machine (OC-SVM)** is a modified support vector machine model that is well-suited for novelty detection (an example of **semi-supervised anomaly detection**). The idea is that the model trains on normal data and is used to detect anomalies when new data is presented to it. While the OC-SVM might seem best suited to semi-supervised anomaly detection, since training on only one class means it's still "partially labeled" when considering the entire dataset, it can also be used for unsupervised anomaly detection, where you pass all the data, get it to learn an optimal boundary, and predict anomalies that way by what lies outside the boundary.

We will perform semi-supervised anomaly detection on the same KDDCUP 1999 dataset used in the Isolation Forest example.

How Does OC-SVM Work?

Before we dive into the code, let's explore what a support vector machine is, and what makes the OC-SVM different. First, let's visualize some data on a 2D plane, as shown in Figure 4-42.

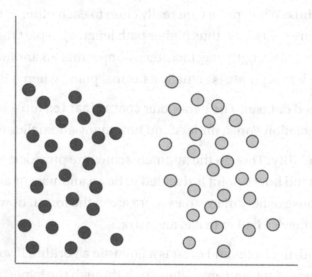

Figure 4-42. *Points in the 2D space that group up in two clusters*

Our goal is to separate the data clumps into two different regions. We can do this by simply drawing a line between the two data clusters, as shown in Figure 4-43.

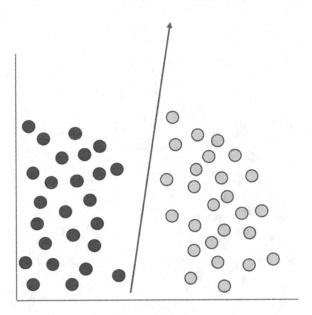

Figure 4-43. *Drawing a line to separate the data clusters into two regions*

When you think about it, there are many ways to draw this line. From a modeling perspective, our goal is to generalize to new data, the distribution of which is similar but can be different to the training data. With the line drawn as appears in Figure 4-43, what happens if we get a point that's still within its region of space but falls on the wrong side of the line?

There is a better way to draw the line and understand it in terms of what are called **support vectors**. They are the whole reason why this model is called a support vector machine, and it's because these support vectors are very influential in determining the decision boundary.

Figure 4-44 depicts what support vectors look like, and what are they exactly? Refer to Figure 4-44.

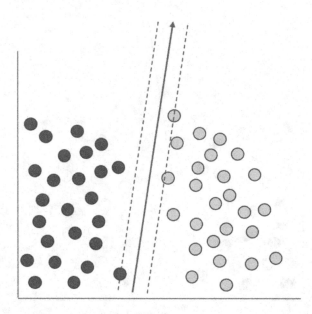

Figure 4-44. *The same line except with support vectors plotted as well*

A **support vector** is a vector parallel to the hyperplane (the line drawn in the middle) that acts as the decision boundary, containing a point that is closest to the hyperplane, and helps establish a margin for the decision boundary. The **hyperplane** is some kind of subspace that is one dimension less than the space it lives in. In the current example, the hyperplane is a line because there are only two dimensions. In 3D space, the hyperplane would be a plane. In four dimensions and beyond, it is just called a "hyperplane."

As shown in Figure 4-44, some space exits between the hyperplane and the support vectors themselves. This is the **margin**, and it is a space that is kept for the sake of accommodating future data points that should belong in their correct regions but stray a bit too far from their distribution. The larger the margin, the more leeway you have in accommodating these straggler data points, allowing you to avoid misclassification.

A more optimal hyperplane margin is shown in Figure 4-45.

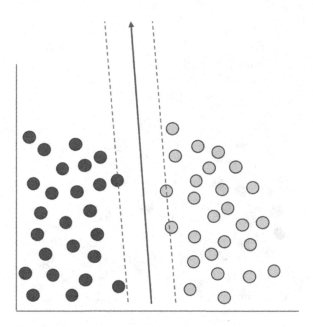

Figure 4-45. *The same data distribution with a more optimal hyperplane and margin thanks to new support vectors*

With how the hyperplane is drawn, the points which their respective support vectors pass through are the closest to the hyperplane. This is a more optimal solution for a hyperplane since the margin for the hyperplane is much larger than in the previous example shown in Figure 4-44.

However, realistically, you will see data distributions that are more like the data distribution shown in Figure 4-46.

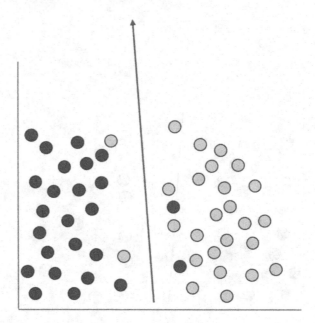

Figure 4-46. *A more realistic, noisy data distribution*

As you can see, there are still two distinct regions where the supermajority of points belong to a certain class. However, some straggler points of the opposite class appear in each of these regions. In this case, the hyperplane drawn in Figure 4-46 is good because it generalizes to new data well. Figure 4-47 presents an alternative possible solution.

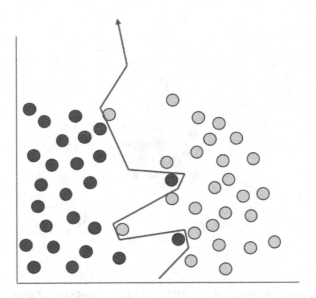

Figure 4-47. *Example of overfitting to the data*

As you can see, we have massively overfit to the data. It has perfect classification performance on this training data, but when generalizing to new data, it will likely perform markedly worse.

The reason the decision boundary is nonlinear here is because support vector machines can transform the data and map it into a different space. In this space, the data may be linearly separable, allowing for a hyperplane to be drawn in this new space. When mapped back to the original space, the decision boundary looks nonlinear.

Figure 4-48 shows an example of how the data distribution itself may not have a straightforward solution in the current space it exists in.

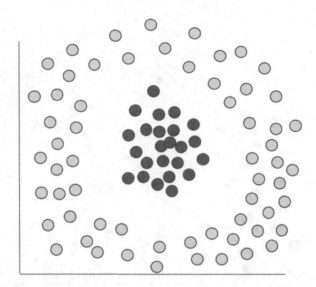

Figure 4-48. *Data distribution that still has clear regions of space where the same class exists within it, but it is not straightforward how to achieve that separation of regions with a linear model*

We cannot draw a line through this data and achieve any good performance. This illustrates the point made earlier about mapping this data into a different space. Assume we used a mathematical function to transform the 2D (x, y) coordinates into 3D space. After applying this transformation, we would see something like Figure 4-49 in the new space.

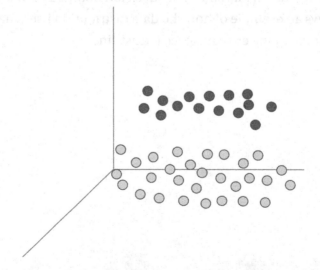

Figure 4-49. *In this new space, we can see that the data is now linearly separable*

Now, there is a clear separation between the two classes, and we can go ahead with separating the data points into two regions, as shown in Figure 4-50.

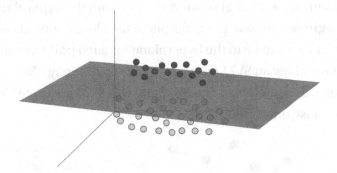

Figure 4-50. *We were able to plot a hyperplane to separate the two regions in this higher-dimensional space*

When we go back to the 2D space, our decision boundary looks like Figure 4-51.

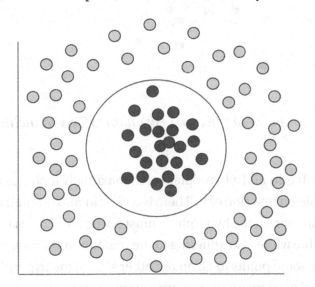

Figure 4-51. *Going back to the original space, our originally linear decision boundary is now nonlinear*

What we just did was use a **kernel** to transform the data into another dimension where there is a clear distinction between the classes of data. This mapping of data is called a **kernel trick**. There are different types of kernels, including the **linear kernel** that we saw in the earlier examples. Other types of kernels include **polynomial kernels**, which map the data to some nth dimension using a polynomial function, and **exponential kernels**, which map the data according to an exponential function.

Finally, the margin is the distance of separation between the closest data point from each class and the hyperplane. As discussed earlier, an **ideal margin**, or **max margin**, involves the maximum separation of support vectors from the hyperplane. A **bad margin** or **suboptimal margin** would have the hyperplane too close to one class or have the distance not be as far as it can be to the hyperplane for each point or support vector.

Now that you know how an SVM operates, what's so different about the OC-SVM? Figure 4-52 illustrates the difference. The SVM and OC-SVM are quite similar but differ in terms of the exact task they perform.

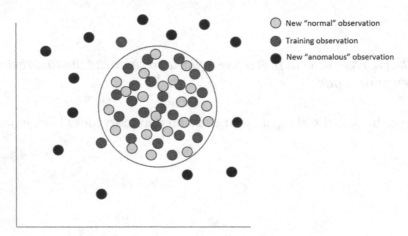

Figure 4-52. An example of prediction task on an OC-SVM including the decision boundary

The OC-SVM's job is to find a hyperplane that maximally includes as many datapoints as possible on one side of it. There is a certain maximum training error that is set as a hyperparameter, so the hyperplane must accommodate as many datapoints on one side as possible while also adhering to the maximum training error allowed and intentionally forcing some points to lie on the other side of the hyperplane.

The trick is to find a balance between hyperparameters so that the OC-SVM maximally contains as many data points as possible while setting the decision boundary in a place where anomalies would clearly fall outside of it. As you will find out shortly, it is a very delicate balance to achieve.

As for the kernel, the OC-SVM commonly uses what is called a **radial basis function (RBF)**, which is a specific type of function mapping that allows the SVM to form nonlinear decision boundaries.

Anomaly Detection with OC-SVM

Once again, we will be using the KDDCUP 1999 dataset. Follow along with the Jupyter notebook available at: https://github.com/apress/beginning-anomaly-detection-python-deep-learning-2e/blob/master/Chapter%204%20Traditional%20Machine%20Learning%20Algorithms/chapter4_ocsvm.ipynb.

Since we covered how to obtain the data in the "Isolation Forest" section, refer to that section if you need to redownload the data.

Let's get right into it by executing the import statements, shown in Figure 4-53.

```
import numpy as np
import pandas as pd
import matplotlib.pyplot as plt
from sklearn.model_selection import train_test_split
from sklearn.preprocessing import LabelEncoder, StandardScaler
from sklearn.svm import OneClassSVM
from sklearn.metrics import precision_score, recall_score, f1_score,
confusion_matrix, ConfusionMatrixDisplay


%matplotlib inline
```

Figure 4-53. *Import statements for the OC-SVM anomaly detection task*

Data Preparation

As in the "Isolation Forest" section, to load the dataframe and define all the columns, execute the code shown in Figure 4-54.

```
columns = ["duration", "protocol_type", "service", "flag", "src_bytes",
"dst_bytes", "land", "wrong_fragment", "urgent",
        "hot", "num_failed_logins", "logged_in", "num_compromised",
"root_shell", "su_attempted", "num_root",
        "num_file_creations", "num_shells", "num_access_files",
"num_outbound_cmds", "is_host_login",
        "is_guest_login", "count", "srv_count", "serror_rate",
"srv_serror_rate", "rerror_rate", "srv_rerror_rate",
        "same_srv_rate", "diff_srv_rate", "srv_diff_host_rate",
"dst_host_count", "dst_host_srv_count",
        "dst_host_same_srv_rate", "dst_host_diff_srv_rate",
"dst_host_same_src_port_rate", "dst_host_srv_diff_host_rate",
        "dst_host_serror_rate", "dst_host_srv_serror_rate",
"dst_host_rerror_rate", "dst_host_srv_rerror_rate", "label"]

df = pd.read_csv("../data/kddcup.data.corrected", sep=",", names=columns,
index_col=None)
```

Figure 4-54. *Defining the columns for the KDDCUP dataset so we can import and reference them properly*

The next few steps are straightforward and repeat what we performed for the Isolation Forest. Execute the code in Figure 4-55.

```
# Filter to only 'http' attacks
df = df[df["service"] == "http"]
df = df.drop("service", axis=1)

# Label of 'normal.' becomes 0, and anything else becomes 1 and is treated
as an anomaly.
df['label'] = df['label'].apply(lambda x: 0 if x=='normal.' else 1)
df['label'].value_counts()
```

Figure 4-55. *Selecting only HTTP data and numerically encoding the label column*

Next, we'll encode the categorical columns. We will also normalize our numerical columns. This is a *very important step* for SVMs as they are very sensitive to data scale! Execute the code shown in Figure 4-56.

```
datatypes = dict(zip(df.dtypes.index, df.dtypes))
encoder_map = {}
for col, datatype in datatypes.items():
    if datatype == 'object':
        encoder = LabelEncoder()
        df[col] = encoder.fit_transform(df[col])
        encoder_map[col] = encoder
    else:
        if col == 'label':
            continue
        scaler = StandardScaler()
        df[col] = scaler.fit_transform(df[col].values.reshape(-1, 1))
        encoder_map[col] = scaler
```

Figure 4-56. *Applying LabelEncoder and StandardScalar where appropriate*

Next, execute the code in Figure 4-57 to derive the training columns and create our training, testing, and validation splits.

```
# Check the variables with highest correlation with 'label'
df2 = df.copy()
label_corr = df2.corr()['label']

# Filter out anything that has null entry or is not weakly correlated
train_cols = label_corr[(~label_corr.isna()) & (np.abs(label_corr) > 0.2)]
train_cols = list(train_cols[:-1].index)
labels = df2['label']

# Conduct a train-test split
x_train, x_test, y_train, y_test = train_test_split(df2[train_cols].values,
labels.values, test_size = 0.15, random_state = 42)

# Additional split of training dataset to create validation split
x_train, x_val, y_train, y_val = train_test_split(x_train, y_train,
test_size=0.2, random_state=42)

print("Shapes")
print(f"x_train:{x_train.shape}\ny_train:{y_train.shape}")
print(f"\nx_val:{x_val.shape}\ny_val:{y_val.shape}")
print(f"\nx_test:{x_test.shape}\ny_test:{y_test.shape}")
```

Figure 4-57. *Creating the train-test-validation splits and printing the output shape*

Once you have executed Figure 4-57, you should see output like that shown earlier in Figure 4-18.

Training

Since the OC-SVM in our example will be training only on normal data (in a semi-supervised setting due to the partial labeling), we need to split the training data into two parts: one comprised of only normal data, intended for training, and another part comprised of normal and anomalous data, intended for testing.

To do that, we will define a function with the code shown in Figure 4-58 to help us with this.

```python
# Split a set into 80% only normal data, and 20% normal data + any
anomalies in set
def split_by_class(x, y):
    # Separate into normal, anomaly
    x_normal = x[y == 0]
    x_anom = x[y==1]

    y_normal = y[y==0]
    y_anom = y[y==1]

    # Split normal into 80-20 split, one for pure training and other for
eval
    x_train_train, x_train_test, y_train_train, y_train_test =
train_test_split(x_normal, y_normal, test_size=0.2, random_state=42)

    # Combine the eval set with the anomalies to test outlier detection
    x_train_test = np.concatenate((x_train_test, x_anom))
    y_train_test = np.concatenate((y_train_test, y_anom))

    # Shuffle the eval set
    random_indices = np.random.choice(list(range(len(x_train_test))),
size=len(x_train_test), replace=False)
    x_train_test = x_train_test[random_indices]
    y_train_test = y_train_test[random_indices]

    return x_train_train, x_train_test, y_train_train, y_train_test
```

Figure 4-58. *A function to help split the data into only normal and a normal + anomaly part*

Let's now split up our training data this way by executing the code in Figure 4-59.

```
### Train on normal data only. The _test splits have normal and anomaly
data both
x_train_train, x_train_test, y_train_train, y_train_test =
split_by_class(x_train, y_train)

print(f"x_train_train: {x_train_train.shape}")
print(f"y_train_train: {y_train_train.shape}")
print(f"x_train_test: {x_train_test.shape}")
print(f"y_train_test: {y_train_test.shape}")
```

Figure 4-59. *Splitting our training set into _train and _test subsets*

You should see output like that shown in Figure 4-60.

```
x_train_train: (336728, 13)
y_train_train: (336728,)
x_train_test: (86973, 13)
y_train_test: (86973,)
```

Figure 4-60. *The new training and testing subsplit sizes. Note, this training data is going to be used only for training purposes*

Before we start training the OC-SVM, it should be known that it does not scale very well with the number of data samples. So, to speed things up, we will use a subset of our training split. Execute Figure 4-61.

```
# nu is a cap on the upper bound of training errors and lower bound of the
fraction of support vectors. We will first try to enter the expected amount
of anomalies
svm = OneClassSVM(nu=0.0065, gamma=0.05)
svm.fit(x_train_train[:50000])
```

Figure 4-61. *Training on the training subsplit with some specific hyperparameter settings (gamma was arbitrarily chosen)*

Now let's make predictions and evaluate it. Note that the OC-SVM predicts

- **1** if the data point is normal (inlier)

- **–1** if the data point is anomalous (outlier)

So we have to map the predictions from either –1, 1 to 0, or 1. To do that, execute the code shown in Figure 4-62.

```
preds = svm.predict(x_train_test)
# -1 is < 0, so it flags as 1. 1 is > 0, so flags as 0.
preds = (preds < 0).astype(int)

pre = precision_score(y_train_test, preds )
rec = recall_score(y_train_test, preds)
f1 = f1_score(y_train_test, preds)

print(f"Precision: {pre}")
print(f"Recall: {rec}")
print(f"F1-Measure: {f1}")
```

Figure 4-62. *Getting predictions and transforming them so that 0 is normal and 1 is an anomaly*

You should see output like that shown in Figure 4-63.

```
Precision: 0.8310991957104558
Recall: 1.0
F1-Measure: 0.9077598828696924
```

Figure 4-63. *Output of executing the code in Figure 4-62*

This score looks really good, but let's see how this does on the testing data by executing the code in Figure 4-64. Judging by how the Isolation Forest went, this likely won't generalize properly.

```
preds = svm.predict(x_test)
preds = (preds < 0).astype(int)
pre = precision_score(y_test, preds )
rec = recall_score(y_test, preds)
f1 = f1_score(y_test, preds)

print(f"Precision: {pre}")
print(f"Recall: {rec}")
print(f"F1-Measure: {f1}")
```

Figure 4-64. *Code to predict on the test set and evaluate the metrics*

Figure 4-65 shows the output of running the code in Figure 4-64.

```
Precision: 0.4809805579036348
Recall: 1.0
F1-Measure: 0.6495433789954338
```

Figure 4-65. *The F1-score dropped quite a bit on the testing split*

As we can see in Figure 4-65, the model most definitely overfit to the training data. Though its recall is perfect, its precision is terrible, indicating that the model went scorched earth and that the decision boundary is not optimal. More specifically, the decision boundary is such that it is allowing too many normal data points to cross to the wrong side of the boundary. This problem was not apparent in the training set, but it is in the test set.

This is fine and to be expected. We still have yet to employ hyperparameter tuning. In particular, the most important hyperparameters to tune are as follows:

- gamma: This parameter is used in the radial basis function (RBF) kernel function. It determines the radius of influence that points have on each other, with smaller gamma translating to larger radii. We want the radius of influence to be such that it is not overfitting but is generalized enough to reliably pick up anomalies in new data. This requires careful tuning of the gamma parameter. If gamma is too large, then the SVM will try and fit the training data more closely, which might lead to overfitting. This makes gamma an important parameter to tune.

- nu: This is an upper bound on the fraction of training errors allowed and a lower bound on the fraction of support vectors. It is a more sensitive hyperparameter than gamma and can influence the output behavior of the model quite a bit, so it is very important to tune this value according to your dataset. You may also need to retune this even if your training data size changes. A very small nu value places a stricter requirement on how many training errors are allowed, while a larger nu value allows for more training errors to slip through when training the model.

Hyperparameter Tuning

Let's define some code for an experimental run, which is captured in Figure 4-66.

```
# Given an svm instance, x, and y data, train and evaluate and return
results
def experiment(svm, x_train, y_train, x_test, y_test):

    # Fit on the training data, predict on the test
    svm.fit(x_train)

    preds = svm.predict(x_test)

    # Predictions are either -1 or 1
    preds = (preds < 0).astype(int)

    pre = precision_score(y_test, preds)
    rec = recall_score(y_test, preds)
    f1 = f1_score(y_test, preds)

    return {'precision': pre, 'recall': rec, 'f1': f1}
```

Figure 4-66. *One experimental run given an instance of an SVM model*

The first parameter we will tune is gamma. We will start with a broad range at first, just arbitrarily chosen. Refer to Figure 4-67.

```
# Perform experimental search for best gamma parameter
validation_results = {}
gamma = [0.005, 0.05, 0.5]
for g in gamma:

    # We are fixing the n_estimators to 50 to be quicker. n_jobs = -1 lets
us train on all cores
    svm = OneClassSVM(nu=0.0065, gamma=g)

    res = experiment(svm, x_train_train[:20000], y_train_train[:20000],
x_val, y_val)
    validation_results[g] = res
```

Figure 4-67. *Grid search for optimal gamma parameters*

Once this is finished, if we run the code in Figure 4-68, we will see the displayed results.

```
# Printing out the results of the validation.
[(f, validation_results[f]['f1']) for f in validation_results.keys()]
```

Figure 4-68. *Code to print out the results of the validation*

From the results, gamma = 0.005 should have given the best F1-measure of 0.639. Let's narrow our search range even more by executing the code in Figure 4-69.

```
# Perform experimental search for gamma with a narrower range. Looks like
smaller gamma is better
validation_results = {}
# Search 1e-5, 5e-5, 1e-4, 1.5e-4, 1e-3, 1.5e-3 and 2e-3
gamma = [1, 5, 10, 15, 20, 100, 150, 200]
for g in gamma:
    g = g / 100000.0

    # We are fixing the n_estimators to 50 to be quicker. n_jobs = -1 lets
us train on all cores
    svm = OneClassSVM(nu=0.0065, gamma=g)

    res = experiment(svm, x_train_train[:20000], y_train_train[:20000],
x_val, y_val)
    validation_results[g] = res
```

Figure 4-69. *Code to run a finer grid search for gamma*

Rerun the code in Figure 4-68 and you should find that gamma = 1e-5 produces the best result of 0.6605 F1-score.

Setting this value for gamma, let's now tune for nu by executing the code in Figure 4-70.

```
# Perform experimental search for nu
validation_results = {}
nu = range(1, 10)
for n in nu:
    n = n / 1000.0

    # We are fixing the n_estimators to 50 to be quicker. n_jobs = -1 lets
us train on all cores
    svm = OneClassSVM(nu=n, gamma=0.00001)

    res = experiment(svm, x_train_train[:20000], y_train_train[:20000],
x_val, y_val)
    validation_results[n] = res
```

Figure 4-70. *Running a broad search over the parameter nu from 0.001 to 0.009*

Rerun Figure 4-68. You will find that nu = 0.001 gave the best result at F1-score. Let's narrow the range of nu some more and run a finer grid search to see if we can improve our performance further. Run the code in Figure 4-71.

```
# Perform experimental search for nu with a finer range
validation_results = {}
nu = range(1, 11)
for n in nu:
    n = n / 10000.0

    # We are fixing the n_estimators to 50 to be quicker. n_jobs = -1 lets
us train on all cores
    svm = OneClassSVM(nu=n, gamma=0.00001)

    res = experiment(svm, x_train_train[:20000], y_train_train[:20000],
x_val, y_val)
    validation_results[n] = res
```

Figure 4-71. *Running a finer grid search for nu*

Once this ends, rerun the code in Figure 4-68. This time, the best result is with nu = 0.0002, with 0.9607 F1-score. Now that we have found the optimal nu setting (any more grid searches would yield diminishing results), let's stick with these settings and train the model.

Evaluation and Summary

Let's now find out how our model performs on the test set. Before that, we must train it with our full training split by executing the code in Figure 4-72. Be aware that when you increase or change the data size, you might need to retune your hyperparameters. In our case, we had to increase gamma from 0.00001 to 0.005.

```
# We increased gamma back to 0.005 as it helped with the fit once we
increased the number of samples.
svm = OneClassSVM(nu=0.0002, gamma=0.005)
svm.fit(x_train_train[:])
```

Figure 4-72. *With the optimal hyperparameter settings, train on the training subsplit*

Now we can evaluate on our test data. Execute Figure 4-73.

```
preds = svm.predict(x_test)
preds = (preds < 0).astype(int)

pre = precision_score(y_test, preds)
rec = recall_score(y_test, preds)
f1 = f1_score(y_test, preds)

print(f"Precision: {pre}")
print(f"Recall: {rec}")
print(f"F1-Measure: {f1}")
```

Figure 4-73. *Evaluating the predictions on the testing split*

The output should look like Figure 4-74.

```
Precision: 0.9709897610921502
Recall: 1.0
F1-Measure: 0.9852813852813853
```

Figure 4-74. *The calculated metrics based on the predictions on the test split*

As we can see in Figure 4-74, the performance is much improved over the untuned model. Also, the F1-measure is really good at 0.985! This is a very strong predictor, as not only did it get every anomaly correct, but its positive predictive power (precision) is 0.971, so there are not a lot of false positives. Let's evaluate the confusion matrix. Refer to Figure 4-75.

```
ConfusionMatrixDisplay(confusion_matrix(y_test, preds)).plot()
```

Figure 4-75. *Code to plot and visualize the confusion matrix*

Figure 4-76 shows the output.

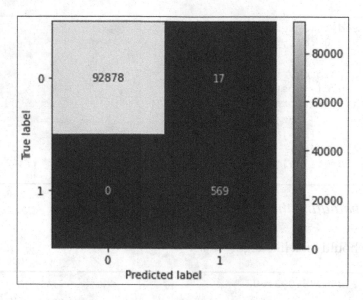

Figure 4-76. *The resulting confusion matrix from the predictions on the test set. There are no false negatives, and very few false positives. This is a very strong result*

You now should have a good understanding of how to implement your own OC-SVM algorithm using scikit-learn, tune it with a holdout-validation and grid search strategy, and use the discovered hyperparameter settings to create a strong anomaly detector.

Like the Isolation Forest model, the OC-SVM model does have some pros and cons that you should consider before you use it in practice.

Pros:

- **High-dimensional data**: The OC-SVM is able to handle higher-dimensional data due to its use of the kernel trick, which maps data into higher dimensions anyway. That being said, it is not immune to the "curse of dimensionality," where you require much more data in higher dimensions to fill out the space adequately for the model to correctly capture a good decision boundary.

- **Unsupervised**: The OC-SVM can be trained in an unsupervised manner. However, if you do have partial labeling, you can use the model in a semi-supervised manner like we did in this section.

- **Nonlinear decision boundaries**: Thanks to kernels like the RBF, the OC-SVM can model nonlinear decision boundaries, as the data is able to first be mapped into a nonlinear, hyperdimensional space before being linearly separated. Back in the original space, the resulting decision boundary becomes nonlinear.

Cons:

- **Sensitive to hyperparameters**: As we experienced, the OC-SVM is very hyperparameter dependent and quite sensitive to these settings. When you increase the number of training samples, the hyperparameter settings may require retuning. However, once it does get going, it's very good.

- **Noisy data**: When the data is noisy, resulting in a bit of an overlap between normal (but noisy) data points and anomalies, the decision boundary becomes hard to model. In these cases, the model will be forced to isolate the noisy normal instances just to target the anomalies.

- **Poor scaling**: The OC-SVM scales quite poorly as the data size increases. There are some ways to get around this, such as by using a kernel mapper to transform the data via an RBF kernel separately, and then using a linear, stochastic gradient descent solver-based SVM (available in sklearn) to arrive at an approximate solution more quickly.

That being said, the OC-SVM is still a very strong algorithm and a great tool to keep in your toolbox. Versions of the OC-SVM, such as the SGD OC-SVM, can offer you alternatives that approximate a similar solution if you are struggling with scaling, but eventually there is still an upper limit for how much data you can train on. So, as long as you properly tune your model and follow the data and any domain knowledge that might help with hyperparameter tuning, the OC-SVM can achieve excellent results depending on the exact nature of the data.

Summary

Having completed this chapter, you now have practiced the fundamental machine learning workflow of data loading, processing, splitting, training, hyperparameter tuning, and evaluation. This same workflow will hold for nearly any kind of machine learning or deep learning modeling task. Beyond that, hopefully you have seen the power that the Isolation Forest and OC-SVM models can offer in terms of anomaly detection, as with the right hyperparameters, they can achieve impressive results.

In Chapter 5, you will be formally introduced to what deep learning is and explore an assortment of techniques to help tune the models for maximizing generalization to new data.

CHAPTER 5

Introduction to Deep Learning

This chapter introduces you to deep learning and all the fundamental, high-level concepts you need to know to implement powerful neural network models. These concepts will apply to the rest of the book and beyond. In the process, you will also implement a simple neural network model in both TensorFlow/Keras and PyTorch to perform supervised anomaly detection and serve as a gateway into learning how to model in these frameworks.

In a nutshell, this chapter covers the following topics:

- What is deep learning?

- The neuron

- Activation functions

- Neural networks

- Loss functions

- Gradient descent and backpropagation

- Loss curve

- Regularization

- Optimizers

- Multilayer perceptron supervised anomaly detection

- Simple neural network: Keras

- Simple neural network: PyTorch

Note   The code examples are provided in Python 3.8. The code repository for this book is available at `https://github.com/apress/beginning-anomaly-detection-python-deep-learning-2e/tree/master`.

The repository also includes a requirements.txt file to check your packages and their versions.

All the notebooks for the rest of this chapter are as follows:

- **Activation Functions**: `https://github.com/apress/beginning-anomaly-detection-python-deep-learning-2e/blob/master/Chapter%205%20Introduction%20to%20Deep%20Learning/chapter5_activation_functions.ipynb`

- **ReLU Nonlinearity with Keras**: `https://github.com/apress/beginning-anomaly-detection-python-deep-learning-2e/blob/master/Chapter%205%20Introduction%20to%20Deep%20Learning/chapter5_relu_keras.ipynb`

- **Gradient Descent**: `https://github.com/apress/beginning-anomaly-detection-python-deep-learning-2e/blob/master/Chapter%205%20Introduction%20to%20Deep%20Learning/chapter5_gradient_descent.ipynb`

- **Overfitting:** `https://github.com/apress/beginning-anomaly-detection-python-deep-learning-2e/blob/master/Chapter%205%20Introduction%20to%20Deep%20Learning/chapter5_overfitting.ipynb`

- **Optimizers**: `https://github.com/apress/beginning-anomaly-detection-python-deep-learning-2e/blob/master/Chapter%205%20Introduction%20to%20Deep%20Learning/chapter5_optimizers.ipynb`

- **Optimizer Comparison**: `https://github.com/apress/beginning-anomaly-detection-python-deep-learning-2e/blob/master/Chapter%205%20Introduction%20to%20Deep%20Learning/chapter5_optimizer_comparison.ipynb`

- **MLP Keras**: https://github.com/apress/beginning-anomaly-detection-python-deep-learning-2e/blob/master/Chapter%205%20Introduction%20to%20Deep%20Learning/chapter5_mlp_keras.ipynb

- **MLP PyTorch**: https://github.com/apress/beginning-anomaly-detection-python-deep-learning-2e/blob/master/Chapter%205%20Introduction%20to%20Deep%20Learning/chapter5_mlp_pytorch.ipynb

These links are also provided in each of the respective sections in which you'll need the notebook. Navigate to "Chapter 5 Introduction to Deep Learning" and then click any of the notebooks you want to try out. The code is provided as .py files as well, though it is the exported version of the notebook.

We will be using JupyterLab to present all of the code examples.

Introduction to Deep Learning

Deep learning is a subset of machine learning that involves the use of neural networks to perform the modeling task. Neural networks can be comprised of hundreds of layers of interconnected neurons, forming a "deep" architecture, hence the term "deep learning."

As described in Chapter 4, "traditional" machine learning algorithms are non-neural network architectures like k-nearest neighbors, support vector machines, random forests, decision trees, and so on. These models are still quite popular, especially in forms such as XGBoost (gradient boosted trees), which remain quite competitive with neural networks on structured, tabular data. Other popular algorithms that remain in use include clustering algorithms and trees. Though deep learning models may receive the spotlight, traditional machine learning algorithms have not fallen out of favor and still have good uses, especially in the cases discussed in Chapter 3.

What Is Deep Learning?

Deep learning is a subfield of machine learning that incorporates the concept of the artificial neuron. What sets deep learning further apart from traditional machine learning in general is its use of many layers of these interconnected artificial neurons to comprise a "deep" network, as opposed to a "shallow" network just a layer or two deep.

The number of layers can range from a handful to up to several hundred. The depth combined with the addition of nonlinearity (covered later in this chapter) enables deep learning models to learn how to perform highly intricate tasks that until recently only humans could perform, such as advanced chatbots, generation of extremely impressive art, and autonomous driving. Most of the impressive groundbreaking technologies widely covered by the media in recent years, such as ChatGPT, deepfakes, voice synthesis, and AI art, are all powered by deep learning. Though deep learning slightly differs from machine learning and is more complex by nature, the basic fundamentals of machine learning all still apply.

In Python, deep learning is implemented primarily with two popular frameworks: PyTorch and TensorFlow. Keras also is popular, but it has been integrated into TensorFlow as of version 2.0. From a beginner's perspective, the TensorFlow 2.0+ Keras module is a great starting point. It abstracts away much of the complexities of the deep learning modeling workflow from you, allowing you to define models in a few lines of code and train and make inferences with just a function each. This feature also makes Keras an excellent tool for rapid prototyping, allowing you to conduct quick experiments or turn to an easier-to-use API when you just need a standard deep learning model for a task.

If you need to delve deeper and define things like custom neural network layers, loss functions, and neural network architectures, or just implement the workflow on a low level, you can use the full functionality of the TensorFlow API beyond Keras.

By contrast, PyTorch is less abstracted than Keras, requiring you to define the model, training loop, and inference loop from scratch. You do not need to reimplement commonly used layers, but PyTorch does not "hold your hand" in the manner that Keras does. However, because of PyTorch's Pythonic API and simpler syntax compared to TensorFlow's API, it remains intuitive and relatively straightforward to use. It is also a favorite among researchers due to its flexibility and ease of use.

Historically, before TensorFlow 2.0 integrated Keras and simplified much of its syntax, TensorFlow was quite syntax heavy and challenging to understand. This posed a barrier both for beginners, since they had a steep learning curve to get started, and for researchers, since they had to deal with all the syntax as well as a harder operating environment in which to debug their models. When PyTorch became available, it quickly grew in popularity. Both frameworks are still quite popular today, though TensorFlow does have a longer history of usage as well as a larger community of support.

Both frameworks are suitable for use in an industrial setting, though TensorFlow might have a slight edge due to its ability to deploy to browsers (via TensorFlow.js) and to mobile devices (TensorFlow Lite). PyTorch is quickly closing the gap with its own deployment capabilities, such as with PyTorch Mobile.

We will cover both PyTorch and Keras in this chapter and provide code for either framework in later chapters on the GitHub repository for those chapters. However, we'll turn to Keras (TensorFlow 2.0+) where applicable due to its brevity.

The Neuron

The artificial neuron is the backbone of the entire deep learning subfield. What exactly is it, and why is it called a neuron? The artificial neuron is loosely derived from its biological counterpart in the brain. A biological neuron's structure is shown in Figure 5-1.

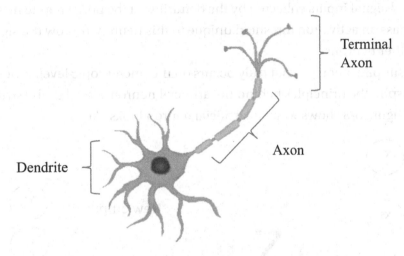

Figure 5-1. *A representation of a biological neuron. The dendrite connects to other terminal axons to form a connection between neurons by which signals can potentially be transmitted*

These neurons form connections between one another where the dendrites seek connections with terminal axons. Once these connections, known as *synapses*, are established, signals can pass through individual neurons based on the input signals it receives from other neurons it connects to. Figure 5-2 shows a connection between two neurons.

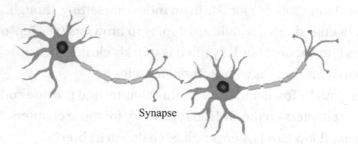

Synapse

Figure 5-2. *A connection between two neurons, known as a synapse*

A synapse is a special junction between two neurons that allows signals to pass through from one neuron to another. Referring to Figure 5-2, when the presynaptic neuron (on the left) fires, it sends a signal outward into the terminal axon, which signal is picked up by the dendrites of postsynaptic neurons (on the right) that have formed synapses with the presynaptic neuron. What determines whether a neuron fires or not? The combined signal inputs collected by the dendrites of the postsynaptic neuron must collectively pass an activation threshold unique to this neuron to allow the signal to pass through to other neurons.

This is a simplification of what truly occurs on the microscopic level, though it is enough to inspire the principles behind the artificial neuron as well as the structures that it builds up. Figure 5-3 shows what an artificial neuron looks like.

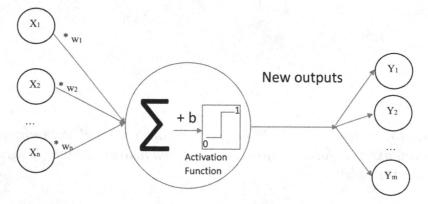

Figure 5-3. *Structure of an artificial neuron. The input, a vector from x_1 to x_n, is multiplied element-wise by a weight vector w_1 to w_n and then summed together. The sum is then offset by a bias term b, and the result passes through an "activation function," which is some mathematical function that delivers an output signal based on the magnitude and sign of the input. An example is a simple step function that outputs 1 if the combined input passes a threshold, or 0 otherwise. These now form the outputs, y_1 to y_m. This y-vector can now serve as the input to another neuron*

As shown in Figure 5-3, the artificial neuron takes in an n-length vector X, multiplies it element-wise by an n-length vector W, adds a bias constant b to each element, and passes it through an activation function. In this case, the activation function is simply 1 if the combined input X•W+b passes a threshold, and 0 otherwise. Here, the • stands for *dot product*. The shorthand version is Wx + b, also known as an *affine transformation* (which is of the general form Ax + b).

Mathematically, as a refresher, Figure 5-4 is what dot-product is.

$$< a, b, d> \bullet < e, f, g > = ae + bf + dg$$

Figure 5-4. *Dot-product between a vector < a, b, d> and a vector < e, f, g>*

Figure 5-5 shows this in a matrix-style format.

$$\begin{bmatrix} a \\ b \\ c \end{bmatrix} \bullet \begin{bmatrix} e \\ f \\ g \end{bmatrix} = ae + bf + cg$$

Figure 5-5. *Figure 5-4 displayed in a different format to resemble matrices*

Mathematically, the entire operation can be summarized as shown in Equation 5-1.

$$y = f\left(\sum_{i=1}^{n} x_i w_i + b \right)$$

Equation 5-1. *Formula representing the artificial neuron's structure. Here, f() is an abstraction for the activation function*

Now that you now know what an artificial neuron looks like and how it operates, let's delve into what activation functions are and why they are used.

Activation Functions

Activation functions are a way to map the input signals into some form of output signal to be interpreted by subsequent neurons. If you do not use an activation function (sometimes called a "linear activation function"), then the output of the affine transformation is just the final output of the neuron.

However, we want to model complex functions that are beyond the scope of linear modeling, so we want to use a nonlinear activation function that maps the inputs into some nonlinear space. Furthermore, most tasks and data are nonlinear, so if we map these inputs into a linear space, we would be severely underfitting. Examples of such nonlinear activation functions include sigmoid and hyperbolic tangent.

Another reason for using nonlinear activation functions is that if we used only a linear activation function, even if we had a deep neural network, it would effectively just become a single layer. A neural network is a sequence of linear transformations. This sequence of linear transformations can itself be rewritten in terms of the original input to be only one linear transformation.

Let's start with the **sigmoid** activation function, a simple way to introduce nonlinearity. If you'd like to know how to generate the figures presented in this section, follow along with the notebook at `https://github.com/apress/beginning-anomaly-detection-python-deep-learning-2e/blob/master/Chapter%205%20Introduction%20to%20Deep%20Learning/chapter5_activation_functions.ipynb`.

The formula for sigmoid is shown in Equation 5-2.

$$f(x) = \frac{1}{1 + e^{-x}}$$

Equation 5-2. *The formula for the sigmoid function*

This is a nonlinear function, meaning that it cannot be adequately modeled by trying to draw a line to hit all of the points. Figure 5-6 shows what the sigmoid graph looks like.

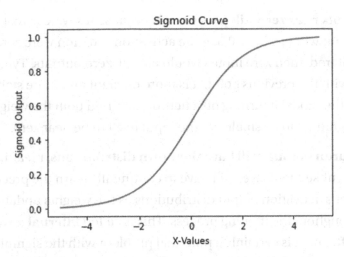

Figure 5-6. *The graph of the sigmoid function. Input values are mapped into a range of (0, 1). The input domain takes any and all x-values, but really large or really small x-values approach 1 or 0, respectively*

As shown in Figure 5-6, the sigmoid activation function maps its inputs into a curved output that ranges from (0, 1), with the function changing most rapidly around its center. Because of this, the sigmoid activation function is well suited for modeling a probability score; for example, "What is the probability of this image being of a dog?"

As a result, sigmoid is appropriate for being used at the very end of the neural network to map the last layer's raw output into a probability score. Sigmoid can also be used as intermediate activation functions to be used within the different non-output layers of the network.

That being said, there are some issues with using sigmoid:

- **Vanishing gradient**: The gradient at the extremes of the graph will be very small. Neural networks update their parameters using the gradient, so tiny gradients would lead to tiny updates, thus slowing down the learning process or even halting it in extreme cases.

- **Not zero-centered**: The output space itself is not zero-centered. This has a few implications for the neural network in general:

  - **Turning "off" neurons** requires negative weights large in magnitude to output to near zero. Weights generally should be smaller in magnitude so that the model is stable. Otherwise, neuron inputs larger in magnitude may either overly activate or overly deactivate the neuron when it shouldn't.

- **Inputs near zero still pass signals** because if x were zero, the output would still be 0.5. If the activation function were zero-centered, then zero inputs would also be zero outputs. Tying in with the previous point, it is more efficient and more stable for the model if turning off a neuron required both the weight's magnitude to be small and the input itself to be near zero.

- **Neuron outputs will have their own distributions**, meaning the subsequent layer will have to continually learn the preceding layer's activation output distributions as the weights update throughout the training process. This is called **internal covariate shift**, and it is certainly a potential problem with the sigmoid activation function.

- **Zig-zag gradient updates**. We update a weight by finding the gradient of the loss function with respect to the same weight. In a neural network with subsequent sigmoid layers, the inputs to each layer will only ever be positive, along with the gradient. And so, the direction of the final gradient step depends entirely on the gradient of the loss function with respect to the output. This leads to a phenomenon where gradient steps tend to zig-zag, leading to inefficient optimization as well as potentially poorer convergence.

  With a zero-centered output space, this issue is largely eliminated because the activation outputs can now be either positive or negative. With the final gradient's sign no longer depending on just the loss gradient with respect to the outputs, gradient steps can be more direct and not oscillate as much.

- **Computationally expensive**: Because of the exponential operation, the sigmoid function itself can be a bit computationally expensive. It may slow down training a bit compared to simpler activation functions (like ReLU, which we will cover shortly).

In practice, sigmoid is not a good activation function to use in your neural network layers, though it is useful to use as a final layer activation function to predict a probability score. You can either have a single output (probability of being an anomaly) or have several output categories that each carry a probability score (e.g., given health metrics, probability of disease X, probability of disease Y).

However, not being zero-centered causes a lot of issues, so let's look at a similar function that is zero-centered: the hyperbolic tangent (**tanh**) function. Its formula is given by Equation 5-3.

$$f(x) = \frac{e^x - e^{-x}}{e^x + e^{-x}}$$

Equation 5-3. *The formula for the hyperbolic tangent function*

The tanh graph is quite similar to the sigmoid graph, but with some differences, as shown in Figure 5-7.

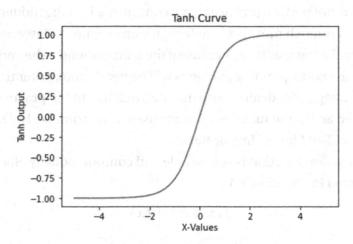

Figure 5-7. *The graph of the tanh function. Input values are mapped into an output range of (–1, 1). Really large or really small x-values approach 1 or –1, respectively*

As you can see in Figure 5-7, the tanh graph is zero-centered and is symmetric about the origin. Compared to sigmoid, tanh carries the following advantages due to being zero-centered:

- **No zig-zag updates**: Gradient updates won't be zig-zagged since activation outputs can now be positive or negative. Subsequent layers now have positive and negative inputs, so layer-wise gradients are no longer only positive.

- **Better distribution**: Being zero-centered means that the distributions of the neuron outputs are more closely centered on a mean of zero, unlike sigmoid. The outputs are now on a more consistent scale compared to sigmoid, though internal covariate shift is still a potential issue. Having a mean potentially near zero means the subsequent layers need not continuously learn an offset, so learning becomes more stable.

- **Inputs near zero have near zero outputs**: Because of this property of tanh, it's far simpler to turn a neuron "off" without needing weights of very large magnitude to transform the input x-values into something very negative just to output something near zero, like in the case of sigmoid.

Though tanh is not perfect, practically speaking, using it as the hidden layer activation function is much more preferable than using sigmoid. However, tanh still has the issue of vanishing gradients, because if the x-inputs tend to be very large in magnitude, then the corresponding gradients will be nearly zero, potentially arresting learning. Some strategies for dealing with this are available, but in practice there are much better-suited activation functions that are used more commonly. One such example is the Rectified Linear Unit, or **ReLU**.

The ReLU activation function is very simple and computationally efficient. Its formula is presented in Equation 5-4.

$$f(x) = \max(0, x)$$

Equation 5-4. *The ReLU activation function. Since there is no exponential function involved, it is very computationally simple as it just involves a direct comparison between x and 0*

The resulting graph of ReLU is shown in Figure 5-8.

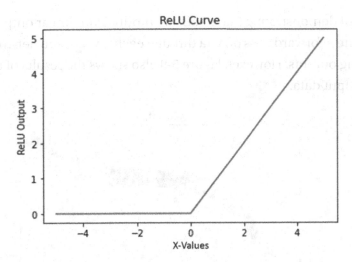

Figure 5-8. *The resulting graph of the ReLU activation function. For x-values above 0, the result is a linear function, but below zero, the output is mapped to zero. ReLU clips the outputs that are negative*

Though it may seem counterintuitive, ReLU also introduces nonlinearities. This has to do with the way it cuts off inputs below a certain threshold, creating pieces of linear sections that can be composed together to form a complex, nonlinear output. When used in tandem with many other ReLU activation functions, with respect to both the breadth (how many neurons in a layer) and depth (how many layers) of the network, the neural network can learn to approximate nonlinear functions in a much more computationally efficient manner than tanh.

To demonstrate this, we will compute a forward pass through a neural network with manually set weights to show how the network, by the end, can piece together a complex, nonlinear function. This function can be used as a decision boundary for classification tasks, or regression for regression tasks. If you would like to follow along with the code, you can find the notebook at `https://github.com/apress/beginning-anomaly-detection-python-deep-learning-2e/blob/master/Chapter%205%20Introduction%20to%20Deep%20Learning/chapter5_relu_keras.ipynb`. You will encounter code that hasn't been formally introduced yet, like the Keras neural network code—Keras modeling is covered later in this chapter.

Let's explore a neural network with two hidden layers that follow the input layer and precede the final layer, the output layer. In total, the network is four layers deep.

For the sake of demonstrating that ReLU can produce nonlinear outputs, we will manually compute a forward pass of data through each layer in the network and observe the corresponding outputs. However, Figure 5-9 also shows the results of a trained network on the input data.

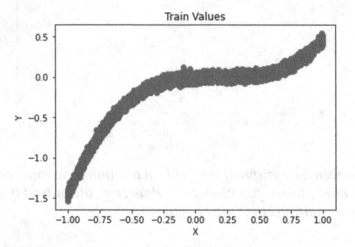

Figure 5-9. *Noisy data that represents the arbitrarily chosen polynomial 0.9x³ – 0.5x² + 0.07x. We added noise to make the task of approximating this function more difficult for a network, as it is forced to generalize and not fit to the noise*

After the first pass, which is summarized by the computation in Equation 5-5, we see output like that shown Figure 5-10.

$$z_1 = x_{train}w_1 + b_1$$

$$ReLU(x) = \max(0,x)$$

$$a_1 = ReLU(z_1)$$

Equation 5-5. *A set of equations that represent the computation for the first pass. Here, z_1 represents the raw output of the first layer, while a_1 represents the output after the ReLU activation function*

Figure 5-10 shows the activation outputs along with the raw outputs.

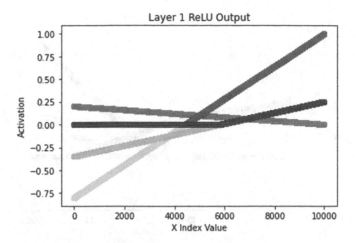

Figure 5-10. *Both the raw layer outputs and the ReLU outputs are shown here. The ReLU outputs are the lines in a darker color. The raw outputs are also graphed here in a lighter color to show how they're clipped off by the ReLU. Anything below zero is set as zero*

As you can see, initially, all of the raw outputs are lines. The ReLU operation sets any value below zero to zero, leaving us with the portions of the lines that are darker. These are the linear snippets that remain, and they're the outputs that become the input to the following layer.

Equation 5-6 shows the computation for the next layer.

$$z_2 = a_1 w_2 + b_2$$

$$a_2 = ReLU(z_2)$$

Equation 5-6. *The computation for the next layer. It takes the post-activation output of the previous layer and uses that as the input*

Recall that a neuron takes a weighted sum of its inputs and offsets it by some bias vector. This means that the raw output for a neuron in the next layer should be some combination of ReLU outputs of the previous layer. Figure 5-11 illustrates this.

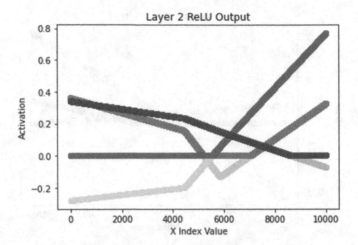

Figure 5-11. *ReLU outputs and raw outputs of the second layer in the network. Notice how it no longer looks linear. If you follow along with the code in the notebook, you can see that layer 2's red output is a weighted combination of 2.3 times layer 1's red output, –0.9 times layer 1's green output, 5.3 times layer 1's blue output, and all of it offset by –0.1*

The output no longer looks linear. Figure 5-12 shows how the third layer outputs look. As a reminder, all of the weights have been manually set just to illustrate how the ReLU function can be composited to create nonlinear outputs.

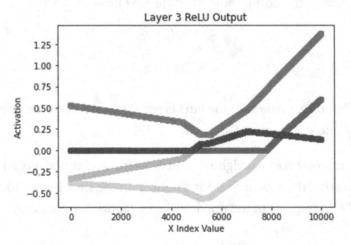

Figure 5-12. *Raw and activation outputs of the third layer*

Finally, Figure 5-13 shows the output layer's raw output. The output layer in this case does not have any activation function.

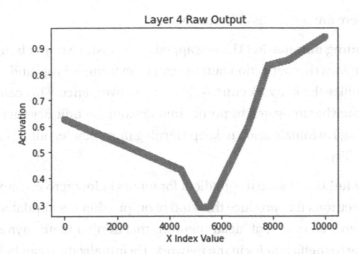

Figure 5-13. *The output of the final layer, which takes three neuron outputs and combines it into one single output*

In a regression context, where the job of the network is to approximate a function given input data, Figure 5-13 shows how it can learn a nonlinear function. Figure 5-13 is not meant to approximate Figure 5-9 and was created as part of an example to show how flexible ReLU can be while remaining very computationally efficient compared to tanh.

This is why ReLU is such a good activation function. Not only is it more computationally efficient to perform the ReLU operation as well as find the gradient, but it is still able to serve as a nonlinearity to allow the network to learn complex functions. It is no wonder, then, that ReLU is commonly used in practical settings. There are some variations of ReLU, like LeakyReLU, that instead of clipping negative values to zero allow a very small signal to pass through. Either way, ReLU is a good baseline to start with when training networks.

Compared to tanh, using ReLU has the following advantages:

- **No gradient traps**. There is no longer the case where gradients on any extreme of the graph become very small, like in tanh. With ReLU, the gradient is simply 1 wherever ReLU was positive and 0 otherwise.

- **Computationally efficient**. There is no exponential to calculate. It's a simple mask that sets anything non-positive to zero.

However, there are some possible drawbacks:

- **Exploding outputs**. ReLU is uncapped in its positive upper bound.
  This means that activation outputs can be extremely large and
  destabilize the network entirely, leading to divergence. This can
  be somewhat mitigated by proper initialization techniques such as
  Kaiming He initialization to keep learning more stable, but it is still a
  possibility.

- **Dying ReLU**. Because the gradient for inputs below zero is simply
  zero, neurons that produce this kind of output don't get updated.
  This can lead to a lot of "dead" neurons that don't update anymore,
  leading to inefficiencies in the network. He initialization can help
  stabilize the network to combat dying ReLU, but it doesn't fully
  solve it.

- **Not zero-centered**. Like with sigmoid, the values and gradient will
  always be positive, meaning there could be zig-zag optimization
  steps taken.

However, modifications of ReLU, such as LeakyReLU, can help because they allow
some negative input through (through max(0.01x, x) instead of max(0, x) like in regular
ReLU), helping mitigate zig-zag gradient steps as well as the dying ReLU problem.

That said, Figure 5-14 shows an example of a simple neural network approximating
the Figure 5-9 true function using the tanh activation function.

Each network was trained for 50 epochs with a batch size of 128. Don't worry if you
don't know what these parameters mean; you simply need to know that both networks
have identical architectures and have been trained under the same setting. The only
difference is that one network uses tanh for its activation function while the other
network uses ReLU.

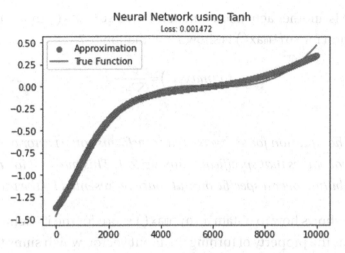

Figure 5-14. *The output approximation of the Figure 5-9 data using tanh as the activation function*

While it's not a perfect fit, it was pretty close. Let's compare this output to a neural network trained using ReLU, shown in Figure 5-15.

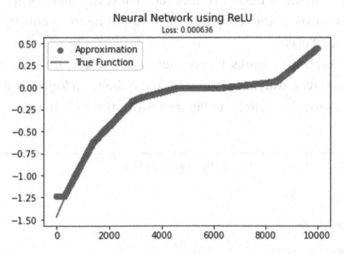

Figure 5-15. *The output approximation of the Figure 5-9 data using ReLU as the activation function. Note how you can see linear sections composed together. It's much rougher than the smooth tanh curves shown in Figure 5-14, but this output is more computationally efficient and quicker to compute*

Finally, there is another activation function called softmax() that is important to know. The equation for softmax() is shown in Equation 5-7.

$$softmax(z_i) = \frac{e^{z_i}}{\sum_j e^{z_j}}$$

Equation 5-7. *The equation for* softmax()*. It transforms some vector of some real values into a vector of real values that specifically sum up to 1. This makes it ideal for creating a probability distribution over a specific output space represented by the vector*

Equation 5-7 shows how to obtain a softmax() score for the ith element of a vector z.

Softmax() has the property of turning the input vector, which sums to some arbitrary number, into an output vector with entries that sum up to 1. This makes the output vector ideal for representing a probability distribution.

In classification tasks, neural networks have an output layer with multiple neurons, each corresponding to a specific class. The first node corresponds to class 0, and the nth neuron corresponds to the n-1 class. Think back to how we categorically encoded strings in Chapter 2—this is exactly the same thing. The output neurons correspond to some categorically encoded class.

To check how softmax() works, let's manually define a vector to represent the raw output of an output layer. You may also come across the term **logits**, which refers to the raw output of the output layer. Refer to Figure 5-16 for the code to run an example of softmax().

```
softmax = lambda z: np.exp(z) / np.exp(z).sum()

z = np.array([-1.2, 2.5, 0.1])

scores = softmax(z)

pred = np.argmax(scores)
print('Raw Output (logits):', z)
print('Probability Scores:', scores)
print('Prediction: ', pred)
```

Figure 5-16. *Code to simulate logits (represented by z),* softmax() *operation to get probability scores, as well as the logic to make a prediction (simply choosing the class with the highest probability)*

As shown in Figure 5-16, the operation argmax() returns the index that holds the highest value in that vector. In terms of softmax() output, it should return the index holding the highest probability score. You should now see something like the output shown in Figure 5-17.

```
Raw Output (logits): [-1.2  2.5  0.1]
Probability Scores: [0.02216479 0.89650602 0.08132919]
Prediction:  1
```

Figure 5-17.   *Output of executing the code in Figure 5-16. The prediction refers to the index of the probability vector with the highest probability value. This index happened to be index 1, corresponding to class 1, whatever that may be*

Softmax() is very heavily used in practice in any classification task that involves the prediction of one specific class given some input. If you want each class to have its own probability distribution, implying that an input can be an example of multiple classes, then you want to use sigmoid over the output layer. This is a task called *multilabel classification*. In this case, whatever class has a probability more than some threshold will be considered to be the prediction.

Now that we have covered the various activation functions, let's dive right into neural networks and their architecture.

Neural Networks

Now that you are familiar with neurons and activation functions, it's time to dig deeper into the layers of neurons alluded to in the previous section.

A **layer** in a neural network is a collection of neurons that each compute some output value using the entire input. The output of a layer is comprised of all the output values computed by the neurons within that layer. A neural network is a sequence of layers of neurons where the output of one layer is the input to the next.

As previously mentioned in the context of ReLU, the first layer of the neural network is the *input layer*, and it takes in the training data as the input. The last layer of the network is the *output layer*, and it outputs values that are used as predictions for whatever task the network is being trained to perform. All layers in between are called *hidden layers*.

Figure 5-18 shows what a neural network looks like in terms of layers.

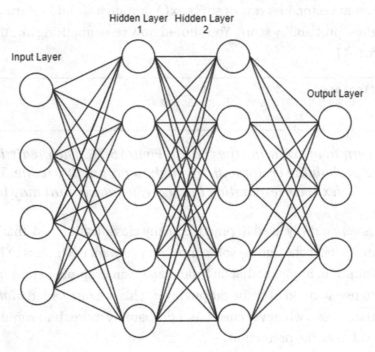

Figure 5-18. *Architecture of a neural network with two hidden layers. Each neuron is fully connected to the previous layer's neurons as well as to the succeeding layer's neurons*

In a mathematical sense, the entire network is a Series of chained matrix multiplication operations with activation functions applied to them. Refer to Equations 5-5 and 5-6 to understand how the layer computations are calculated mathematically. We use the notation z_L to denote layer L's affine transformation output, and a_l to denote the activation function output using z_L as the input. The value a_L is now the input to layer L+1, becoming z_{L+1} and then a_{L+1}.

When data flows from the leftmost layer to the rightmost layer, this is called the **forward pass**. In other words, the forward pass is the process through which the final output is calculated from some given input.

The **backward pass** involves moving from the rightmost layer to the leftmost layer one layer at a time, using the backpropagation algorithm to compute the gradients. **Backpropagation**, short for "backward propagation of errors," computes the gradients with respect to the learnable parameters of each layer by using chain-rule in such a manner as to avoid recomputing all intermediate terms. Since the processes are so interrelated, these two terms are used interchangeably at times. However, the distinction is that the backward pass is strictly the process of going backward through the network layer by layer, whereas backpropagation (used in the backward pass) is the process of computing the gradients.

Once all of the gradients have been computed, the optimization step of updating the learnable parameters takes place. We will go over this later in the chapter in the section "Gradient Descent and Backpropagation."

Let's now look at how the data propagates through the network during the forward pass to result in the output prediction by revisiting Figure 5-18 in the context of classification. Each of the three output nodes, respectively, corresponds to output classes 0, 1, and 2, standing in for "dog," "cat," and "bird." Let's say that the input data, though only four nodes wide in this case, is a highly compressed representation of images of either dogs, cats, or birds.

Assume for the sake of illustration in the following figures that the activation functions used are sigmoid, that the neurons that maximally fire are black (activation output is close to 1), and that the neurons that minimally fire are white (activation output is close to 0). Neurons that fire more strongly will propagate that signal to the next layer, while neurons that fire very weakly will not propagate a signal to the next layer.

As for the weights themselves (represented by the lines [edges] interconnecting neurons between the two layers), edges corresponding with very strong weights are depicted with thicker lines, while edges corresponding to small weights are depicted with thinner lines. How "strong" a weight on an edge is, is in terms of the magnitude. An edge with a thick line can have a weight that is strongly negative or strongly positive.

Remember that a neuron is fully connected with every neuron in the preceding layer and that it will have a different weight assigned to each of these connections. Two neurons are considered to have a strong connection if the activation of the former very likely activates the latter, which is represented by a higher weight on the edge connecting these two neurons. The opposite case is also true, and you will see a lower weight on the edge connecting two neurons where the firing of the former rarely leads to the firing of the latter.

Let's say some arbitrary neuron rarely ever fires except when the input data represents a dog. In this case, let's also assume that the neuron fires very strongly. Then, the weights connecting this neuron to the next layer may likely be large in magnitude because of the significant influence of this neuron's activation on the output prediction.

Unfortunately, how neural networks come to "understand" the input data and interpret it will differ across architectures and even across different instances of the same architecture. Because there are so many connections, explainability of neural networks is extremely poor, and they're often regarded as a black box for this reason. When networks make a decision, we do not know exactly why, only that after a forward pass through a network with up to millions of finely tuned parameters, we got a certain prediction. Compare this to a decision tree that can tell us why it made the decision it did. Furthermore, it is important not to anthropomorphize neural networks, as they're basically large mathematical functions that map inputs to outputs.

With that said, take the following illustrations with a grain of salt—they are meant to be gross simplifications of what truly goes on in a neural network. The reality of understanding how neural networks make decisions is far more complex and is an active area of research.

Figure 5-19 shows the initial data transformation in layer 1, the input layer.

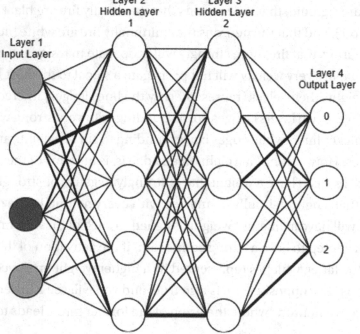

Figure 5-19. *The input data has just been processed by layer 1, the input layer, and the signal has not propagated to layer 2 yet. The edges represented as thicker lines indicate that in the process of training this network, these connections have been reinforced to have higher weights because the activation of the former neuron was important to correctly determine the output class*

Referencing Figure 5-19, we can see in the input layer that one neuron has not activated much, while two neurons have mildly activated (with an activation of around 0.5). One neuron has a higher activation at around 0.8. These signals will propagate to the next layer, and the subsequent neurons will fire based on the weighted combination of the inputs along with the post-sigmoid output.

As shown in Figure 5-20, the next layer's neurons activate according to their individual weights and post-sigmoid activations. The outputs of layer 2 will now be the inputs to layer 3, which will fire depending on their own individual weights and biases, as shown in Figure 5-21.

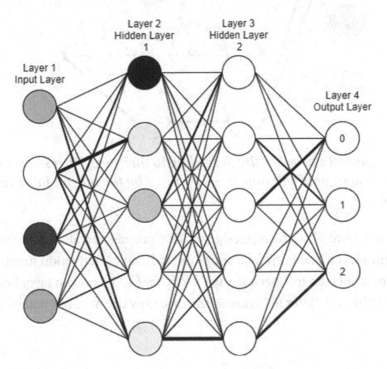

Figure 5-20. *Signals have propagated to the next layer, layer 2, and have fired according to the weighted sum of the inputs (weights can be positive or negative, so a bold edge can be a strongly negative connection) as well as the post-sigmoid activation output*

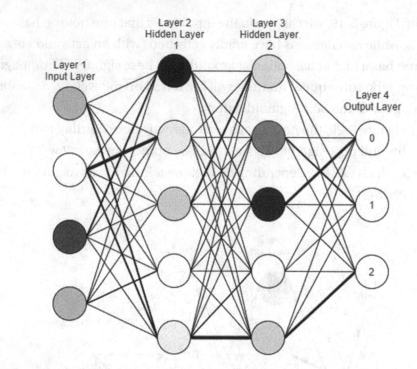

Figure 5-21. *Layer 3's neurons fire according to their weighted sums and sigmoid activations. These signals will now be interpreted by the output layer to generate probability scores*

As shown in Figure 5-21, the input signals have propagated all the way to just before the output layer. The output layer has the softmax() activation function, so its activations should all sum to 1 because, with softmax(), the output layer becomes a probability distribution. Refer to Figure 5-22 to see how the network makes its final predictions.

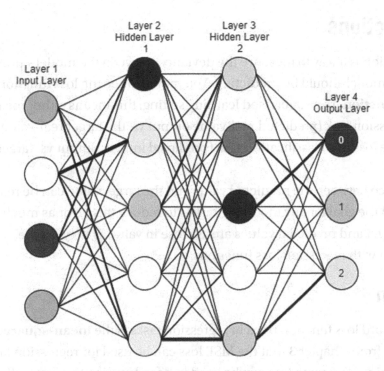

Figure 5-22. *The final predictions are made and the output layer outputs probability scores for each class. The neuron representing class zero has the highest probability of approximately 0.75. The other two neurons have a much lower probability that together sums to 0.25. The way to interpret the prediction made by the network is to take the neuron's index that has the highest probability (by using* argmax()*). In this case, the output neuron at index zero has the highest probability, and so the model's prediction for the input will be class zero*

As Figure 5-22 shows, this is the sequence of activations that resulted in the output prediction of zero given the input. After reversing the categorical encoding, we find that the label associated with the encoded value of zero is "dog."

Now that you have a better idea of how neural networks operate, let's cover the concept of the loss function, which determines how far off the neural network's predictions are compared to the true correct values that we want the network to predict.

Loss Functions

The loss function is a way to measure the deviance between the model's predictions and what the model should be predicting. You may also see the loss function referred to as a **cost function**. In a supervised learning setting, this means either measuring the loss in a regression task (predicted real-value output vs. the target real-value output) or measuring the loss in a classification task (predicted integer output vs. target integer output).

A difference between the predicted value and the target value can be referred to as an "error," and the goal for any modeling task is to reduce this error as much as possible so that the target and predicted values are as close in value as they can be.

Let's begin with regression loss functions.

Regression

A straightforward loss function used in regression tasks is the **mean-squared error (MSE)**. Recall from Chapter 3 that the MSE loss can be used for regression tasks in machine learning. The same loss can be used in deep learning tasks as well. As a reminder, the MSE loss equation is shown in Equation 5-8.

$$MSE = \frac{1}{N}\sum_{i=1}^{N}\left(y_i - \widehat{y}_i\right)^2$$

Equation 5-8. $\widehat{y}_i$ *is the prediction made on the ith data sample, x_i. y_i is the corresponding true output we want the model to target. If the model made perfect predictions, the MSE would be 0*

Also recall from Chapter 3 that other regression loss functions exist, such as **mean absolute error (MAE)** (Equation 5-9), **mean absolute percentage error (MAPE)** (Equation 5-10), and more, though what loss function you choose to use may depend on the nature of the task and the threshold of tolerance for deviances from the true outputs.

$$MAE = \frac{1}{N}\sum_{i=1}^{N}\left|y_i - \widehat{y}_i\right|$$

Equation 5-9. *Mean absolute error finds the average magnitude of deviance of the prediction from the true value*

$$MAPE = \frac{1}{N}\sum_{i=1}^{N}\left|\frac{y_i - \widehat{y_i}}{y_i}\right|$$

Equation 5-10. *Mean absolute percentage error finds the deviance of the prediction from the true value in terms of percentages and averages these out*

Which loss/cost function you should use depends on the exact nature of the task. If you want to penalize prediction errors more harshly, then the MSE loss function is good choice due to the squaring component of the loss. If you don't want the loss to be as sensitive to outliers, then MAE loss may be a good error metric because it measures only the direct magnitude of errors.

If you simply want to view the errors in relative terms, then MAPE may be a good choice, since it measures errors in terms of the target value. This can be more appropriate for time-series forecasting tasks, for example, where magnitude error or squared error differences may incur a higher loss than a percentage error. For example, if you predicted a stock price would be 10,100 but it was actually 10,000, your absolute error would be 100 and your squared error would be 10,000. However, your percentage error would be 0.01 (1% error).

Classification

With classification tasks, you're dealing with discrete integer labels, so it's not as simple as finding the numerical deviation between the target and the prediction. Furthermore, the model outputs real-value probabilities that the input example belongs to a certain class. We want to incorporate this probability prediction into the loss because it provides an indication of the strength of the prediction. Though both 0.71 and 0.99 round up to 1, the latter is a much stronger prediction than the former.

Once again, we can bring back the **cross-entropy loss/log loss** function from Chapter 3. Given a binary classification task (there are only two possible labels, either 0 or 1), we can incorporate an equation like Equation 5-11. Since this loss function only deals with two classes, it can also be referred to as **binary cross-entropy loss**.

$$Log\ Loss = -\frac{1}{N}\sum_{i=1}^{N} y_i * \log\left(\widehat{y_i}\right) + \left(1 - y_i\right) * \log\left(1 - \widehat{y_i}\right)$$

Equation 5-11. *In this loss function, y_i represents the true label (a discrete value either 0 or 1) corresponding to x_i, the ith input value, while $\widehat{y_i}$ represents the probability score output by the model (a real value between 0 and 1) for the input x_i*

However, we can have more than just two classes. For several classes, we want to extend the binary cross-entropy loss into a generalized **categorical cross-entropy loss** function that is capable of dealing with multiple classes.

Recall from Chapter 2 that **one-hot encoding** creates vector of all 0s except for a single 1 at a particular index. This index value where the encoding holds a 1 is the integer class label.

For example, if there were three classes total in a classification task, then the one-hot encoding for class 0 is [1, 0, 0], the one-hot encoding for class 1 is [0, 1, 0], and the one-hot encoding for class 2 is [0, 0, 1].

With this in mind, let's visit the formula for categorical cross-entropy. There will be two summations: the outer one loops through each individual logits (z) and ground truth (y) pair. The logits is of the same dimension as the one-hot encoded y vector. The index where the logits is maximum is the class prediction of the model. Now refer to Equation 5-12.

$$CE = -\frac{1}{N}\sum_{i=1}^{N}\sum_{j=1}^{C} y_{i,j} \log\left(softmax(z_i)_j\right)$$

Equation 5-12. *The categorical cross-entropy loss. Here, y_i is the ith ground truth/target value that corresponds with x_i, the ith input value being passed in. z_i refers to the output logits of the model (no* softmax() *is applied) that was produced by the neural network given x_i. C refers to the total number of classes and is also the length of each y_i vector*

Equation 5-13 shows the categorical cross-entropy loss without one-hot encoded true labels. In this case, the true labels are just an array of integer labels. This is a simplified version of Equation 5-12 that does not require additional one-hot encoding processing of the ground truths.

$$CE = -\frac{1}{N}\sum_{i=1}^{N}\log\left(softmax\left(z_i\right)_{y_i}\right)$$

Equation 5-13. *A simpler, equivalent version of Equation 5-12. All that is required is the score at the index of the y-label of the logits after* softmax()

This is equivalent to Equation 5-12 because, looking at it closer, the one-hot encoding of the y vector sets the summation terms to 0 anywhere other than the index containing 1, so these terms do not contribute to the loss function anyway.

Now that you know some of the basic loss functions that we will use in regression and classification tasks, let's get into how exactly neural networks update the weights and biases using the loss and calculated gradients.

Gradient Descent and Backpropagation

Recall that neural networks update their weights using gradients, which are calculated through backpropagation during the backward pass. So, what are these gradients, and how exactly are they computed? To follow the code used to generate the graphs in this section, refer to the notebook at `https://github.com/apress/beginning-anomaly-detection-python-deep-learning-2e/blob/master/Chapter%205%20Introduction%20to%20Deep%20Learning/chapter5_gradient_descent.ipynb`.

If you remember from calculus class, the **derivative** is the instantaneous rate of change of a function at a specific point with respect to a single variable. Take a simple graph like the parabola with its equation of y=x². Its derivative is simply $\frac{dy}{dx} = 2x$. Refer to Figure 5-23 to see this graphically.

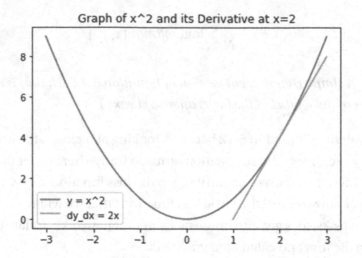

Figure 5-23. *The graph of y=x² and its derivative function, $\frac{dy}{dx} = 2x$. The instantaneous slope at x=y points in the direction of the maximum since the slope is 4. Thus, if we want to find the minimum, we must simply go in the opposite direction, or –4. This would lead us to the minimum at x=0*

The slope has an interesting property in that it always points toward the direction on the original graph that would increase the y-value. For example, at x=2, the instantaneous slope (value of the derivative) is 4. This means that at x=2, at that instant, the y-value is increasing by 4 for a 1 increase in x. As we keep increasing x, we will find that y keeps increasing as well.

If we look at a different graph, y = x³ – 5x², captured in Figure 5-24, we can see another example of how the derivative points toward a direction that would maximize the y-value.

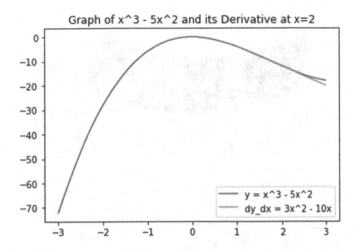

Figure 5-24. *The equation $y = x^3 - 5x^2$ and its derivative, $\frac{dy}{dx} = 3x^2 - 10x$ at x=2. As we can see, the derivative at this instant is –8. If we traveled in this direction (left), we would see an increase in the y-value until it hits the local maximum at x=0*

This time, our derivative at x=2 is –8. This can be interpreted as either an instantaneous rate of change of y decreasing by 8 for every increase of 1 in x, or y increasing by 8 for every decrease of 1 in x. Remember that slope is change in y over change in x, so the negative can apply to either the numerator or the denominator.

In this case, the derivative "points" at the local maximum of the graph at x=0. This also happens to be the global maximum of the graph.

This definition works fine in a two-dimensional setting, but machine learning problems tend to deal with variables that can go up to even hundreds of dimensions. In a three-dimensional setting and beyond, we can extend the concept of a derivative and call it the **gradient**, which instead of a scalar value, is now a vector. For $y = x^2 + z^2$, the

gradient would be $\nabla y = \left(\frac{\partial y}{\partial x}, \frac{\partial y}{\partial z} \right)$, where the ∂ symbol refers to the partial derivative.

The partial derivative is just the derivative of y, which is multivariate, with respect to just one of the variables. Now, as we can see in Figure 5-25 (3D example of $y = x^2 + z^2$), the same principles apply as we saw with the derivative in the 2D example.

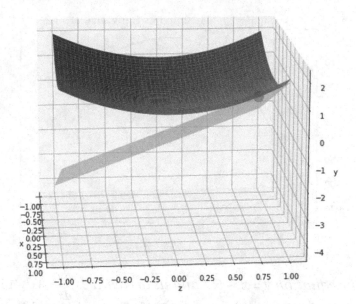

Figure 5-25. *A three-dimensional graph depicting the y function as well as a point at x=0.75, z=0.75, and the tangent plane*

Since the gradient is just the higher-dimensional extension of the derivative, it stands to reason that the direction opposite of the gradient should point toward the minimum. Let's assume that we have initialized our x and z values to be 0.75 both (where the red dot is located in Figure 5-25). Hypothetically, if we wanted to find the x and z values that would lead to the minimum y-value in this graph (x=0, z=0), then starting at x=0.75 and z=0.75, we should take a step in the opposite direction of the gradient. Not too large a step, but a small one so that we don't overshoot the minimum. We can repeat this process of taking small steps many times until we get close to the minimum to find the values of x and z that effectively "minimize" y.

We can formulate this general "optimization step" as follows, where the gradient is with respect to x:

$$x = x - step\_size * gradient$$

The **step_size** parameter controls how large of an update step to take at each iteration. In this case, this step size can also be referred to as the **learning rate**. We will cover more details about the learning rate in the "Optimizers" section.

Having discussed the foundational theory, let's see if we can apply this to a simple y=mx+b line and iteratively solve for m and b given some data.

First, let's set up the problem. We want to approximate the line y = 4x + 3 using the approximation $\hat{y}$ = mx+b with m = 1 and b = 0 as the initial starting values. The loss we will be using is MSE. We need to find the derivative of the MSE loss function with respect to m and with respect to b.

Let's go over how we can calculate this. First, we must begin with what the MSE loss function is defined to be. We will generalize the loss function to L(y, $\hat{y}$), as shown in Equation 5-14.

$$L(y,\hat{y}) = \frac{\sum\limits_{i}^{N}(y_i - \hat{y}_i)^2}{N}$$

Equation 5-14. *The equation for MSE loss referred to as L(y, $\hat{y}$)*

The loss function in Equation 5-14 will serve as a feedback mechanism for how "bad" our predictions are using m and b. To achieve perfect predictions, we'd need to minimize the loss function so that, in the best case, it equals zero. Intuitively, what exactly does this mean? We can visualize the loss curve by taking a grid of m, b values, computing the corresponding loss, and plotting it. Refer to Figure 5-26.

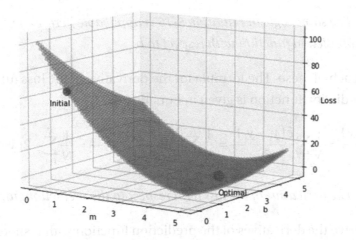

Figure 5-26. *The loss surface of our problem setup visualized with the initial starting point on this surface given by m = 1 and b = 0, and the optimal target point given by m = 4 and b = 3*

As we can see in Figure 5-26, the parameter settings for m and for b are quite far off from what is optimal. If we took the gradient of the loss with respect to m and b, and took a step in the opposite direction of the gradient, m and b would be updated to be closer to the optimal m and b.

Now what would the gradient even look like? And how do we find the gradient of the loss function with respect to our variables? First, let's use chain-rule to help us out. The gradients of the loss function with respect to m and b are given in Equation 5-15 and Equation 5-16.

$$\frac{\partial L(y,\hat{y})}{\partial m} = \frac{\partial L(y,\hat{y})}{\partial \hat{y}} * \frac{\partial \hat{y}}{\partial m}$$

Equation 5-15. *The derivative of the loss function with respect to m can be obtained through chain-rule. We first find the derivative of the loss with respect to our predictions, then get the derivative of our prediction function with respect to m*

$$\frac{\partial L(y,\hat{y})}{\partial b} = \frac{\partial L(y,\hat{y})}{\partial \hat{y}} * \frac{\partial \hat{y}}{\partial b}$$

Equation 5-16. *The same logic in Equation 5-15 is applicable here, except we take the derivative of the prediction function with respect to b*

Let's derive each of these. The formula for the derivative of the loss function with respect to the prediction function is given in Equation 5-17.

$$\frac{\partial L(y,\hat{y})}{\partial \hat{y}} = \frac{1}{N}\sum_{i=1}^{N}\frac{\partial(y_i - \hat{y}_i)^2}{\partial \hat{y}_i} = \frac{1}{N}\sum_{i=1}^{N}\frac{2(y_i - \hat{y}_i)*-\partial\hat{y}_i}{\partial \hat{y}_i} = \frac{1}{N}\sum_{i=1}^{N}-2(y_i - \hat{y}_i)$$

Equation 5-17. *The derivative of the loss with respect to the prediction function*

Next, let's derive the derivatives of the prediction function with respect to m and b. Refer to Equations 5-18 and 5-19.

$$\frac{\partial \hat{y}}{\partial m} = \frac{\partial(mx + b)}{\partial m} = x$$

Equation 5-18. *When you have several variables but need to take the derivative with respect to only one, treat the other variables as a constant. This leaves us with a simple result of just x*

$$\frac{\partial \hat{y}}{\partial b} = \frac{\partial (mx+b)}{\partial b} = 1$$

Equation 5-19. *Similarly, we achieve a simple result of just 1 by following the logic in Equation 5-18 but taking the derivative with respect to b this time*

Putting it all together, we achieve Equations 5-20 and 5-21.

$$\frac{\partial L(y,\hat{y})}{\partial m} = \frac{1}{N}\sum_{i=1}^{N} -2(y_i - \hat{y}_i) * x$$

Equation 5-20. *Plugging in the earlier derivations, we get the final derivative equation for the derivative of the loss with respect to m*

$$\frac{\partial L(y,\hat{y})}{\partial b} = \frac{1}{N}\sum_{i=1}^{N} -2(y_i - \hat{y}_i) * 1$$

Equation 5-21. *Similarly, we obtain the derivative of the loss with respect to b*

This type of derivation is the same principle that powers neural networks. If you have many nested layers, you simply need to apply chain-rule until you can get the derivative that you need. This is how we can obtain all of the gradients that we need. With this in mind, let's minimize the loss curve using simple **gradient descent**. Gradient descent is the process of iteratively updating some learnable parameters by taking small steps in the opposite direction of computed gradients with respect to these parameters. Our update steps for m and b are shown in Equation 5-22 and Equation 5-23, respectively.

$$m = m - step\_size * \frac{\partial L(y,\hat{y})}{\partial m}$$

Equation 5-22. *The update step to update m toward the optimal m*

$$b = b - step\_size * \frac{\partial L(y,\hat{y})}{b}$$

Equation 5-23. *The update step to update b toward the optimal b*

With our starting values of m=1 and b = 0, our initial predictions look like Figure 5-27.

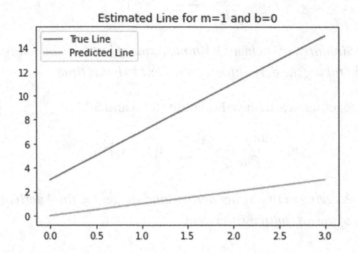

Figure 5-27. *A comparison between the true line we want to approximate and the predicted line using m=1 and b=0*

With a step size of 0.001 and 1,000 iterations (both arbitrarily chosen), we will run gradient descent and plot the results. Be sure to follow along with the code to find out how to implement gradient descent from scratch as well as visualize all these plots. After running this, we obtained values for m and b to be approximately m=4.34 and b = 2.28 after rounding. Refer to Figure 5-28 to see the predicted line this time.

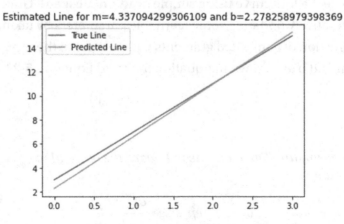

Figure 5-28. *Plotting the approximation of the true line after running gradient descent. It is clear that while it's not the most optimal, the learned parameters are pretty close to the true values of m=4 and b=3*

Let's now look at how the learned m and b parameters appear on the loss curve. Refer to Figure 5-29.

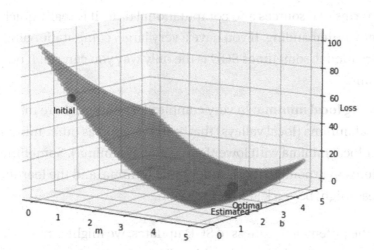

Figure 5-29. *The estimated parameters of m and b result in a loss that is very close to the optimal loss. Gradient descent worked, even if the resulting parameters are not fully optimal*

From Figure 5-29, we can gather that gradient descent is a simple yet powerful mechanism to allow us to learn parameters to approximate various functions. It is this same premise that powers deep learning models, even extremely capable models like ChatGPT.

There are some variations of gradient descent to be aware of:

- **Stochastic gradient descent**: Performing a gradient update after computing the gradient for each individual data sample

- **Mini-batch gradient descent**: Performing a gradient update after computing the averaged gradients for a mini-batch of N data samples

- **Batch gradient descent:** Performing a gradient update after computing the average gradient for the entire batch of data

There are certain observations to be made for each.

Stochastic gradient descent (SGD) can be thought of as having a mini-batch size of 1, although mini-batch gradient descent is sometimes referred to as SGD. The pros and cons of stochastic gradient descent are list next.

Pros:

- **Computationally efficient**: You only need a sample size of one, so in terms of resources and computational time, it is really quick and not as demanding. If you have a very large, computationally hungry model, sometimes SGD is the only way you can run it on your machine.

- **Escaping local minima**: In very complex loss curves, there may be local minima (local valleys) that aren't as good as other minima (other local minima with lower loss or global minima). Sometimes, gradients may point to these suboptimal minima, and the learning process might be stuck at a bad solution.

  With the potential noisiness of SGD updates, we might arrive at a parameter setting that escapes this valley, allowing the new gradients to point toward a deeper/more optimal minimum. This way, the learning process eventually learns a better set of parameters that results in a lower loss than if learning stopped at the earlier, shallower minimum.

Cons:

- **Unstable**: Updating variables based on feedback from just one sample might introduce a lot of noisiness and lead to suboptimal optimization steps.

  Assume for a second you're a tourist in Italy and need to find the Leaning Tower of Pisa. The true location is somewhere northwest of you. You plan to keep asking tourists after traversing some arbitrary distance (taking a gradient descent step). You ask four people: three point north (correct direction), while the other points southeast. The directions received from these people represent individual gradient directions toward the minimum. If you followed SGD, you'd take three "steps" north, and then take a step southeast. However, if you used mini-batch or even batched gradient descent, you'd consider the average direction, north.

- **Poorer convergence**: Because of the potentially noisy and unstable nature of SGD optimization, you might have trouble converging more closely to the optimal solution; the parameter values might keep oscillating around the minimum on the loss curve due to the noisy updates.

Mini-batch gradient descent considers mini-batches (commonly just referred to as having a batch size of N) of N data samples at a time when performing gradient updates. This helps to average out the noise and allows for more accurate update steps toward the minimum.

Pros:

- **More stable**: Averaging out the noise helps to take more accurate update steps and potentially avoid unintended oscillation due to noise.

- **Computationally efficient**: Due to how neural networks are constructed, we can easily take advantage of the matrix form to perform the computations on batches of data at a time, improving throughput and taking advantage of the GPU's parallel computation architectures. This allows mini-batch GD to be efficient and fast if you have the hardware.

Con:

- **Batch-size hyperparameter**: The mini-batch size becomes a hyperparameter to tune. Too small, and you may make the optimization more susceptible to the noisiness of the data. Too large, and you might average out the gradients too much, potentially leading to the parameters getting stuck in some local minimum.

Mini-batch gradient descent is usually the optimal learning strategy when incorporating gradient descent learning, and it is by far the most prevalent in practice. It can lead to better convergence to more optimal solutions, as SGD is too vulnerable to noise and tends to oscillate too much, while batch gradient descent dulls out the signal too much and may lead to suboptimal solutions that are unable to escape local minima.

With a better, more thorough understanding of gradient descent and the mechanism behind the training process, let's move on to how we can interpret the loss curves of neural networks.

Loss Curve

Loss curves can tell us a lot about a model's training process as well as its ability to generalize to new data. Recall that we can split the data into training, validation, and testing sets. The training set is obviously what we train the model on. More specifically, it is what we pass through the network, compute the loss and perform gradient descent with.

One **epoch** represents one full pass of the entire training dataset through the network. Regardless of what batch size you use (1 being SGD, the full number of training samples, and N being batch gradient descent, with mini-batch gradient descent using a batch size anywhere in between), once the model has seen every sample of data in the training set, you have completed one epoch of training.

The validation set can serve as a blind "test" set that you evaluate your neural network on after one epoch of training. Really, it is up to the implementation how the validation set is used, but ideally, it will give you a good indication of the training process and what the model's generalization capabilities look like.

The test set is something you use for model evaluation after your training process has been completed.

The loss curve shows the current training loss over the number of iterations, optimization steps, or epochs, depending on how you display it. The goal is to simply display the loss as the training progresses.

For the code used to generate the graphs in this section, be sure to refer to the accompanying notebook at https://github.com/apress/beginning-anomaly-detection-python-deep-learning-2e/blob/master/Chapter%205%20Introduction%20 to%20Deep%20Learning/chapter5_overfitting.ipynb.

Figure 5-30 shows an example training and validation loss curve that was obtained during the training process. A convolutional neural network was trained on the MNIST dataset for 64 epochs (to make overfitting more likely to occur, a higher number of epochs was chosen).

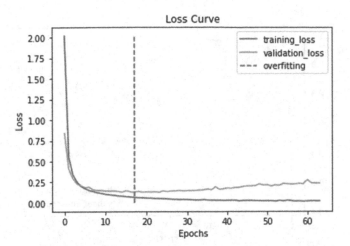

Figure 5-30. *An example of a neural network loss curve showing both the training loss and the validation loss. When the validation loss reaches its lowest point and starts ticking upward again, you have crossed the line into overfitting, which is shown by the red dashed vertical line on epoch 17*

As you can see in Figure 5-30, both training and validation loss start off high and then decline over time. At some point, both training loss and validation loss are at their lowest, which is the best time to stop the training of the network so that you can ensure the model doesn't continue to overfit but instead generalizes as best as it can.

The accuracy curve, plotting the training accuracy and validation accuracy throughout the training process, is shown in Figure 5-31.

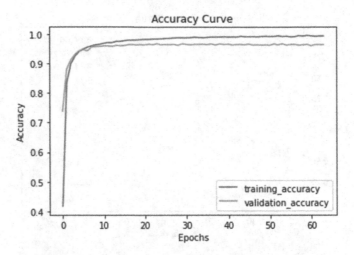

Figure 5-31. *Graph showing the training accuracy against the validation accuracy. Gains in validation accuracy stop after some point, while training accuracy keeps getting better. This is yet another sign of overfitting*

If the training is allowed to continue, the training loss decreases, but the validation loss starts to increase. This is a clear sign the model is overfitting. If we were to pick an epoch to stop at, we could simply take the epoch with the lowest validation loss, which happens to be epoch 17 in this particular training run.

If you rerun the code for this section, your epoch where overfitting begins may differ, but you should see similar results to the graphs shown here. The graph in Figure 5-30 shows the approximate point at which the validation loss starts to tick upward again. The graph in Figure 5-30 shows a similar trend, where we get a disparity between the training accuracy and validation accuracy as we keep training the model because the training accuracy keeps improving while the validation accuracy either stagnates or very slightly starts to decrease.

So how can we combat overfitting? We can employ a bunch of regularization techniques combined with early stopping to ensure we get the model to generalize as best as it can.

Regularization

There are a variety of techniques that we can use to combat overfitting:

- **L1/L2 penalty**: We can apply a penalty by using either the L1 norm (sum of absolute value of weights) or the L2 norm (sum of squares of the weights), incentivizing them to become smaller in magnitude to minimize the penalty.

- **Dropout**: Given the probability of retaining a node, p, we randomly deactivate certain nodes within a layer and prevent them from passing a signal to the next layer. This process repeats with each forward pass, so no neuron is guaranteed to stay active across successive forward passes.

    These deactivated neurons also do not contribute toward the backward pass, so they do not take part in the training process whatsoever if they are dropped. Since the following layer can no longer rely on any one neuron to be present during the forward pass, it must learn to form connections with as many other neurons as it can. This helps avoid overfitting and promotes generalization because single neurons will no longer dominate the connections with the following layer and such connections cannot take place as easily.

- **Early stopping**: As discussed in the previous section on loss curves, early stopping halts training at the point of best generalization performance (using the blind validation data that the model never trains on) so that the model won't have overfit on the data. To simply stop overfitting just as it starts to occur is a form of regularization to ensure your model generalizes.

- **More training data**: Sometimes, your training data cannot sufficiently capture the possible input space compared to the test data or any other new data. In these cases, adding more, diverse training samples can help the model properly learn the task and generalize better.

As an example of this, if you trained a model only on clear images of dogs and cats but then evaluated the model on a test dataset that also included partially obscured images, the model would not generalize well, and would have overfit to the training data. Adding images with occlusion in the training data can help the model properly learn the task and generalize better to the new data.

You can use any combination of these techniques to help your model generalize well to new data, although you don't necessarily want to go too hard on the regularization because that might actually lead to worse performance overall.

A model that is overregularized (such as too harsh of an L1/L2 penalty combined with too high a dropout rate) can fail to learn the training task properly, consequently doing worse on the testing set since training went poorly.

Both Keras and PyTorch allow you to implement these techniques. The parameter regularizers are something supported by the frameworks themselves as you construct the neural network architecture. We will go over this in the examples of a supervised neural network later this chapter.

Dropout is also supported by both frameworks. Early stopping is easily supported in Keras, though in PyTorch, you will have to manually code this as part of the training loop. We will implement this in our PyTorch example.

Optimizers

There are many optimizer techniques that go beyond simple gradient descent. We previously covered stochastic gradient descent, but we can make many modification to SGD to improve the optimization process and help the neural networks converge to better solutions.

In the context of optimization, **convergence** means to iteratively arrive at an optimal set of weights that minimizes loss to a certain extent. We cannot find the globally optimal set of weights to find the globally optimal minimum, because iterative techniques won't ever exactly arrive at the best solution, but they can get very close to it.

So, when a model converges to a solution, it has arrived at some local minimum on the loss curve, but not necessarily the best global minimum. The opposite scenario is when a model **diverges** or starts to balloon up the loss instead of minimizing it. Diverging models are a sign of instability in your network or training setup and may indicate poor hyperparameter settings (too high a learning rate can lead to large optimization steps).

Covering the optimization methods would get deeply mathematical very quickly, so we will just go over the high-level intuition behind each optimizer as well as its pros and cons. If you are interested in the code to generate the visualizations in this section, refer to the notebook at `https://github.com/apress/beginning-anomaly-detection-python-deep-learning-2e/blob/master/Chapter%205%20Introduction%20to%20Deep%20Learning/chapter5_optimizers.ipynb`.

Let's formally go over the generalized SGD algorithm. Before we get into the code itself, let's go over the variables and functions:

- `compute_gradient()`: An abstract function that takes the loss, the data, and the learnable parameters of the model and returns the corresponding gradients.

- `parameters`: A list of the weights and biases of some arbitrary neural network.

- `gradients`: A list of the gradients of the weights and biases in the variable `parameters`. The gradients correspond one-to-one to the parameters as the order is preserved.

- `lr`: The step size used in the gradient update.

Figure 5-24 demonstrates the variations of the SGD algorithm on the function in shown in Equation 5-24.

$$y = \sin\frac{(4\pi x)}{3.75x} + (x-1)^4$$

Equation 5-24. *A variation of the Gramacy & Lee (2012) function (Source: `https://www.sfu.ca/~ssurjano/grlee12.html`)*

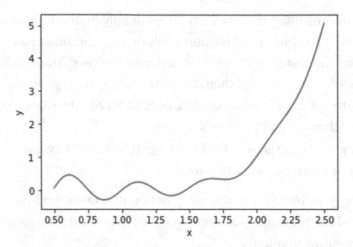

Figure 5-32. *The function we will apply SGD and its variations on. This function can be thought of as the loss function we are trying to minimize by finding the optimal x. There are a couple local minima in the graph that must be overcome in order to reach the global minima (lowest point on the graph). A poor optimizer will get stuck in a local minima that is suboptimal, which would result in the model achieving subpar performance since it would not minimize the loss as well as it could*

Refer to Figure 5-33 to see the template code for SGD's update algorithm.

```
for batch in training_data:

    gradients = compute_gradient(loss, batch, parameters)
    for p in range(len(parameters)):
        parameters[p] = parameters[p] - lr * gradients[p]
```

Figure 5-33. *SGD training loop including the gradient update step. The gradients are computed and the update step is executed. The base SGD algorithm is very simple*

Utilizing PyTorch and a starting x-value of 2.25, we iteratively use base SGD over 1,000 iterations to see if we can converge to a good solution. The resulting optimization history is shown in Figure 5-34.

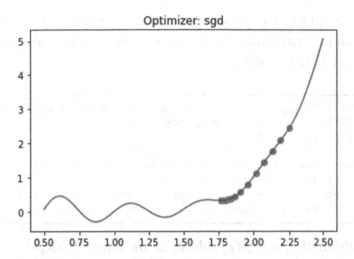

Figure 5-34. *Base SGD's optimization performance over 1,000 iterations. The optimizer got stuck in the local minimum*

As we can see in Figure 5-34, SGD got stuck in a very shallow local minimum, having reached the minimum loss in 52 iterations. How can we improve upon this and keep the optimizer going toward a better minimum? This is where the concept of **momentum** comes in. Imagine the red dots as a ball rolling down the hill. Wouldn't it be great if the ball accrued some momentum that kept it going forward, allowing it to overcome that tiny hill and keep going until it fell into a deeper minimum?

To achieve this, we can use the concept of **velocity**, which is a decaying average of accumulated gradients added to the gradient update. This gives the optimizer some inertia that helps overcome shallow valleys in the curvature as well as dampen any oscillating behavior.

Figure 5-35 shows the SGD with momentum training loop algorithm. There are two additional variables to consider:

- `velocities`: An initial vector of all zeros that, like the variable `gradients`, corresponds one-to-one with each parameter by index. This variable holds the decaying average of the accumulated gradient updates to be added to the new gradient update to add inertia to the optimization process.

- `alpha`: A coefficient applied to the old velocity when adding it to the update step. It controls the rate of decay of the accumulated velocity. A high `alpha` value adds much more weight to past gradients and adds a lot more inertia. An `alpha` closer to zero makes SGD with momentum behave more like plain SGD, with `alpha` equal to zero being vanilla SGD exactly.

```
#initial velocity vector of all  zeros corresponding to each parameter
velocities = [0 for f in range(len(parameters))]
for batch in training_data:

    gradients = compute_gradient(loss, batch, parameters)
    for p in range(len(parameters)):
        velocities[p] = alpha * velocities[p] - lr * gradients[p]
        parameters[p] = parameters[p] + velocities[p]
```

Figure 5-35. *Adding the concept of velocity to the base SGD algorithm with the decay of accumulated gradient updates controlled by alpha*

Revisiting the prior function in Figure 5-34, Figure 5-36 shows how this simple modification has allowed the optimizer to overcome the minima.

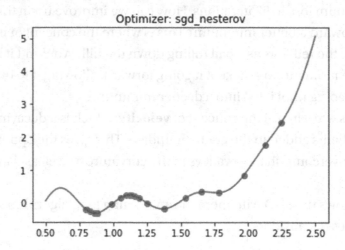

Figure 5-36. *Performance of SGD with momentum optimizer. The momentum coefficient, alpha, was set to 0.9*

The optimizer reached the lowest loss in 190 iterations and has hit the global minimum. However, as we can see, quite a bit of oscillation is still going on, likely because there is a lot of inertia retained that needs to be dissipated. There is yet another modification we could make here. Wouldn't it be nice if the red ball were able to see where it was heading and decelerate if needed? This way it could start "braking" before reaching some minimum so that it could converge much sooner.

This is where **Nesterov momentum** comes into play. The idea with Nesterov momentum is that when computing the gradients, we will not use the active parameters for that current iteration, but rather a gradient-updated "future" version of those parameters. In a sense, the optimizer "looks ahead" into the future, allowing it to compute a different, final gradient update with this "future" outcome in mind. This will result in acceleration/deceleration as well as more accurate gradient updates. Now, the optimizer won't oscillate as much because there won't be as much accumulated velocity that needs to be shed—the optimizer would decelerate in advance before hitting the bottom of a steeper valley, for example.

The code for SGD with Nesterov (also referred to as Nesterov Accelerated Gradient, or NAG) is shown in Figure 5-37.

```
#initial velocity vector of all  zeros corresponding to each parameter
velocities = [0 for f in range(len(parameters))]
for batch in training_data:
    #look-ahead update
    for p in range(len(parameters)):
        parameters[p] = parameters[p] + alpha*velocities[p]

    gradients = compute_gradient(loss, batch, parameters)
    for p in range(len(parameters)):
        velocities[p] = alpha * velocities[p] - lr * gradients[p]
        parameters[p] = parameters[p] + velocities[p]
```

Figure 5-37. *Before we compute the gradients, we update the parameters with the existing accumulated velocity. In this sense, the optimizer "looks ahead" to see where the momentum would carry it, and with the new gradients, it can apply any correction such as acceleration/deceleration, resulting in more accurate parameter updates*

Figure 5-38 shows the results of this modification to SGD with momentum.

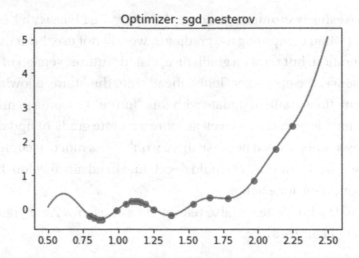

Figure 5-38. *Compared to Figure 5-36, SGD with Nesterov momentum looks like it had far fewer oscillations. At the global minimum, it looks like the "red ball" was able to slow down. Over the hump just before the global minimum, it looks like it slowed down at the peak but accelerated toward the valley once it got over. Using the "ball on a hill" analogy, this looks much more realistic, direct, and iteration-efficient compared to the prior versions of SGD*

As we can see in Figure 5-38, the optimizer looked like it converged much sooner with less oscillation. Indeed, the lowest loss was achieved in 42 iterations.

Although this looks like a huge improvement over SGD and SGD with momentum and is a pretty good optimizer, it isn't perfect. One issue is the fact that the learning rate remains constant. We could manually adjust this or use a learning rate scheduler that automatically changes learning rate over time, but then the learning rate would become yet another hyperparameter to tune.

It would be great to have an optimizer that could dynamically update the learning rate, wouldn't it? There are optimizers like RMSProp and Adadelta that deal with some decaying average of the square of accumulated gradients to dynamically adapt the learning rate per parameter and follow the topology of the loss curve better. They are also well suited for higher-dimensional problems as well as sparse data in general, making optimizers like RMSProp a strong optimizer for situations like training Long-Short Term Memory (LSTM) models for natural language processing (NLP) problems such as sentiment analysis or text generation.

Figure 5-39 shows the RMSProp code. There are a few new variables to consider:

- r: The squared gradient accumulation variable. Similar to velocity, it holds the exponentially decaying average of the squared gradients.

- rho: The decay coefficient.

- eps: A very small number, like 1e-6 or 1e-7, added for numerical stability.

```
r = [0 for f in range(len(parameters))]
for batch in training_data:

        gradients = compute_gradient(loss, batch, parameters)
        for p in range(len(parameters)):
                r[p] = rho*r[p] + (1-rho)*gradients[p]**2
                update = - lr * gradients[p] / (r[p] + eps)**0.5
                parameters[p] = parameters[p] + update
```

Figure 5-39. *The RMSProp update equation. The update step coefficient of the gradients will differ across iterations, effectively making the step size (learning rate) dynamic. You can set some initial learning rate, but throughout training, it will vary*

The function that we used to demonstrate SGD and its variations on is too simple to properly showcase the perks of RMSProp, so we will do a full comparison of all the optimizers we discussed along with the Adam optimizer.

The Adam optimizer can be thought of as a "best-of-both-worlds" type optimizer that incorporates momentum as well as accumulated squared gradients like in RMSProp and Adadelta.

Although RMSProp can be modified to include momentum (and even Nesterov momentum), the Adam optimizer incorporates bias correction (allowing for more stable learning, especially earlier in the training process). The Adam optimizer is also more robust to hyperparameter choices, making it a great go-to optimizer across various different training tasks. The Adam optimizer has also been found to typically outperform RMSProp. With these advantages, the Adam optimizer is used pretty widely in practice.

However, there is no best overall optimizer, and what optimizer performs best can vary based on the training task. Each optimizer also has pros and cons. For example, the Adam optimizer is much more memory intensive due to how many variables it uses. In comparison, SGD is more memory efficient and is computationally lighter, though it may not be the best-performing optimizer. It may be worth trying out different optimizers with different hyperparameter settings if you have the computational power to spare and if performance is a very high priority.

The code for the Adam optimizer is show in Figure 5-40. It introduces several new parameters to consider:

- s: The first-order moment array. Similar to the velocities vector, s stores the velocities and can be considered akin to momentum.

- r: The second-order moment array. Like in RMSProp, this holds the squared gradients and computes an exponentially decaying average on it.

- beta1: The exponentially decaying average coefficient used for the first-order moment s vector.

- beta2: The exponentially decaying average coefficient used for the second-order moment r vector.

- eps: A very small number added for numerical stability, typically something like 1e-6 or 1e-7.

- t: The iteration counter. Keeps count of what iteration of forward pass it is.

```
s = [0 for f in range(len(parameters))]
r = [0 for f in range(len(parameters))]
t = 0
for batch in training_data:

        gradients = compute_gradient(loss, batch, parameters)
        t = t + 1
        for p in range(len(parameters)):
                s[p] = beta1*s[p] + (1-beta1)*gradients[p]
                r[p] = beta2*r[p] + (1-beta2)*gradients[p]**2
                s_hat = s[p] / (1 - beta1**t)
                r_hat = r[p] / (1 - beta2**t)
                update = - lr * s_hat / (r_hat**0.5 + eps)
                parameters[p] = parameters[p] + update
```

Figure 5-40. *The pseudocode for the Adam optimizer. It is more resource heavy due to having to store the first-order momentum as well as the squared gradients. We must also keep track of the iteration number*

As you can see in Figure 5-40, the Adam optimizer is a bit more computationally involved than the other optimizer algorithms previously introduced. It keeps track of both the first-order moment (the momentum) and the second-order moment (the accumulation of the squared gradients we saw in RMSProp). The Adam optimizer can also be modified to incorporate Nesterov momentum.

Having covered SGD, SGD + momentum, NAG, RMSProp, and the Adam optimizer, let's see how each of these optimizers performs on a simple neural network trained on the MNIST dataset. All of the optimizers will have a learning rate of 1e-3, be trained on a batch size of 256, and be trained for 64 epochs.

For more details on the experimental setup, be sure to refer to the accompanying Jupyter notebook on GitHub at `https://github.com/apress/beginning-anomaly-detection-python-deep-learning-2e/blob/master/Chapter%205%20Introduction%20to%20Deep%20Learning/chapter5_optimizer_comparison.ipynb`.

Refer to Figure 5-41 to see the results.

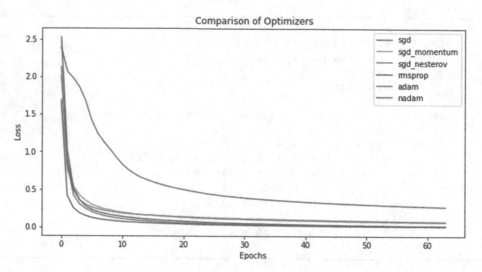

Figure 5-41. *The training results of all of the optimizer setups. This time, RMSProp performed the best out of all the optimizers, reaching convergence the soonest as well as achieving the best solution. SGD performed the worst by far, with Adam and Nadam eventually reaching a similarly optimal solution as RMSProp did*

It is important to keep in mind that this is a simple toy example and not necessarily indicative of real-world performance in all cases. These problems are far too complex, but as a general rule of thumb, the Adam optimizer is a relatively good pick to experiment with due to its robustness to the initial learning rate as well as its relatively quick rate of convergence.

In this case, RMSProp converged the soonest and to the best solution. Again, this isn't always indicative of real-world performance, so if you can afford the computational costs, trying out several optimizers is a good strategy (via cross-validation, for example, to ensure experimentally sound results) to further boost your model's performance.

Now that we have gone over all the major components of deep learning modeling, let's begin creating our first supervised multilayer perceptron model for anomaly detection.

Multilayer Perceptron Supervised Anomaly Detection

Let's run our own supervised anomaly detection model by implementing a multilayer perceptron both in Keras and in PyTorch. The dataset we will perform this task on is the KDDCUP 1999 dataset, just like in Chapter 4.

Much of the data preparation code will look exactly the same as in Chapter 4. Additionally, this data preparation code will be shared between both the Keras model and the PyTorch model, so be sure to execute these blocks first within this section before continuing on to the Keras and/or PyTorch sections. We will also go over the neural network architecture shared by both models in this section.

This section's code is the same as that used across the Keras and PyTorch notebooks, so you can refer to either:

- **Keras notebook**: https://github.com/apress/beginning-anomaly-detection-python-deep-learning-2e/blob/master/Chapter%205%20Introduction%20to%20Deep%20Learning/chapter5_mlp_keras.ipynb

- **PyTorch notebook**: https://github.com/apress/beginning-anomaly-detection-python-deep-learning-2e/blob/master/Chapter%205%20Introduction%20to%20Deep%20Learning/chapter5_mlp_pytorch.ipynb

Let's begin with Figure 5-42, the imports.

```
import numpy as np
import pandas as pd
import matplotlib.pyplot as plt
from sklearn.model_selection import train_test_split
from sklearn.preprocessing import LabelEncoder
from sklearn.metrics import precision_score, recall_score, f1_score,
confusion_matrix, roc_auc_score

import seaborn as sns
print('seaborn: ', sns.__version__)
```

***Figure 5-42.** The necessary import statements. You should have at least seaborn version 0.12.2*

The version of seaborn used in the examples is 0.12.2. Seaborn is a plotting library that builds on top of matplotlib to provide more advanced functionality.

We will once again define all the columns and load the dataframe. Refer to Figure 5-43.

```
columns = ["duration", "protocol_type", "service", "flag", "src_bytes",
"dst_bytes", "land", "wrong_fragment", "urgent",
        "hot", "num_failed_logins", "logged_in", "num_compromised",
"root_shell", "su_attempted", "num_root",
        "num_file_creations", "num_shells", "num_access_files",
"num_outbound_cmds", "is_host_login",
        "is_guest_login", "count", "srv_count", "serror_rate",
"srv_serror_rate", "rerror_rate", "srv_rerror_rate",
        "same_srv_rate", "diff_srv_rate", "srv_diff_host_rate",
"dst_host_count", "dst_host_srv_count",
        "dst_host_same_srv_rate", "dst_host_diff_srv_rate",
"dst_host_same_src_port_rate", "dst_host_srv_diff_host_rate",
        "dst_host_serror_rate", "dst_host_srv_serror_rate",
"dst_host_rerror_rate", "dst_host_srv_rerror_rate", "label"]

df = pd.read_csv("../data/kddcup.data.corrected", sep=",", names=columns,
index_col=None)
```

Figure 5-43. *Defining the columns and loading the KDDCUP 1999 data as a dataframe*

Make sure to have the data in the right path, much like in Chapter 4. Next, we will filter the dataframe to only include HTTP attacks. Refer to Figure 5-44.

```
# Filter to only 'http' attacks
df = df[df["service"] == "http"]
df = df.drop("service", axis=1)
```

Figure 5-44. *Code to filter the dataframe to only include HTTP attacks. The row count should be 623,091 rows with 41 columns*

Once this is done, we must apply our label map so that anything labeled "normal." will be counted as a normal data point, and anything labeled otherwise will be an anomaly. Run the code in Figure 5-45.

```
# Filter to only 'http' attacks
df = df[df["service"] == "http"]
df = df.drop("service", axis=1)
```

Figure 5-45. *Applying the numerical encoding and checking values counts*

Next, we must encode all other categorical columns. Run the code in Figure 5-46.

```
datatypes = dict(zip(df.dtypes.index, df.dtypes))

encoder_map = {}
for col, datatype in datatypes.items():
    if datatype == 'object':
        encoder = LabelEncoder()
        df[col] = encoder.fit_transform(df[col])
        encoder_map[col] = encoder
```

Figure 5-46. *Encoding the non-numeric columns into numeric equivalents*

Just as before, we will include only the columns with the highest correlation with the label column. Run the code in Figure 5-47.

```
# Check the variables with highest correlation with 'label'
df2 = df.copy()
label_corr = df2.corr()['label']

# Filter out anything that has null entry or is not weakly
correlated
train_cols = label_corr[(~label_corr.isna()) &
(np.abs(label_corr) > 0.2)]
train_cols = list(train_cols[:-1].index)
train_cols
```

Figure 5-47. *Filtering out columns with null entries or no linear correlation while preserving columns at least weakly correlated. We also index* `train_cols[:-1]` *to remove label as its own correlation*

With this, we can create our train-test-val splits. Refer to Figure 5-48 to create the testing split and initial training split.

```
labels = df2['label']
# Conduct a train-test split
x_train, x_test, y_train, y_test =
train_test_split(df2[train_cols].values, labels.values,
test_size = 0.15, random_state = 42)
```

Figure 5-48. *Creating the testing split and the initial training split. The* `random_state` *has been set at 42 for reproducibility*

Finally, we can create our validation set. Refer to Figure 5-49.

```
# Additional split of training dataset to create validation
split
x_train, x_val, y_train, y_val = train_test_split(x_train,
y_train, test_size=0.2, random_state=42)
```

Figure 5-49. *Creating the validation split and the final training split*

We can print out the shapes just to be sure the processing was done correctly. Refer to Figure 5-50.

```
print("Shapes")
print(f"x_train:{x_train.shape}\ny_train:{y_train.shape}")
print(f"\nx_val:{x_val.shape}\ny_val:{y_val.shape}")
print(f"\nx_test:{x_test.shape}\ny_test:{y_test.shape}")
```

Figure 5-50. *Code to print out the shapes of all of the data splits we have created*

If all went well, we should see output like that shown in Figure 5-51.

```
Shapes
x_train:(423701, 13)
y_train:(423701,)

x_val:(105926, 13)
y_val:(105926,)

x_test:(93464, 13)
y_test:(93464,)
```

Figure 5-51. *The output shapes of the training, validation, and test sets*

With that, we have now prepared the datasets for further modeling with either PyTorch or Keras. Before we proceed, let's go over the architecture of the neural network we will be implementing. Refer to Figure 5-52.

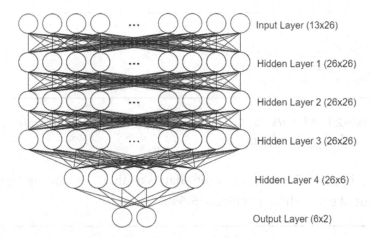

Input Layer (13x26)

Hidden Layer 1 (26x26)

Hidden Layer 2 (26x26)

Hidden Layer 3 (26x26)

Hidden Layer 4 (26x6)

Output Layer (6x2)

Figure 5-52. *The neural network architecture we will be implementing. The number of layers and the number of units were arbitrarily chosen—feel free to experiment with fewer or more layers and/or neuron counts. Each hidden layer, including the input layer, has the ReLU activation function. The output layer has the* softmax() *activation function*

In your implementations of the neural network, feel free to experiment with different activation functions, the number of hidden layers, and the number of neurons for each layer to get a sense of what works and what doesn't. The architecture shown in Figure 5-52 was arbitrarily chosen, so don't take it as a model example of what to do for every task. What architecture works will differ per task, along with how deep it really needs to be and how many neurons each layer should have.

With that information in mind, let's begin with TensorFlow and Keras.

Simple Neural Network: Keras

Be sure to follow along with the Jupyter notebook for this section, available at https://github.com/apress/beginning-anomaly-detection-python-deep-learning-2e/blob/master/Chapter%205%20Introduction%20to%20Deep%20Learning/chapter5_mlp_keras.ipynb. The examples in the notebook were run on TensorFlow 2.7.0.

Before we get started with the modeling, we need to transform the y-sets into a one-hot format. This is necessary because Keras expects the output predictions to be post-softmax() scores. To do this, Keras provides a utility function called to_categorical() that we can use to transform an array of integers into a one-hot output. Refer to Figure 5-53.

```
from tensorflow.keras.utils import to_categorical

y_train =  to_categorical(y_train)
y_test =  to_categorical(y_test)
y_val =  to_categorical(y_val)
```

Figure 5-53. *Using the Keras* `to_categorical()` *function to transform the y-sets into one-hot encoded equivalents*

After running Figure 5-53, let's check the shapes again by rerunning Figure 5-51. You should see output like that shown in Figure 5-54.

```
Shapes
x_train:(423701, 13)
y_train:(423701, 2)

x_val:(105926, 13)
y_val:(105926, 2)

x_test:(93464, 13)
y_test:(93464, 2)
```

Figure 5-54. *Looking at the shapes of the y-sets after applying* `to_categorical()`. *AThey are one-hot encoded over the two possible output classes*

Next, let's import all of the components of Keras that we will use. Refer to Figure 5-55.

```
from tensorflow.keras.models import Model
from tensorflow.keras.layers import Input, Dense
from tensorflow.keras.regularizers import L2
from tensorflow.keras.callbacks import EarlyStopping
```

Figure 5-55. *The* `import` *statements required*

Now let's define our neural network structure and explain what each layer type means. Refer to Figure 5-56.

```
# The input layer requires you to specify the dimensionality of the x-
features (and not the number of samples)
input_layer = Input(shape=(13))
h1 = Dense(26, activation='relu', kernel_initializer = 'he_uniform',
kernel_regularizer = L2(l2=1e-5))(input_layer)
h2 = Dense(26, activation='relu', kernel_initializer = 'he_uniform',
kernel_regularizer = L2(l2=1e-5))(h1)
h3 = Dense(26, activation='relu', kernel_initializer = 'he_uniform',
kernel_regularizer = L2(l2=1e-5))(h2)
h4 = Dense(6, activation='relu', kernel_initializer = 'he_uniform',
kernel_regularizer = L2(l2=1e-5))(h3)
output_layer = Dense(2, activation='softmax', kernel_regularizer =
L2(l2=1e-5))(h4)

# Creating a model by specifying the input layer and output layer
model = Model(input_layer, output_layer)
```

Figure 5-56. *Implementing the neural network structure described in Figure 5-52*

Let's now go over what each of the layer types means:

- **Input**: The input layer specifies the dimensionality of the x-features.
  That is, what is the dimensionality of each individual x data point?
  In this case, we selected 13 features to pass in, so this would be 13.
  However, it's possible to have multidimensional x data points, such as
  with images that have a length, width, and color channel data.

  You can find out more information about this layer at https://www.
  tensorflow.org/versions/r2.7/api_docs/python/tf/keras/
  layers/InputLayer.

- **Dense**: This is a neural network layer comprised of many neurons
  and is the basic layer discussed earlier in the chapter. There are
  many possible arguments to this class, but the most important and
  commonly used parameters are these two:

  - units: The number of neurons in this layer

  - activation: The activation function to use

Many other possible parameters, such as kernel_regularizer in Figure 5-56, let you customize the behavior of a Dense layer. If you wanted to specify how to initialize the weights and biases, you could do so, as well as determine whether to use a bias at all. More information on the possible parameters can be found at https://www.tensorflow.org/versions/r2.7/api_docs/python/tf/keras/layers/Dense.

The code in Figure 5-56 also specifies the kernel_initializer parameter to be he_uniform. This parameter is used to initialize the weights, and it can have a noticeable effect on the resulting stability and performance of the model. By default, Keras uses a method known as 'glorot_uniform', which is more suited for stability in networks using sigmoid or tanh functions. (It is named after Xavier Glorot. You may see this method referred to as Glorot initialization or Xavier initialization.)

For ReLU networks, an initialization method known as Kaiming Normal or He Normal (named after Kaiming He) is more optimal for model stability and proper learning. In PyTorch, the default is Kaiming Uniform.

Finally, we create the model by using the Model() class and passing in the input and output layers. This is what is called the **functional API**, when we define each layer separately like we did in Figure 5-56, each calling upon the previous layer, and then create a model using the input and output layers. There is another method of creating a model called the **sequential API**, where we pass in a list of layers to a class named Sequential().

Which method you choose is up to you. You can find more information on the functional API at https://www.tensorflow.org/guide/keras/functional_api, and find more information on the sequential API at https://www.tensorflow.org/guide/keras/sequential_model.

Now that we have defined our model, let's define our early stopping mechanism and compile our model. Refer to Figure 5-57.

```
es = EarlyStopping(patience=5, min_delta=1e-3, monitor='val_loss',
restore_best_weights=True)

callbacks = [es]

model.compile(optimizer='adam', loss='categorical_crossentropy',
metrics='accuracy')
```

Figure 5-57. *Defining the early stopping callback and adding it to a list of callbacks to be used in the training process. Then, we compile the model with the Adam optimizer (at default parameter settings) and categorical cross-entropy loss*

In Keras, callbacks are a set of functions that can be called throughout the training process. Figure 5-57 shows the `EarlyStopping()` callback function. If the target metric, such as validation loss, does not decrease or remain within some threshold within a certain number of epochs, `EarlyStopping` stops the training and can optionally restore the best-performing epoch's weights. This ties back to the discussion earlier in the chapter of the loss curve and the optimal stopping point during the training process.

The following list explains the parameters of `EarlyStopping`:

- `patience`: The number of epochs with no improvement after which training will be stopped. The model will keep training as long as the monitored metric keeps improving, even if the improvements are small. If the metric stops improving for a number of epochs specified by `patience`, then the training process will stop early.

- `min_delta`: The threshold for considering whether an improvement is significant. If the absolute change in the monitored metric between epochs is less than `min_delta`, it is not considered an improvement. If the metric fails to improve significantly for a `patience` number of epochs, training will stop early.

- `monitor`: The metric to monitor for early stopping purposes. Setting it to `'val_loss'` to monitor validation loss can help stop overfitting, since if validation loss stops decreasing over a `patience` number of epochs, we will early stop.

- `restore_best_weights`: If set to `True`, the model's weights will be reverted to the state of the epoch with the best value of the monitored metric. This means that even if the model's performance deteriorated in later epochs due to overfitting, you would still retain the best model state.

Other callback functions in Keras include `ReduceLROnPlateau`, which reduces the learning rate if the target metric doesn't really improve. This can address oscillating behavior caused by step sizes that are too large, allowing the optimizer to take smaller and more precise steps and lead to an even better convergence.

Now that you're familiar with the early stopping mechanism, you're ready to start training the model. Refer to Figure 5-58.

```
epochs = 20
batch_size = 128

history = model.fit(x_train, y_train, validation_data=(x_val, y_val),
epochs=epochs, batch_size=batch_size, callbacks=callbacks)
```

Figure 5-58. *Training the model and saving the training history to a variable named history. This step is where we can specify the number of epochs, the minibatch size, the validation data, and the callbacks to use during training. The number of epochs was arbitrarily set at 20, which is higher than it needs to be for the purpose of demonstrating how early stopping can halt training well before the 20 epochs*

Because the weights are randomly initialized, your training may end at a different epoch than in the example. Regardless, you should see something like Figure 5-59 as the training output.

```
Epoch 1/20
3311/3311 [==============================] - 16s 5ms/step - loss: 0.3778 -
accuracy: 0.9962 - val_loss: 0.0052 - val_accuracy: 0.9999
...
Epoch 6/20
3311/3311 [==============================] - 15s 4ms/step - loss: 0.0073 -
accuracy: 0.9972 - val_loss: 0.0039 - val_accuracy: 0.9999
Epoch 7/20
3311/3311 [==============================] - 14s 4ms/step - loss: 0.0090 -
accuracy: 0.9978 - val_loss: 0.0062 - val_accuracy: 0.9987
Epoch 8/20
3311/3311 [==============================] - 15s 4ms/step - loss: 0.0063 -
accuracy: 0.9976 - val_loss: 0.0082 - val_accuracy: 0.9964
Epoch 9/20
3311/3311 [==============================] - 15s 4ms/step - loss: 0.0069 -
accuracy: 0.9978 - val_loss: 0.0061 - val_accuracy: 0.9982
Epoch 10/20
3311/3311 [==============================] - 15s 5ms/step - loss: 0.0057 -
accuracy: 0.9989 - val_loss: 0.0053 - val_accuracy: 0.9986
Epoch 11/20
3311/3311 [==============================] - 15s 5ms/step - loss: 0.0047 -
accuracy: 0.9989 - val_loss: 0.0064 - val_accuracy: 0.9979
```

Figure 5-59. *The training output of the code in Figure 5-58. Notice that the early stopper kicked in at Epoch 11 and halted the training. This is because the best validation loss was achieved in Epoch 6, but the validation loss did not "improve" over the following three epochs, so the training was stopped and the weights were restored to Epoch 6 as it achieves the best validation loss (lowest loss is best performance)*

Keep in mind that because the weights are randomly initialized, you may see a different set of accuracies being achieved by the model. Feel free to redefine the model starting from Figure 5-56 and rerun the code with a different set of initialized weights, as the performance may fluctuate. With the `'he_uniform'` initialization though, training performance should be consistent.

Let's evaluate this trained model on the test set to see what accuracy we achieve. Refer to Figure 5-60.

```
model.evaluate(x_test, y_test)
```

Figure 5-60. *The evaluate method built into the model. Pass in the test set to automatically score the model's prediction and output the loss and the accuracy*

After evaluating, you should see output like that shown in Figure 5-61.

```
2921/2921 [==============================] - 8s 3ms/step - loss: 0.0032 -
accuracy: 0.9999
[0.003189682262018323, 0.9999144077301025]
```

Figure 5-61. *We have achieved a loss of 0.0032 and an accuracy at 0.9999. The list returned by the method below gives us the full, precise values for the loss and the accuracy in order. The accuracy on the test set is actually 99.9914%*

Remember that accuracy is not the best metric to evaluate performance with if your classes are imbalanced. Let's get the model's predictions for the test set and compute the precision, recall, and F1-score metrics. Refer to Figure 5-62.

```
preds = model.predict(x_test)

# One hot to the original label encodings
y_true = y_test.argmax(axis=1)

# Derive the label predictions from the probability scores
y_preds = preds.argmax(axis=1)

# Compute precision, recall, f1 scores
precision = precision_score(y_true, y_preds)
recall = recall_score(y_true, y_preds)
f1_measure = f1_score(y_true, y_preds)

print(f" Precision: {precision}")
print(f"Recall: {recall}")
print(f"F1-Measure: {f1_measure}")
```

Figure 5-62. *Making predictions and computing precision, recall, and f1-scores*

You should see output similar to Figure 5-63 if all went well with the training process.

```
Precision: 0.9895287958115183
Recall: 0.9964850615114236
F1-Measure: 0.9929947460595447
```

Figure 5-63. *The precision, recall, and f1-score outputs. The model did very well with a really high precision (near perfect anomaly predictions) and a really high recall (captured almost all the anomalies in the test set)*

While we are at it, let's also compute the AUC score, which is given by:

```
roc_auc_score(y_true, y_preds)
```

The output result should be something similar to 0.9982102362296339, which is a very strong anomaly detector. To end this modeling experiment, let's compute and visualize the confusion matrix. Refer to Figure 5-64 for the code.

```
cm = confusion_matrix(y_true, y_preds)
plt.title("Confusion Matrix")
ax = sns.heatmap(cm, annot=True, fmt='0.0f')
ax.invert_yaxis()
ax.invert_xaxis()
ax.set_xlabel('Predicted Label')
ax.set_ylabel('True Label')
```

Figure 5-64. *Code to compute and plot the confusion matrix given the predictions and the true y-labels for the test set that were computed in Figure 5-62*

The confusion matrix result should look similar to Figure 5-65.

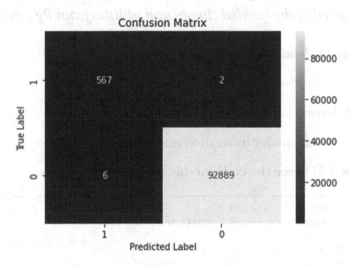

Figure 5-65. *The confusion matrix result of the model's prediction performance on the test set. There are 6 false positives (predicted 1, but truly 0) and 2 false negatives (predicted 0, but truly 1)*

With that, you now know how to implement a neural network from scratch, train it, and evaluate it with TensorFlow and Keras. Let's now do the same but in PyTorch.

Simple Neural Network: PyTorch

Be sure to follow along with the notebook at https://github.com/apress/beginning-anomaly-detection-python-deep-learning-2e/blob/master/Chapter%205%20Introduction%20to%20Deep%20Learning/chapter5_mlp_pytorch.ipynb. We are running torch version 1.9.0+cu111.

We do not need to perform any one-hot encoding with PyTorch. However, we will be constructing dataloaders using PyTorch's APIs so that we can easily batch our dataset and sample minibatches during training. Refer to Figure 5-66 for the code to import all the necessary components.

```python
import torch

print(f"Torch version: {torch.__version__}")
from torch.nn import Linear, CrossEntropyLoss
from torch.optim import Adam
from torch.utils.data import DataLoader, Dataset, TensorDataset
```

Figure 5-66. *Importing the needed classes and utilities from PyTorch*

Let's create our dataloaders. This is a three-step process:

1. Convert our datasets into tensors.

2. Create a TensorDataset from the tensors.

3. Create the DataLoader using the TensorDataset.

Refer to Figure 5-67 to see the code for this.

```python
x_train = torch.Tensor(x_train)
y_train = torch.Tensor(y_train).type(torch.long)

x_val = torch.Tensor(x_val)
y_val = torch.Tensor(y_val).type(torch.long)

x_test = torch.Tensor(x_test)
y_test = torch.Tensor(y_test).type(torch.long)

train_set = TensorDataset(x_train, y_train)
val_set = TensorDataset(x_val, y_val)
test_set = TensorDataset(x_test, y_test)

train_dataloader = DataLoader(train_set, batch_size = 128)
val_dataloader = DataLoader(val_set, batch_size = 128)
test_dataloader = DataLoader(test_set, batch_size = 128)
```

Figure 5-67. *Creating the DataLoader objects to be used in the training and evaluation processes*

Now that we have our datasets prepared, let's look at how to build a model in PyTorch, which is relatively straightforward. We will construct a class for our custom model and define all the layers within. Refer to Figure 5-68.

```
class model(torch.nn.Module):
    def __init__(self):
        super().__init__()

        self.h1 = Linear(13, 26)
        self.h2 = Linear(26, 26)
        self.h3 = Linear(26, 26)
        self.h4 = Linear(26, 6)
        self.h5 = Linear(6, 2)

    def forward(self, x):
        x = self.h1(x).relu()
        x = self.h2(x).relu()
        x = self.h3(x).relu()
        x = self.h4(x).relu()
        x = self.h5(x)

        return x

mlp = model()
```

Figure 5-68. *Model building in PyTorch. We can first define the layers as private variables (not necessary, but it's a standard way of doing it) and then, in the* forward() *function, determine how the input data is processed through all the layers*

The layers in PyTorch are more straightforward in what parameters they take. For the Linear layer's documentation, refer to https://pytorch.org/docs/stable/generated/torch.nn.Linear.html.

Following this, we must define both what our optimizer is and what our loss function is. We will be choosing the Adam optimizer and the Categorical CrossEntropy Loss (just called CrossEntropyLoss in PyTorch). Refer to Figure 5-69.

```
optim = Adam(mlp.parameters(), lr=1e-3)
criterion = CrossEntropyLoss()
```

Figure 5-69. *Defining the optimizer and the loss function. The optimizer as the first parameter takes in the model's learnable parameters, which we can get by using* model_name.parameters(). *We pass in the learning rate to be 1e-3*

Now we can define our training loop, which has early stopping incorporated into it. The parameter `patience` works just as it does in the Keras `EarlyStopping` callback function, and the `tolerance` variable is the equivalent to `min_delta` in `EarlyStopping`. We manually save the best path using PyTorch save/load functionality so that we can retain the best-performing model's weights.

This is a long training loop, so we will split it up into two figures: Figure 5-70 and Figure 5-71. Mind the tabbing—the validation loop should be in the same indent level as the actual training `for` loop. For best consistency, refer to the GitHub notebook.

```python
epochs = 20
tolerance = 1e-3
lambda_ = 1e-5
best_loss = np.inf
patience = 5
early_stop_counter = 0
best_save_path = 'ch5_mlp_pytorch_best.pt'
mlp.cuda()
for e in range(epochs):
    for i, (data, labels) in enumerate(train_dataloader):
        optim.zero_grad()

        data = data.cuda()
        labels = labels.cuda()

        preds = mlp.forward(data)
        loss = criterion(preds, labels)

        l2_norm = sum(p.pow(2.0).sum() for p in mlp.parameters())
        loss = loss + lambda_ * l2_norm

        loss.backward()
        optim.step()
        predictions = preds.detach().cpu().numpy().argmax(axis=1)
        y_true = labels.detach().cpu().numpy()
        acc = np.mean(y_true == predictions)
        print(f'\rEpoch {e} / {epochs}: {i}/{len(train_dataloader)} | loss:
{loss.item()} acc: {acc}'.ljust(200, ' '), end= '')
```

Figure 5-70. *The first half of the training script. This contains all of the hyperparameter settings for the early stop algorithm as well as the training loop itself, such as the number of epochs*

```python
    # Validation

with torch.no_grad():
    hold = np.array([])
    loss_hold = []
    for i, (data, labels) in enumerate(val_dataloader):

        data = data.cuda()
        labels = labels.cuda()

        preds = mlp.forward(data)

        loss = criterion(preds, labels)
        l2_norm = sum(p.pow(2.0).sum() for p in mlp.parameters())
        loss = loss + lambda_ * l2_norm

        predictions = preds.detach().cpu().numpy().argmax(axis=1)
        y_true = labels.detach().cpu().numpy()

        hold = np.concatenate((y_true == predictions, hold))
        loss_hold.append(loss.item())

    val_acc = np.mean(hold)
    val_loss = np.mean(loss_hold)
    print(f'\rEpoch {e} / {epochs}: {i}/{len(val_dataloader)} | loss:
{loss.item()} acc: {acc} val_loss: {val_loss} val_acc:
{val_acc}'.ljust(200, ' '), end= '\n')

if val_loss < best_loss:
    best_loss = val_loss
    early_stop_counter = 0

    torch.save(mlp.state_dict(), best_save_path)

elif abs(best_loss - val_loss) <= tolerance:
    pass
else:
    early_stop_counter += 1

    if early_stop_counter >= patience:
        print(f'\rEpoch {e} / {epochs}: {i}/{len(val_dataloader)} |
loss: {loss.item()} acc: {acc} val_loss: {val_loss} val_acc: {val_acc}
Early Stopping'.ljust(200, ' '), end= '\n')
        mlp.load_state_dict(torch.load(best_save_path))
        break
```

Figure 5-71. *The second half of the training loop, containing the validation logic as well as the early stopper logic. The idea is for the validation loss to either be the best or be within an absolute deviation within the threshold from the best validation loss. If it is not, we increment the counter until we hit the patience limit, after which point we early stop*

Figures 5-70 and 5-71 contain the entire training loop. Executing the code should start the training process and the training output. Once done, you should see output resembling Figure 5-72.

```
Epoch 0 / 20: 827/828 | loss: 0.0007202776032499969 acc: 1.0 val_loss:
0.003406143917886041 val_acc: 0.9998583917074184
Epoch 1 / 20: 827/828 | loss: 0.00105832249391824 acc: 1.0 val_loss:
0.002974088849828259 val_acc: 0.9998678322602571
Epoch 2 / 20: 827/828 | loss: 0.000773684645537287 acc: 1.0 val_loss:
0.002741797632033878 val_acc: 0.9998772728130959
Epoch 3 / 20: 827/828 | loss: 0.0006962550687603652 acc: 1.0 val_loss:
0.002940932255303532 val_acc: 0.9998961539187735
Epoch 4 / 20: 827/828 | loss: 0.0007242353749461472 acc: 1.0 val_loss:
0.002827661038488307 val_acc: 0.9998961539187735
Epoch 5 / 20: 827/828 | loss: 0.0006321458495222032 acc: 1.0 val_loss:
0.0025162437367185517 val_acc: 0.9998961539187735
Epoch 6 / 20: 827/828 | loss: 0.000612203439231962 acc: 1.0 val_loss:
0.009361552663939298 val_acc: 0.9962804221815229
Epoch 7 / 20: 827/828 | loss: 0.0008201689925044775 acc: 1.0 val_loss:
0.0072782886366259784 val_acc: 0.9964220304741046
Epoch 8 / 20: 827/828 | loss: 0.0010553736938163638 acc: 1.0 val_loss:
0.005277823925902392 val_acc: 0.9998961539187735
Epoch 9 / 20: 827/828 | loss: 0.0012233523884788156 acc: 1.0 val_loss:
0.007399877606849479 val_acc: 0.9973849668636595
Epoch 10 / 20: 827/828 | loss: 0.0013556014746427536 acc: 1.0 val_loss:
0.007573924820076063 val_acc: 0.9967713309291392
Epoch 10 / 20: 827/828 | loss: 0.0013556014746427536 acc: 1.0 val_loss:
0.007573924820076063 val_acc: 0.9967713309291392 Early Stopping
```

Figure 5-72. *The training output of the code in Figure 5-70 and 5-71, the training script. We early stopped because the validation loss was no longer improving after hitting its lowest point, indicating overfitting. The best weights were restored, so we don't keep the overfitted weights*

Once we have our trained model, we can begin the evaluation loop to get the loss and accuracy in a similar manner to the Keras model evaluate function. Refer to Figure 5-73 to see the code.

```
with torch.no_grad():
    hold = np.array([])
    loss_hold = []
    for i, (data, labels) in enumerate(test_dataloader):

        data = data.cuda()
        labels = labels.cuda()

        preds = mlp.forward(data)

        loss = criterion(preds, labels)

        predictions = preds.detach().cpu().numpy().argmax(axis=1)
        y_true = labels.detach().cpu().numpy()

        hold = np.concatenate((y_true == predictions, hold))
        loss_hold.append(loss.item())
        print(f'\rEvaluating {i}/{len(test_dataloader)} | loss:
{loss.item()} acc: {np.mean(y_true==predictions)}'.ljust(200, ' '), end=
'')

    test_acc = np.mean(hold)
    test_loss = np.mean(loss_hold)
    print(f'\rEvaluating {i}/{len(test_dataloader)} | loss: {test_loss}
acc: {test_acc}'.ljust(200, ' '), end= '')
```

Figure 5-73. *The evaluation code, which is identical to the validation loop*

The output of executing the code in Figure 5-73 should look somewhat like Figure 5-74.

```
Evaluating 730/731 | loss: 0.00024468251649472625 acc: 0.9999679020799452
```

Figure 5-74. *Output of executing the evaluation code in Figure 5-73. The model achieved a very close convergence—closer than the Keras model in this case*

Let's now look at the precision, recall, and related metrics. Execute the code in Figure 5-75.

```
preds = mlp.forward(x_test.cuda())

# Detach from Cuda and derive the label predictions from the probability
scores
y_preds = preds.detach().cpu().numpy().argmax(axis=1)

# Original labels became a tensor, so convert back to a numpy array
y_true = y_test.numpy()

# Compute precision, recall, f1 scores
precision = precision_score(y_true, y_preds)
recall = recall_score(y_true, y_preds)
f1_measure = f1_score(y_true, y_preds)

print(f"Precision: {precision}")
print(f"Recall: {recall}")
print(f"F1-Measure: {f1_measure}")
```

Figure 5-75. *Computing the precision, recall, and F1-measure scores*

You should see output like that shown in Figure 5-76.

```
Precision: 1.0
Recall: 0.9947275922671354
F1-Measure: 0.9973568281938325
```

Figure 5-76. *The model achieved perfect precision, meaning all of its anomaly predictions were correct. Its recall was near perfect, meaning that of all the true anomalies, the model almost got all of them*

To compute the AUC score, execute

```
roc_auc_score(y_true, y_preds)
```

You should see a result somewhat like 0.9973637961335677, which indicates a very strong predictor.

Finally, let's observe the confusion matrix. Refer to Figure 5-77 for the code.

```
cm = confusion_matrix(y_true, y_preds)
plt.title("Confusion Matrix")
ax = sns.heatmap(cm, annot=True, fmt='0.0f')
ax.invert_yaxis()
ax.invert_xaxis()
ax.set_xlabel('Predicted Label')
ax.set_ylabel('True Label')
```

Figure 5-77. *Code to compute and plot the confusion matrix*

The result is as depicted in Figure 5-78.

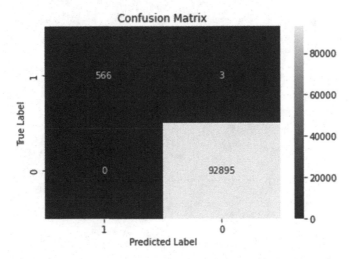

Figure 5-78. *The resulting confusion matrix from the code in Figure 5-77. The model performed almost perfectly, with the exception of three false negatives*

As we can see from Figure 5-78, this was a very strong anomaly detector model. Now you know how to implement, train, and evaluate a neural network in PyTorch.

Summary

This chapter gave you a solid understanding of what deep learning is, how it works, and how you can implement neural networks in both TensorFlow/Keras and PyTorch. Using this knowledge, you should be able to follow along with the rest of the book as we delve into the implementation of various deep learning architectures to perform anomaly detection.

In Chapter 6, you will be introduced to the concept of an autoencoder, different variations of the autoencoder, and how they all can be applied to the task of anomaly detection.

CHAPTER 6

Autoencoders

In this chapter, you will learn about autoencoder neural networks and the different types of autoencoders. You will also learn how autoencoders can be used to detect anomalies and how you can implement anomaly detection using autoencoders.

In a nutshell, this chapter covers the following topics:

- What are autoencoders?

- Simple autoencoders

- Sparse autoencoders

- Deep autoencoders

- Convolutional autoencoders

- Denoising autoencoders

- Variational autoencoders

Note Code examples are provided in Python 3.8. The code repository for this book is available at `https://github.com/apress/beginning-anomaly-detection-python-deep-learning-2e/tree/master`.

The repository also includes a requirements.txt file to check your packages and their versions.

All the notebooks for the rest of this chapter are as follows:

- **Simple, Sparse, and Deep Autoencoders**: `https://github.com/apress/beginning-anomaly-detection-python-deep-learning-2e/blob/master/Chapter%206%20Autoencoders/chapter6_autoencoder.ipynb`

- **Convolutional Autoencoders**: https://github.com/apress/ beginning-anomaly-detection-python-deep-learning-2e/ blob/master/Chapter%206%20Autoencoders/chapter6_ cnnautoencoder.ipynb

- **Denoising Autoencoders**: https://github.com/apress/ beginning-anomaly-detection-python-deep-learning-2e/ blob/master/Chapter%206%20Autoencoders/chapter6_ denoisingautoencoder.ipynb

- **Variational Autoencoders**: https://github.com/apress/ beginning-anomaly-detection-python-deep-learning-2e/ blob/master/Chapter%206%20Autoencoders/chapter6_ variationalautoencoder.ipynb

These links are also provided in each of the respective sections in which you'll need the notebook.

Navigate to "Chapter 6 Autoencoders" and then click any of the notebooks you want to try out. The code is provided as .py files as well, though it is the exported version of the notebook.

We will be using JupyterLab to present all of the code examples.

What Are Autoencoders?

Chapter 5 introduced the basic functioning of neural networks. The basic concept is that a neural network computes a weighted calculation of inputs to produce outputs. The inputs are in the input layer, the outputs are in the output layer, and there are one or more hidden layers in between the input and the output layers. Backpropagation is a technique used to train the network while trying to adjust the weights until the error is minimized. Autoencoders use this property of a neural network in a special way to accomplish some very efficient methods of training networks to learn normal behavior, thus helping to detect anomalies when they occur. Figure 6-1 shows a typical neural network.

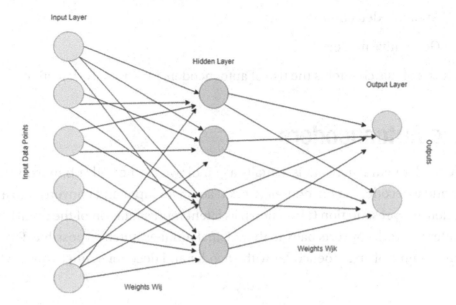

Figure 6-1. *Typical neural network*

Autoencoders are neural networks that have the ability to discover low-dimensional representations of high-dimensional data and reconstruct the input from the output. Autoencoders are made up of two pieces of the neural network: an encoder and a decoder. The encoder reduces the dimensionality of a high-dimensional dataset to a low-dimensional dataset, whereas the decoder expands the low-dimensional data to high-dimensional data. The goal of such a process is to try to reconstruct the original input. If the neural network is good, then there is a good chance of reconstructing the original input from the encoded data. This inherent principle is critical in building an anomaly detection module.

Note that autoencoders are not very useful if you have training samples containing few dimensions/features at each input point. Autoencoders perform well for five or more dimensions. If you have just one dimension/feature, you are just doing a linear transformation, which is not useful.

Autoencoders are incredibly useful in many use cases. The following are some popular applications of autoencoders:

- Training deep learning networks
- Compression
- Classification

- Anomaly detection

- Generative models

The focus of this chapter is the use of autoencoders for anomaly detection.

Simple Autoencoders

An autoencoder neural network is actually a pair of two connected subnetworks: an encoder and a decoder. An encoder network takes in an input and converts it to a smaller, dense representation (also known as latent representation of the input), which the decoder network converts back to the original input as much as possible. Figure 6-2 shows an example of an autoencoder with encoder and decoder subnetworks.

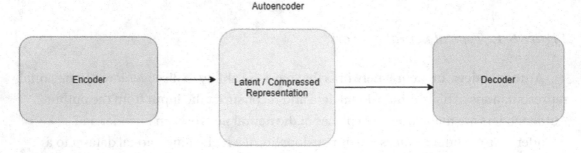

Figure 6-2. *Depiction of an autoencoder*

Autoencoders use a data compression logic where the compression and decompression functions that are implemented by neural networks are lossy and are mostly unsupervised without much intervention. Figure 6-3 shows an expanded view of an autoencoder

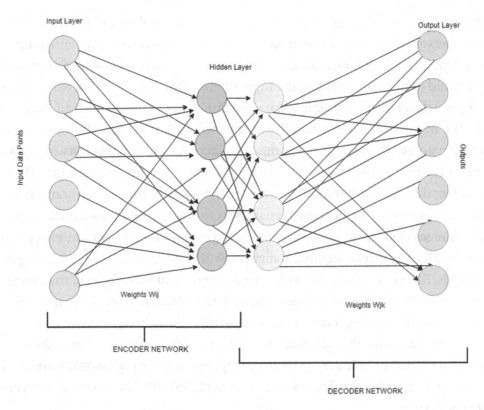

Figure 6-3. *Expanded view of an autoencoder*

The entire network is usually trained as a whole. The loss function is usually either the mean-squared error or cross-entropy between the output and the input, known as the *reconstruction loss*, which penalizes the network for creating outputs different from the input. Because the encoding (which is simply the output of the hidden layer in the middle) has far fewer units than the input, the encoder must choose to discard information. The encoder learns to preserve as much of the relevant information as possible in the limited encoding and intelligently discard irrelevant parts. The decoder learns to take the encoding and properly reconstruct it into the input. If you are processing an image, then the output is an image. If the input is an audio file, the output is an audio file. If the input is some feature-engineered dataset, the output is a dataset too. We will use a credit card transaction example to illustrate autoencoders in this chapter.

Why even bother learning the presentation of the original input only to reconstruct the output as well as possible? The answer is that when you have input with many features, generating a compressed representation via the hidden layers of the neural network could help in compressing the input of the training sample. So, when the neural network goes through all the training data and fine-tunes the weights of all the hidden layer nodes, what will happen is that the weights will truly represent the kind of input that we typically see it- free of noise or other artifacts. As a result, if you try to input some other type of data, such as data with some noise, the autoencoder network will be able to detect the noise and remove at least some portion of the noise when generating the output. This is fantastic because it enables us to potentially remove noise from, for example, images of cats and dogs. As a real-world example, security monitoring cameras often capture hazy, unclear pictures at night or during adverse weather, resulting in noisy images. An autoencoder network can help remove at least some of that noise.

The logic behind denoising autoencoders is that by training on clean, normal images, it is possible to detect and remove noise from noisy images because this noise is not some salient characteristic of the data. For the code, you can find the notebook on GitHub at `https://github.com/apress/beginning-anomaly-detection-python-deep-learning-2e/blob/master/Chapter%206%20Autoencoders/chapter6_autoencoder.ipynb`.

Figure 6-4 shows the basic code to import all necessary packages in Jupyter Notebook.

```
import tensorflow.keras as keras
from tensorflow.keras import optimizers
from tensorflow.keras import losses
from tensorflow.keras.models import Sequential, Model
from tensorflow.keras.layers import Dense, Input, Dropout, Embedding, LSTM
from tensorflow.keras.optimizers import RMSprop, Adam, Nadam
from tensorflow.keras.preprocessing import sequence
from tensorflow.keras.callbacks import TensorBoard
from tensorflow.keras import regularizers

import sklearn
from sklearn.preprocessing import StandardScaler
from sklearn.model_selection import train_test_split
from sklearn.metrics import confusion_matrix, roc_auc_score
from sklearn.preprocessing import MinMaxScaler
from sklearn.metrics import classification_report

import seaborn as sns
import pandas as pd
import numpy as np
import matplotlib

import matplotlib.pyplot as plt
import matplotlib.gridspec as gridspec
%matplotlib inline

import tensorflow
import sys
print("Python: ", sys.version)

print("pandas: ", pd.__version__)
print("numpy: ", np.__version__)
print("seaborn: ", sns.__version__)
print("matplotlib: ", matplotlib.__version__)
print("sklearn: ", sklearn.__version__)
print("Tensorflow: ", tensorflow.__version__)
```

Figure 6-4. Importing packages in Jupyter Notebook

The outputs are as follows:

```
Python:  3.8.12 (default, Oct 12 2021, 03:01:40) [MSC v.1916 64 bit
(AMD64)]
pandas:  2.0.0
numpy:  1.22.2
seaborn:  0.11.2
matplotlib:  3.7.1
sklearn:  1.2.2
Tensorflow:  2.7.0
```

Shown below is code to visualize the results via a confusion matrix, chart for the anomalies and the chart for the errors (difference between predicted and truth) while training. Shown below (Figure 6-5) is `Visualization` helper class.

```
class Visualization:
    labels = ["Normal", "Anomaly"]

    def draw_confusion_matrix(self, y, ypred):
        matrix = confusion_matrix(y, ypred)

        plt.figure(figsize=(10, 8))
        colors=[ "orange","green"]
        sns.heatmap(matrix, xticklabels=self.labels,
yticklabels=self.labels, cmap=colors, annot=True, fmt="d")
        plt.title("Confusion Matrix")
        plt.ylabel('Actual')
        plt.xlabel('Predicted')
        plt.show()


    def draw_anomaly(self, y, error, threshold):
        groupsDF = pd.DataFrame({'error': error,
                                 'true': y}).groupby('true')

        figure, axes = plt.subplots(figsize=(12, 8))

        for name, group in groupsDF:
            axes.plot(group.index, group.error, marker='x' if name == 1
else 'o', linestyle='',
                      color='r' if name == 1 else 'g', label="Anomaly" if
name == 1 else "Normal")

        axes.hlines(threshold, axes.get_xlim()[0], axes.get_xlim()[1],
colors="b", zorder=100, label='Threshold')
        axes.legend()

        plt.title("Anomalies")
        plt.ylabel("Error")
        plt.xlabel("Data")
        plt.show()

    def draw_error(self, error, threshold):
            plt.plot(error, marker='o', ms=3.5, linestyle='',
                     label='Point')

            plt.hlines(threshold, xmin=0, xmax=len(error)-1, colors="b",
zorder=100, label='Threshold')
            plt.legend()
            plt.title("Reconstruction error")
            plt.ylabel("Error")
            plt.xlabel("Data")
            plt.show()
```

Figure 6-5. *Visualization helpers*

We will use the example of credit card data to detect whether a transaction is normal/expected or abnormal/anomalous. Figure 6-6 shows the data being loaded into a Pandas DataFrame. You can download this dataset at `https://www.kaggle.com/datasets/mlg-ulb/creditcardfraud`

```
filePath = '../data/creditcard.csv'
df = pd.read_csv(filepath_or_buffer=filePath, header=0, sep=',')
print(df.shape[0])
df.head()
```

Figure 6-6. *The pandas dataframe*

The dataset is also hosted at `https://github.com/apress/beginning-anomaly-detection-python-deep-learning-2e/blob/master/data/creditcard.csv.zip`. Simply extract the .zip file locally to obtain creditcard.csv.

You should see output like that shown in Table 6-1 after executing the code in Figure 6-6.

Table 6-1. *Truncated Output of df.head() from Code in Figure 6-6*

	Time	V1	V2	...	V27	V28	Amount	Class
0	0.0	-1.359807	-0.072781	...	0.133558	-0.021053	149.62	0
1	0.0	1.191857	0.266151	...	-0.008983	0.014724	2.69	0
2	1.0	-1.358354	-1.340163	...	-0.055353	-0.059752	378.66	0
3	1.0	-0.966272	-0.185226	...	0.062723	0.061458	123.50	0
4	2.0	-1.158233	0.877737	...	0.219422	0.215153	69.99	0

We will collect 20,000 normal records and 400 abnormal records. You can pick different ratios, but in general, using more normal data examples than abnormal data examples is better, as you want to teach your autoencoder what normal data looks like. Too much abnormal data in training will train the autoencoder to learn that the anomalies are actually normal, which goes against our goal. Figure 6-7 shows sampling the Dataframe choosing a majority of normal data.

```
df['Amount'] =
StandardScaler().fit_transform(df['Amount'].values.reshape(-1, 1))
df0 = df.query('Class == 0').sample(20000)
df1 = df.query('Class == 1').sample(400)
df = pd.concat([df0, df1])
```

Figure 6-7. *Sampling the Dataframe choosing a majority of normal data*

Split the dataframe into training and testing data (80-20 split), as shown in Figure 6-8.

```
x_train, x_test, y_train, y_test =
train_test_split(df.drop(labels=['Time', 'Class'], axis = 1) ,
                                              df['Class'],
test_size=0.2, random_state=42)
print(x_train.shape, 'train samples')
print(x_test.shape, 'test samples')
```

Figure 6-8. *Splitting the data into test and train using 20% as holdout test data*

You should see printed output like this:

```
(16320, 29) train samples
(4080, 29) test samples
```

Now it's time to create a simple neural network model with just an encoder layer and a decoder layer. We will encode the 29 columns of the input credit card data set into 12 features using the encoder. The decoder expands the 12 features back into the 29 features. Figure 6-9 shows the code to create the neural network.

```
#simple autoencoder
logfilename = "simpleautoencoder"

encoding_dim = 12
input_dim = x_train.shape[1]

inputArray = Input(shape=(input_dim,))
encoded = Dense(encoding_dim, activation='relu')(inputArray)

decoded = Dense(input_dim, activation=None)(encoded)

autoencoder = Model(inputArray, decoded)
autoencoder.summary()
```

Figure 6-9. *Creating the simple autoencoder neural network*

You should see output like the following:

```
Model: "model"
_____
 Layer (type)                Output Shape              Param #
=================================================================
 input_1 (InputLayer)        [(None, 29)]              0

 dense (Dense)               (None, 12)                360

 dense_1 (Dense)             (None, 29)                377

=================================================================
Total params: 737
Trainable params: 737
Non-trainable params: 0
```

If you look at the code above you will see we used the following activation function:

- **ReLU**: The Rectified Linear Unit is the most commonly used activation function in deep learning models. The function returns 0 if it receives any negative input, but for any positive value xx it returns that value. So it can be written as

  $f(x) = max(0, x)$.

Several other activation functions are available, and you can refer to the Keras documentation to look at the options: `https://keras.io/activations/`.

Now, compile the model using RMSprop as the optimizer and mean-squared error for the loss computation. The RMSprop optimizer is similar to the gradient descent algorithm with momentum. A metric function is similar to a loss function, except that the results from evaluating a metric are not used when training the model. You may use as a metric function any of the loss functions listed at `https://keras.io/losses/`. Figure 6-10 shows the code to compile the model using mean absolute error (MAE) and accuracy as metrics.

```
autoencoder.compile(optimizer=RMSprop(),
                    loss='mean_squared_error',
                    metrics=['mae', 'accuracy'])
```

Figure 6-10. *Compiling the model*

Now, we can start training the model using the training dataset using the testing dataset to validate the model at every step. We chose 32 as the batch size and 20 epochs. Figure 6-11 shows the code to train the model, which is the most time-consuming part of the code.

```
batch_size = 32
epochs = 20

history = autoencoder.fit(x_train, x_train,
                    batch_size=batch_size,
                    epochs=epochs,
                    verbose=1,
                    shuffle=True,
                    validation_data=(x_test, x_test),

callbacks=[TensorBoard(log_dir='./logs/{0}'.format(logfilename))])
```

Figure 6-11. *Training the model*

As you see the training process outputs the loss, accuracy as well as the validation loss and validation accuracy at each epoch. Shown below (Figure 6-12) is output of the training step.

```
Epoch 1/20
510/510 [==============================] - 6s 6ms/step - loss: 1.5464 -
mae: 0.6881 - accuracy: 0.0986 - val_loss: 1.5224 - val_mae: 0.6708 -
val_accuracy: 0.2414

...

Epoch 14/20
510/510 [==============================] - 2s 5ms/step - loss: 1.4690 -
mae: 0.6569 - accuracy: 0.4714 - val_loss: 1.4596 - val_mae: 0.6467 -
val_accuracy: 0.4596
Epoch 15/20
510/510 [==============================] - 3s 5ms/step - loss: 1.4689 -
mae: 0.6569 - accuracy: 0.4711 - val_loss: 1.4595 - val_mae: 0.6467 -
val_accuracy: 0.4615
Epoch 16/20
510/510 [==============================] - 3s 5ms/step - loss: 1.4690 -
mae: 0.6569 - accuracy: 0.4700 - val_loss: 1.4595 - val_mae: 0.6467 -
val_accuracy: 0.4623
Epoch 17/20
510/510 [==============================] - 3s 5ms/step - loss: 1.4689 -
mae: 0.6569 - accuracy: 0.4718 - val_loss: 1.4596 - val_mae: 0.6467 -
val_accuracy: 0.4551
Epoch 18/20
510/510 [==============================] - 3s 6ms/step - loss: 1.4688 -
mae: 0.6568 - accuracy: 0.4710 - val_loss: 1.4594 - val_mae: 0.6467 -
val_accuracy: 0.4610
Epoch 19/20
510/510 [==============================] - 3s 5ms/step - loss: 1.4688 -
mae: 0.6568 - accuracy: 0.4709 - val_loss: 1.4597 - val_mae: 0.6467 -
val_accuracy: 0.4556
Epoch 20/20
510/510 [==============================] - 3s 6ms/step - loss: 1.4688 -
mae: 0.6568 - accuracy: 0.4715 - val_loss: 1.4593 - val_mae: 0.6466 -
val_accuracy: 0.4608
```

Figure 6-12. *The progress of training phase*

You may have noticed the TensorBoard callback. TensorBoard is a utility tool that is typically automatically installed alongside TensorFlow. It allows you to visualize different metrics related to the training process of each model. Here, with the path specified to log_dir, TensorBoard creates a folder named logs within the same folder. To start TensorBoard, execute the following in a command prompt window:

```
tensorboard -logdir=log_file_path
```

If you are using Conda, make sure you're in the correct Conda environment. If you have Jupyter Notebook or JupyterLab, you can open a terminal instance and type the following, making sure log_file_path points to the logs folder:

```
tensorboard --logdir=./logs
```

Once the command prompt output tells you that TensorBoard is being hosted at some URL (such as `http://localhost:6006/`), navigate to that website to access the TensorBoard UI, where you will find the outputs of your training runs.

Figure 6-13 shows the graph of the model as visualized by TensorBoard (in the Graphs tab).

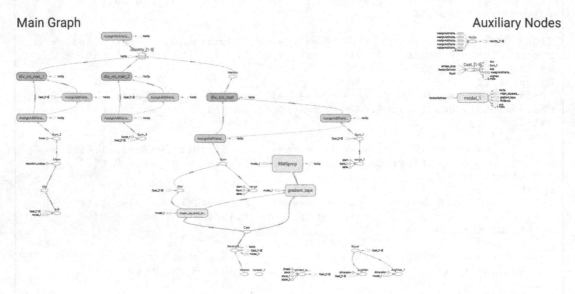

Figure 6-13. *Model graph shown in TensorBoard. It is large and very detailed, so if you want to analyze the computational graph of your model, view your model's graph in your TensorBoard UI*

Figure 6-14 shows the plotting of the accuracy during the training process through the epochs of training.

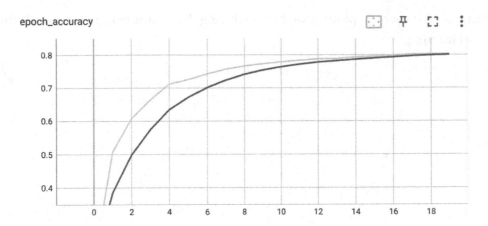

Figure 6-14. *Plotting of accuracy shown in TensorBoard*

Figure 6-15 shows the plotting of the mean absolute error during the training process through the epochs of training.

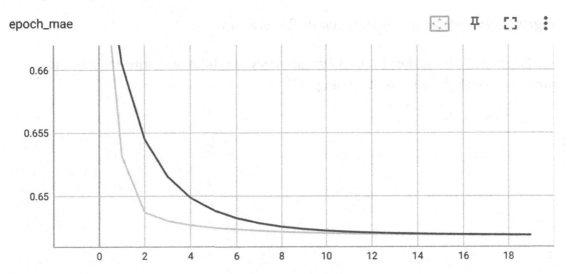

Figure 6-15. *Plotting of MAE shown in TensorBoard*

Figure 6-16 shows the plotting of the loss during the training process through the epochs of training.

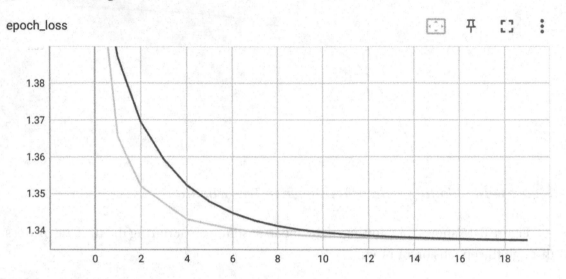

Figure 6-16. *Plotting of loss shown in TensorBoard*

Figure 6-17 shows the plotting of the accuracy of validation during the training process through the epochs of training.

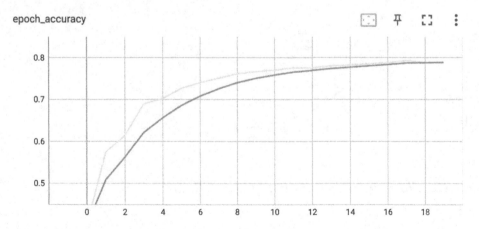

Figure 6-17. *Plotting of validation accuracy shown in TensorBoard*

Figure 6-18 shows the plotting of the loss of validation during the training process through the epochs of training.

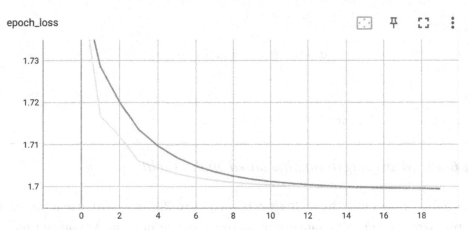

Figure 6-18. *Plotting of validation loss shown in TensorBoard*

Now that the training process is complete, let's evaluate the model for loss and accuracy by running the code shown in Figure 6-19.

```
score = autoencoder.evaluate(x_test, x_test, verbose=1)
print('Test loss:', score[0])
print('Test accuracy:', score[1])
```

Figure 6-19. *Code to evaluate the model*

You should see output like this, with pretty good accuracy of 0.79:

```
128/128 [==============================] - 0s 3ms/step - loss: 1.6993 -
mae: 0.6686 - accuracy: 0.7892
Test loss: 1.6993268728256226
Test accuracy: 0.6685927510261536
```

The next step is to calculate the errors, detect anomalies, then plot the anomalies and errors. We chose threshold of 10. Figure 6-20 shows the code to measure anomalies based on threshold.

```
threshold=10.00
y_pred = autoencoder.predict(x_test)
y_dist = np.linalg.norm(x_test - y_pred, axis=-1)
z = zip(y_dist >= threshold, y_dist)
y_label=[]
error = []
for idx, (is_anomaly, y_dist) in enumerate(z):
    if is_anomaly:
        y_label.append(1)
    else:
        y_label.append(0)
    error.append(y_dist)
```

Figure 6-20. *Measuring anomalies based on threshold*

Let's delve deeper into the code shown in Figure 6-20, as we will use it throughout the chapter when we classify data points as anomalies or normal. As you can see, this is based on a special parameter called the threshold. We are simply looking at error (difference between actual and predicted) and comparing it to the threshold. First calculate the precision and recall for `threshold=10`. Figure 6-21a shows the code to show the precision and recall.

```
print(classification_report(y_test,y_label))
```

	precision	recall	f1-score	support
0	0.99	0.99	0.99	3987
1	0.57	0.62	0.59	93
accuracy			0.98	4080
macro avg	0.78	0.81	0.79	4080
weighted avg	0.98	0.98	0.98	4080

Figure 6-21a. *Code to show the precision and recall*

Let's use the code shown in Figure 6-20 to calculate the precision and recall for `threshold=1.0`, and `threshold=5.0`. Figures 6-21b, and 6-21c show the output for the respective thresholds. Note: your exact precision recall values may not match what is depicted due to the random initialization of the neural network weights. You are welcome to also try `threshold=15.0`.

```
print(classification_report(y_test,y_label))
```

	precision	recall	f1-score	support
0	0.00	0.00	0.00	3987
1	0.02	1.00	0.04	93
accuracy			0.02	4080
macro avg	0.01	0.50	0.02	4080
weighted avg	0.00	0.02	0.00	4080

Figure 6-21b. *Code to show the precision and recall for* threshold=1.0

```
print(classification_report(y_test,y_label))
```

	precision	recall	f1-score	support
0	1.00	0.77	0.87	3987
1	0.09	0.96	0.16	93
accuracy			0.77	4080
macro avg	0.54	0.86	0.51	4080
weighted avg	0.98	0.77	0.85	4080

Figure 6-21c. *Code to show the precision and recall for* threshold=5.0

If you observe the three classification reports, you can see that the precision and recall columns are not good (very low values for row 0 and row 1) for `threshold=1` and `threshold=5`. They look better for `threshold=10`. We want as high a precision and recall as we can get for both row 0 and row 1. In fact, `threshold=10` performed the best, with better precision and recall than for `threshold=1` or `threshold=5`. However, being so simple a model, the performance leaves something to be desired.

Picking a threshold is a matter of experimentation in this and other models and changes as per the data being trained on.

Figure 6-21d shows the code for computing the area undercurve (AUC) score (0.0 to 1.0) and the output of 0.806.

```
roc_auc_score(y_test, y_label)
```

```
0.806310023706077
```

Figure 6-21d. *Code to show AUC*

We can now visualize the confusion matrix to see how well we did with the model, as shown in Figure 6-22.

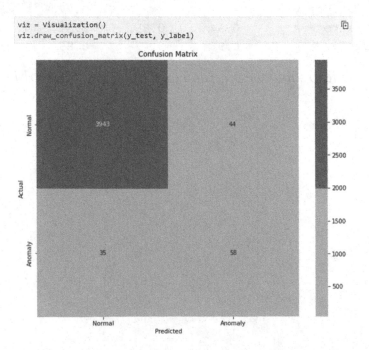

Figure 6-22. *Confusion matrix*

Now, using the predictions of the labels (normal or anomaly), we can plot the anomalies in comparison to the normal data points. Figure 6-23 shows the anomalies based on threshold.

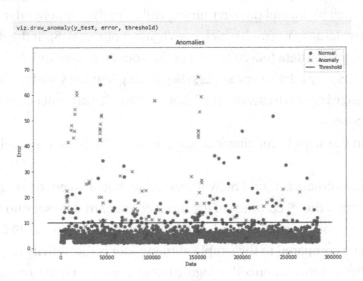

Figure 6-23. *Anomalies based on threshold*

Sparse Autoencoders

In the example of simple autoencoders in the previous section, the representations were constrained only by the size of the hidden layer (12). In such a situation, what typically happens is that the hidden layer is learning an approximation of principal component analysis (PCA). But another way to constrain the representations to be compact is to add a sparsity constraint on the activity of the hidden representations, so fewer units would fire at a given time. In Keras, this can be done by adding an `activity_regularizer` to our `Dense` layer.

The difference between the simple autoencoders and sparse autoencoders is mostly due to the regularization term being added to the loss during training.

In this section, we will use the same credit card dataset as in the simple autoencoder example. We will use the example of credit card data to detect whether a transaction is normal/expected or abnormal/anomalous. Shown below is the data being loaded into Pandas Dataframe.

Then, we will collect 20,000 normal records and 400 abnormal records. You can pick different ratios, but in general, using more normal data examples than abnormal data examples is better, as you want to teach your autoencoder what normal data looks like. Too much abnormal data in training will train the autoencoder to learn that the anomalies are actually normal, which goes against our goal. Split the dataframe into training and testing data (80-20 split). For the code, you can find the notebook on GitHub at `https://github.com/apress/beginning-anomaly-detection-python-deep-learning-2e/blob/master/Chapter%206%20Autoencoders/chapter6_autoencoder.ipynb`.

Simply rerun the steps in the previous section but with the new model definition in Figure 6-24.

Now it's time to create a neural network model with just an encoder and a decoder layer. We will encode the 29 columns of the input credit card data set into 12 features using the encoder. The decoder expands the 12 features back into the 29 features. The key difference compared to the simple autoencoder is the activity regularizer to accommodate the sparse autoencoder. Figure 6-24 shows the code to create the neural network.

```
#sparse autoencoder
logfilename = "sparseautoencoder"

encoding_dim = 12
input_dim = x_train.shape[1]

inputArray = Input(shape=(input_dim,))
encoded = Dense(encoding_dim, activation='relu',
               activity_regularizer=regularizers.l1(10e-5))(inputArray)

decoded = Dense(input_dim, activation='softmax')(encoded)

autoencoder = Model(inputArray, decoded)
autoencoder.summary()
```

Figure 6-24. *Code to create the neural network*

The model output should look somewhat like the following:

```
Model: "model_1"
_____
Layer (type)                 Output Shape              Param #
=================================================================
input_2 (InputLayer)         [(None, 29)]              0

dense_2 (Dense)              (None, 12)                360

dense_3 (Dense)              (None, 29)                377

=================================================================
Total params: 737
Trainable params: 737
Non-trainable params: 0
```

Figure 6-25 shows the graph of the model as visualized by TensorBoard.

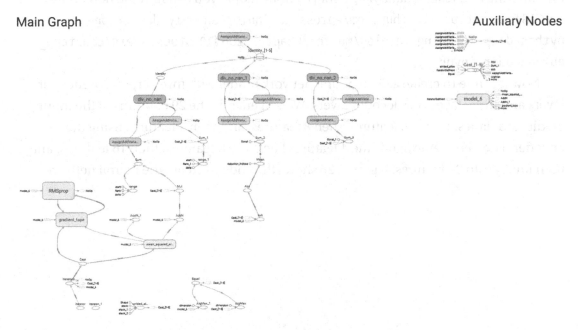

Figure 6-25. *Model graph created by TensorBoard*

Deep Autoencoders

Instead of limiting ourselves to a single layer as encoder or decoder, we could use a stack of layers. It's not a good idea to use too many hidden layers (to avoid possible overfitting), and how many layers is optimal depends on the use case, so you have to experiment to determine the optimal number of layers and the compressions.

The only thing that really changes is the number of layers.

We will use the example of credit card data to detect whether a transaction is normal/expected or abnormal/anomalous. Shown below is the data being loaded into Pandas Dataframe.

We will collect 20,000 normal records and 400 abnormal records. Again, you can pick different ratios, but in general, using more normal data examples than abnormal data examples is better, as you want to teach your autoencoder what normal data looks like. Too much abnormal data in training will train the autoencoder to learn that the anomalies are actually normal, which goes against our goal. Split the dataframe into training and testing data (80-20 split). For the code, you can find the notebook on GitHub at `https://github.com/apress/beginning-anomaly-detection-python-deep-learning-2e/blob/master/Chapter%206%20Autoencoders/chapter6_autoencoder.ipynb`.

Now it's time to create a deep neural network model with three layers for encoder layers and three layers for decoder layers. We will encode the 29 columns of the input credit card data set into 16 features, then 8 features, and then 4 features using the encoder. The decoder expands the 4 features back into 8 features, then 16 features, and then finally into 29 features. Figure 6-26 shows the code to create the neural network.

```
#deep autoencoder
logfilename = "deepautoencoder"

encoding_dim = 16
input_dim = x_train.shape[1]

inputArray = Input(shape=(input_dim,))
encoded = Dense(encoding_dim, activation='relu')(inputArray)
encoded = Dense(8, activation='relu')(encoded)
encoded = Dense(4, activation='relu')(encoded)

decoded = Dense(8, activation='relu')(encoded)
decoded = Dense(encoding_dim, activation='relu')(decoded)
decoded = Dense(input_dim, activation='softmax')(decoded)

autoencoder = Model(inputArray, decoded)
autoencoder.summary()
```

Figure 6-26. *Code to create the neural network*

Figure 6-27 shows the graph of the model as visualized by TensorBoard.

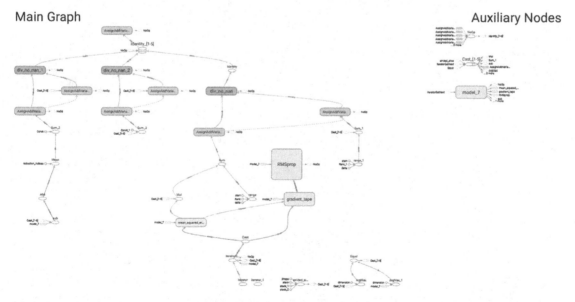

Figure 6-27. *Model graph shown in TensorBoard*

Convolutional Autoencoders

Whenever our inputs are images, it makes sense to use convolutional neural networks (convnets or CNNs) as encoders and decoders. In practical settings, autoencoders applied to images are always convolutional autoencoders, as they simply perform much better.

Let's implement an encoder that consists of a stack of `Conv2D` and `MaxPooling2D` layers (max pooling being used for spatial downsampling), and a decoder that consists of a stack of `Conv2D` and `UpSampling2D` layers.

For the code, you can find the notebook on GitHub at `https://github.com/apress/beginning-anomaly-detection-python-deep-learning-2e/blob/master/Chapter%20 6%20Autoencoders/chapter6_cnnautoencoder.ipynb`.

Figure 6-28 shows the basic code to import all necessary packages. Also note the versions of various packages we are using.

```
import tensorflow.keras as keras
from tensorflow.keras import optimizers
from tensorflow.keras import losses
from tensorflow.keras.models import Sequential, Model
from tensorflow.keras.layers import Dense, Input, Dropout, Embedding, LSTM
from tensorflow.keras.optimizers import RMSprop, Adam, Nadam
from tensorflow.keras.preprocessing import sequence
from tensorflow.keras.callbacks import TensorBoard
from tensorflow.keras import regularizers

import sklearn
from sklearn.preprocessing import StandardScaler
from sklearn.model_selection import train_test_split
from sklearn.metrics import confusion_matrix, roc_auc_score
from sklearn.preprocessing import MinMaxScaler

import seaborn as sns
import pandas as pd
import numpy as np
import matplotlib

import matplotlib.pyplot as plt
import matplotlib.gridspec as gridspec
%matplotlib inline

import tensorflow
import sys
print("Python: ", sys.version)

print("pandas: ", pd.__version__)
print("numpy: ", np.__version__)
print("seaborn: ", sns.__version__)
print("matplotlib: ", matplotlib.__version__)
print("sklearn: ", sklearn.__version__)
print("Tensorflow: ", tensorflow.__version__)
```

Figure 6-28. *Importing packages in Jupyter Notebook*

We will use the MNIST images dataset for this purpose. MNIST contains images for the digits 0 to 9 and is used for many different use cases. Figure 6-29 show the code to load the MNIST data.

```
from keras.datasets import mnist
import numpy as np
(x_train, _), (x_test, _) = mnist.load_data()
```

Figure 6-29. *Code to load MNIST data*

Split the dataset into training and testing subsets and reshape the MNIST data to 28×28 images, as shown in Figure 6-30.

```
from keras.datasets import mnist
import numpy as np

(x_train, _), (x_test, y_test) = mnist.load_data()

x_train = x_train.astype('float32') / 255.
x_test = x_test.astype('float32') / 255.
x_train = np.reshape(x_train, (len(x_train), 28, 28, 1))  # adapt this if
using `channels_first` image data format
x_test = np.reshape(x_test, (len(x_test), 28, 28, 1))   # adapt this if
using `channels_first` image data format
```

Figure 6-30. *Code to transform the images from MNIST*

Create a CNN (convolutional neural network) model with Convolutions and MaxPool layers. Figure 6-31 shows the code to create the neural network.

```
from keras.layers import Input, Dense, Conv2D, MaxPooling2D, UpSampling2D
from keras.models import Model
from keras import backend as K

#cnn autoencoder
logfilename = "cnnautoencoder2"

input_img = Input(shape=(28, 28, 1))  # adapt this if using
`channels_first` image data format

x = Conv2D(16, (3, 3), activation='relu', padding='same')(input_img)
x = MaxPooling2D((2, 2), padding='same')(x)
x = Conv2D(8, (3, 3), activation='relu', padding='same')(x)
x = MaxPooling2D((2, 2), padding='same')(x)
x = Conv2D(8, (3, 3), activation='relu', padding='same')(x)
encoded = MaxPooling2D((2, 2), padding='same')(x)

# at this point the representation is (4, 4, 8) i.e. 128-dimensional
x = Conv2D(8, (3, 3), activation='relu', padding='same')(encoded)
x = UpSampling2D((2, 2))(x)
x = Conv2D(8, (3, 3), activation='relu', padding='same')(x)
x = UpSampling2D((2, 2))(x)
x = Conv2D(16, (3, 3), activation='relu')(x)
x = UpSampling2D((2, 2))(x)
decoded = Conv2D(1, (3, 3), activation='sigmoid', padding='same')(x)

autoencoder = Model(input_img, decoded)

autoencoder.summary()
```

Figure 6-31. *Code to create the neural network*

Compile the model using RMSprop as the optimizer and mean-squared error for the loss computation. The RMSprop optimizer is similar to the gradient descent algorithm with momentum. Figure 6-32 show the code to compile the model.

```
autoencoder.compile(optimizer=RMSprop(),
                    loss='mean_squared_error',
                    metrics=['mae', 'accuracy'])
```

Figure 6-32. *Code to compile the model*

Now, we can start training the model using the training dataset using the testing dataset to validate the model at every step. We chose 32 as the batch size and 20 epochs. As you see the training process outputs the loss, accuracy as well as the validation loss and validation accuracy at each epoch. Shown below (Figure 6-33) is model being trained.

```
batch_size = 32
epochs = 20

history = autoencoder.fit(x_train, x_train,
                    batch_size=batch_size,
                    epochs=epochs,
                    verbose=1,
                    shuffle=True,
                    validation_data=(x_test, x_test),

callbacks=[TensorBoard(log_dir='./logs/{0}'.format(logfilename))])
```

Figure 6-33. *Model being trained*

Now that the training process is complete, let's evaluate the model for loss and accuracy, as shown in Figure 6-34.

```
score = autoencoder.evaluate(x_test, x_test, verbose=1)
print('Test loss:', score[0])
print('Test accuracy:', score[1])
```

Figure 6-34. *Code to evaluate the model*

The output should look like this, with pretty good accuracy of 0.81:

```
313/313 [==============================] - 2s 5ms/step - loss: 0.0110 -
mae: 0.0353 - accuracy: 0.8126
Test loss: 0.011009062640368938
Test accuracy: 0.035254064947366714
```

The next step is to use the model to generate the output images for the testing subset. This will show us how well the reconstruction phase is going. Figure 6-35a shows the code to predict based on the model.

```
decoded_imgs = autoencoder.predict(x_test)

n = 10
plt.figure(figsize=(20, 4))
for i in range(1, n):
    # display original
    ax = plt.subplot(2, n, i)
    plt.imshow(x_test[i].reshape(28, 28))
    plt.gray()
    ax.get_xaxis().set_visible(False)
    ax.get_yaxis().set_visible(False)

    # display reconstruction
    ax = plt.subplot(2, n, i + n)
    plt.imshow(decoded_imgs[i].reshape(28, 28))
    plt.gray()
    ax.get_xaxis().set_visible(False)
    ax.get_yaxis().set_visible(False)
plt.show()
```

Figure 6-35a. *Code to predict based on the model*

Figure 6-35b shows the original images (top row) and the reconstructions (bottom row).

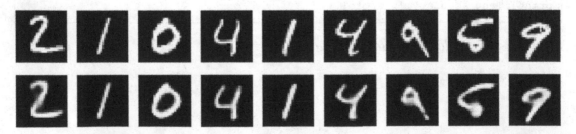

Figure 6-35b. *Output reconstructions of the model generated by Figure 6-35a*

We can also see how the encoder phase is working by displaying the test subset images in this phase. Figure 6-36a and Figure 6-36b shows the encoded images themselves.

```
encoder = Model(input_img, encoded)
encoded_imgs = encoder.predict(x_test)
n = 10
plt.figure(figsize=(20, 8))
for i in range(1, n):
    ax = plt.subplot(1, n, i)
    plt.imshow(encoded_imgs[i].reshape(4, 4 * 8).T)
    plt.gray()
    ax.get_xaxis().set_visible(False)
    ax.get_yaxis().set_visible(False)
plt.show()
```

Figure 6-36a. *Code to display encoded images*

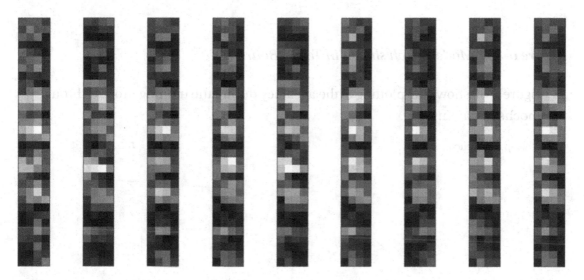

Figure 6-36b. *Displayed encoded images*

Figure 6-37 shows a graph of the model created by TensorBoard.

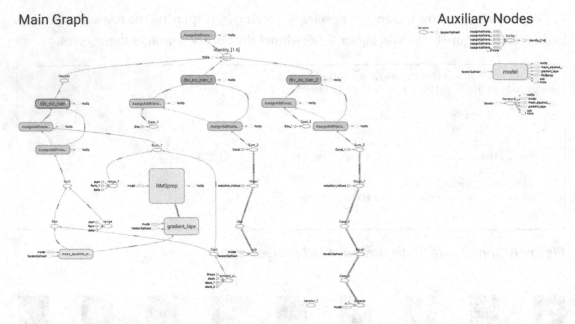

Figure 6-37. *Model graph shown in TensorBoard*

Figure 6-38 shows the plotting of the accuracy during the training process through the epochs of training.

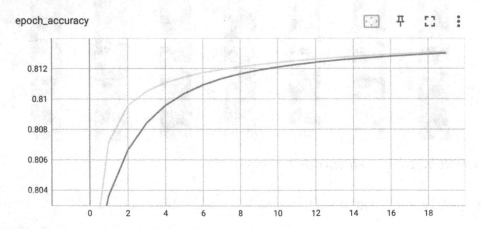

Figure 6-38. *Plotting of accuracy shown in TensorBoard*

Figure 6-39 shows the plotting of the loss during the training process through the epochs of training.

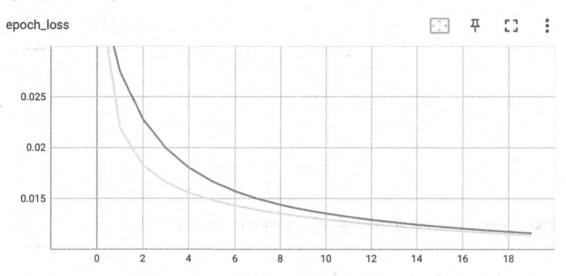

Figure 6-39. *Plotting of loss shown in TensorBoard*

Figure 6-40 shows the plotting of the accuracy of validation during the training process through the epochs of training.

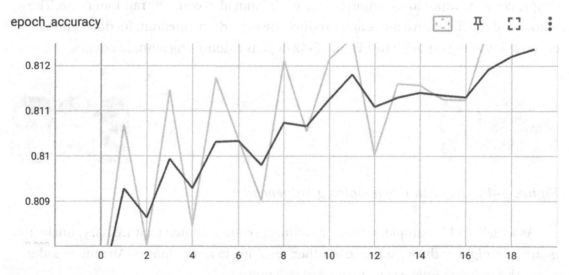

Figure 6-40. *Plotting of validation accuracy shown in TensorBoard*

Figure 6-41 shows the plotting of the loss of validation during the training process through the epochs of training.

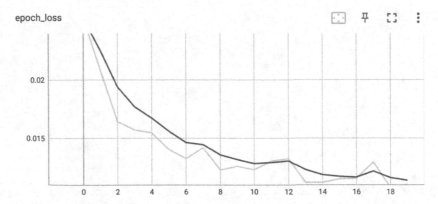

Figure 6-41. *Plotting of validation loss shown in TensorBoard*

Denoising Autoencoders

We can force the autoencoder to learn useful features by adding random noise to its inputs and making it recover the original noise-free data. This way the autoencoder can't simply copy the input to its output because the input also contains random noise. The autoencoder will remove noise and produce the underlying meaningful data. This is called a *denoising autoencoder.* Figure 6-42 depicts a denoising autoencoder.

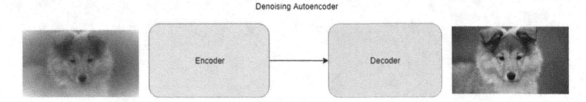

Figure 6-42. *Depiction of denoising autoencoder*

As a real-world example, security monitoring cameras often capture hazy, unclear pictures at night or during adverse weather, resulting in noisy images. An autoencoder network can help remove at least some of that noise.

The logic behind denoising autoencoders is that by training on clean, normal images, it is possible to detect and remove noise from noisy images because this noise is not some salient characteristic of the data.

For the code, you can find the notebook on GitHub at https://github.com/apress/
beginning-anomaly-detection-python-deep-learning-2e/blob/master/Chapter%20
6%20Autoencoders/chapter6_denoisingautoencoder.ipynb

Figure 6-43 shows the basic code to import all necessary packages. Also note the
versions of various packages we are using.

```
import tensorflow.keras as keras
from tensorflow.keras import optimizers
from tensorflow.keras import losses
from tensorflow.keras.models import Sequential, Model
from tensorflow.keras.layers import Dense, Input, Dropout, Embedding, LSTM
from tensorflow.keras.optimizers import RMSprop, Adam, Nadam
from tensorflow.keras.preprocessing import sequence
from tensorflow.keras.callbacks import TensorBoard
from tensorflow.keras import regularizers

import sklearn
from sklearn.preprocessing import StandardScaler
from sklearn.model_selection import train_test_split
from sklearn.metrics import confusion_matrix, roc_auc_score
from sklearn.preprocessing import MinMaxScaler

import seaborn as sns
import pandas as pd
import numpy as np
import matplotlib

import matplotlib.pyplot as plt
import matplotlib.gridspec as gridspec
%matplotlib inline

import tensorflow
import sys
print("Python: ", sys.version)

print("pandas: ", pd.__version__)
print("numpy: ", np.__version__)
print("seaborn: ", sns.__version__)
print("matplotlib: ", matplotlib.__version__)
print("sklearn: ", sklearn.__version__)
print("Tensorflow: ", tensorflow.__version__)
```

Figure 6-43. *Code to import packages*

We will use MNIST images dataset for this purpose. MNIST contains images for the
digits 0 to 9 and is used for many different use cases. Figure 6-44 shows the code to load
MNIST images.

```
from keras.datasets import mnist
import numpy as np
(x_train, _), (x_test, _) = mnist.load_data()
```

Figure 6-44. *Code to load MNIST images*

Split the dataset into training and testing subsets and reshape the MNIST data to 28×28 images, as shown in Figure 6-45.

```
from keras.datasets import mnist
import numpy as np

(x_train, _), (x_test, y_test) = mnist.load_data()

x_train = x_train.astype('float32') / 255.
x_test = x_test.astype('float32') / 255.
x_train = np.reshape(x_train, (len(x_train), 28, 28, 1))  # adapt this if
using `channels_first` image data format
x_test = np.reshape(x_test, (len(x_test), 28, 28, 1))  # adapt this if
using `channels_first` image data format

noise_factor = 0.3
x_train_noisy = x_train + noise_factor * np.random.normal(loc=0.0,
scale=1.0, size=x_train.shape)
x_test_noisy = x_test + noise_factor * np.random.normal(loc=0.0,
scale=1.0, size=x_test.shape)

x_train_noisy = np.clip(x_train_noisy, 0., 1.)
x_test_noisy = np.clip(x_test_noisy, 0., 1.)

print(x_train_noisy.shape)
print(x_test_noisy.shape)
print(y_test.shape)
```

Figure 6-45. *Code to load and reshape images*

The output should look as follows:

```
(60000, 28, 28, 1)
(10000, 28, 28, 1)
(10000,)
```

Figure 6-46a shows the code to display the images.

```
n = 11
plt.figure(figsize=(20, 2))
for i in range(1, n):
    ax = plt.subplot(1, n, i)
    plt.imshow(x_test_noisy[i].reshape(28, 28))
    plt.gray()
    ax.get_xaxis().set_visible(False)
    ax.get_yaxis().set_visible(False)
plt.show()
```

Figure 6-46a. *Code to display the images*

Figure 6-46b. *The displayed output of Figure 6-46a*

Create a CNN model with Convolutions and MaxPool layers. Figure 6-47 shows the code to create the neural network.

```
from keras.layers import Input, Dense, Conv2D, MaxPooling2D, UpSampling2D
from keras.models import Model
from keras import backend as K

#cnn autoencoder
logfilename = "DenoisingAutoencoder2"

input_img = Input(shape=(28, 28, 1))  # adapt this if using
`channels_first` image data format

x = Conv2D(16, (3, 3), activation='relu', padding='same')(input_img)
x = MaxPooling2D((2, 2), padding='same')(x)
x = Conv2D(8, (3, 3), activation='relu', padding='same')(x)
x = MaxPooling2D((2, 2), padding='same')(x)
x = Conv2D(8, (3, 3), activation='relu', padding='same')(x)
encoded = MaxPooling2D((2, 2), padding='same')(x)

# at this point the representation is (4, 4, 8) i.e. 128-dimensional

x = Conv2D(8, (3, 3), activation='relu', padding='same')(encoded)
x = UpSampling2D((2, 2))(x)
x = Conv2D(8, (3, 3), activation='relu', padding='same')(x)
x = UpSampling2D((2, 2))(x)
x = Conv2D(16, (3, 3), activation='relu')(x)
x = UpSampling2D((2, 2))(x)
decoded = Conv2D(1, (3, 3), activation='sigmoid', padding='same')(x)

autoencoder = Model(input_img, decoded)


autoencoder.summary()
```

Figure 6-47. *Code to create the neural network*

Compile the model using RMSprop as the optimizer and mean-squared error for the loss computation. The RMSprop optimizer is similar to the gradient descent algorithm with momentum. Figure 6-48 shows the code to compile the model.

```
#autoencoder.compile(optimizer='adadelta', loss='binary_crossentropy')
autoencoder.compile(optimizer=RMSprop(),
                    loss='mean_squared_error',
                    metrics=['accuracy'])
```

Figure 6-48. *Code to compile the model*

Now, we can start training the model using the training dataset using the testing dataset to validate the model at every step. We chose 32 as the batch size and 20 epochs. As you see the training process outputs the loss, accuracy as well as the validation loss and validation accuracy at each epoch. Shown below (Figure 6-49) is code to start training of the model.

```
batch_size = 32
epochs = 20

history = autoencoder.fit(x_train_noisy, x_train,
                    batch_size=batch_size,
                    epochs=epochs,
                    verbose=1,
                    shuffle=True,
                    validation_data=(x_test_noisy, x_test),

callbacks=[TensorBoard(log_dir='./logs/{0}'.format(logfilename))])
```

Figure 6-49. *Code to start training of the model*

Now that the training process is complete, let's evaluate the model for loss and accuracy, as shown in Figure 6-50.

```
score = autoencoder.evaluate(x_test, x_test, verbose=1)
print('Test loss:', score[0])
print('Test accuracy:', score[1])
```

Figure 6-50. *Code to evaluate the model*

The output should look like this, with pretty good accuracy of 0.81, indicating that most of the pixels are able to be reconstructed:

```
313/313 [==============================] - 1s 4ms/step - loss: 0.0115 -
accuracy: 0.8123
Test loss: 0.011469859629869461
Test accuracy: 0.8122942447662354
```

The next step is to use the model to generate the output images for the testing subset. This will show us how well the reconstruction phase is going. Figure 6-51a shows the code to display denoised images.

```
decoded_imgs = autoencoder.predict(x_test_noisy)

n = 10
plt.figure(figsize=(20, 4))
for i in range(1, n):
    # display original
    ax = plt.subplot(2, n, i)
    plt.imshow(x_test_noisy[i].reshape(28, 28))
    plt.gray()
    ax.get_xaxis().set_visible(False)
    ax.get_yaxis().set_visible(False)

    # display reconstruction
    ax = plt.subplot(2, n, i + n)
    plt.imshow(decoded_imgs[i].reshape(28, 28))
    plt.gray()
    ax.get_xaxis().set_visible(False)
    ax.get_yaxis().set_visible(False)
plt.show()
```

Figure 6-51a. *Code to display denoised images*

The denoised images are displayed in Figure 6-51b.

Figure 6-51b. *The displayed denoised images*

We can also see how the encoder phase is working by displaying the test subset images in this phase. Figure 6-52a shows the code to display encoded images.

```
encoder = Model(input_img, encoded)
encoded_imgs = encoder.predict(x_test_noisy)
n = 10
plt.figure(figsize=(20, 8))
for i in range(1, n):
    ax = plt.subplot(1, n, i)
    plt.imshow(encoded_imgs[i].reshape(4, 4 * 8).T)
    plt.gray()
    ax.get_xaxis().set_visible(False)
    ax.get_yaxis().set_visible(False)
plt.show()
```

Figure 6-52a. *Code to display encoded images*

The encoded images are shown in Figure 6-52b.

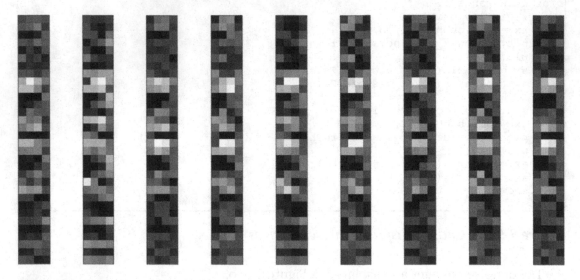

Figure 6-52b. *Visualizing the encoded images*

Figure 6-53 shows the graph of the model as visualized by TensorBoard.

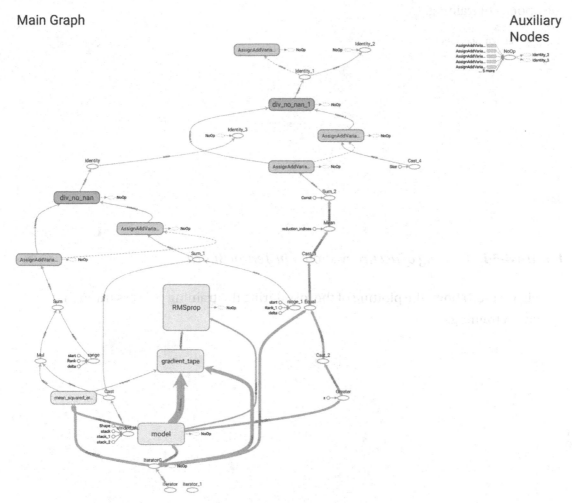

Figure 6-53. *Model graph shown in TensorBoard*

Figure 6-54 shows the plotting of the accuracy during the training process through the epochs of training.

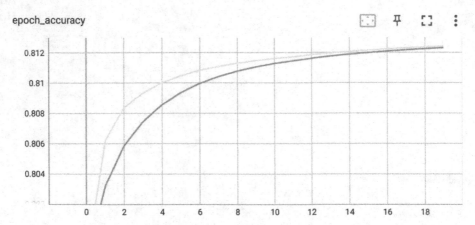

Figure 6-54. *Plotting of accuracy shown in TensorBoard*

Figure 6-55 shows the plotting of the loss during the training process through the epochs of training.

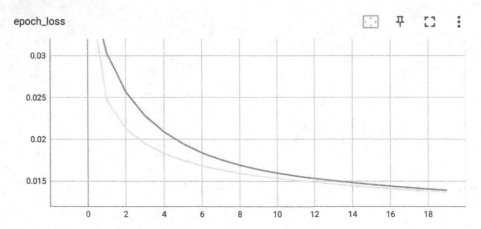

Figure 6-55. *Plotting of loss shown in TensorBoard*

Figure 6-56 shows the plotting of the accuracy of validation during the training process through the epochs of training.

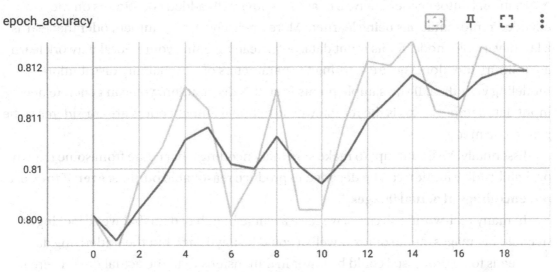

Figure 6-56. *Plotting of validation accuracy shown in TensorBoard*

Figure 6-57 shows the plotting of the loss of validation during the training process through the epochs of training.

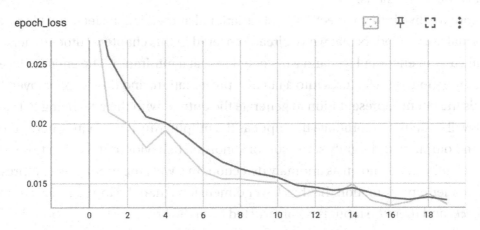

Figure 6-57. *Plotting of validation loss shown in TensorBoard*

Variational Autoencoders

A variational autoencoder is a type of autoencoder with added constraints on the encoded representations being learned. More precisely, it is an autoencoder that learns a latent variable model for its input data. So instead of letting your neural network learn an arbitrary function, you are learning the parameters of a probability distribution modeling your data. If you sample points from this distribution, you can generate new input data samples. This is the reason why variational autoencoders are considered to be generative models.

Essentially, VAEs attempt to make sure that encodings that come from some known probability distribution can be decoded to produce reasonable outputs, **even if they are not encodings of actual images**.

In many real-world use cases, we have a whole bunch of data that we're looking at (it could be images, audio or text...well, it could be anything), but the underlying data that needs to be processed could be lower in dimensions than the actual data. Therefore, many of the machine learning models involve some sort of dimensionality reduction. One very popular technique is singular value decomposition or principal component analysis. Similarly, in deep learning space, variational autoencoders do the task of reducing the dimensions.

Before we dive into the mechanics of variational autoencoders, let's quickly recap the normal autoencoders that we've already covered in this chapter. Autoencoders basically use an encoder layer and a decoder layer, at a minimum. The encoder layer reduces the input data features into a latent representation, and the decoder layer expands the latent representation to generate the output, with the goal being to train the model well enough to reproduce the input as the output. Any discrepancy between the input and output could signify some sort of abnormal behavior or deviation from what is normal, otherwise known as anomaly detection. In a way, the input gets compressed into a smaller representation that has fewer dimensions than the input, known as the *bottleneck*, and then the input is reconstructed from the bottleneck.

By contrast, with a variational autoencoder, instead of mapping the input to a fixed vector, we map the input to a distribution, so the big difference is that the bottleneck vector seen in the normal order in quarters is replaced with the mean vector and a standard deviation vector by looking at the distributions and then taking the sampled latent vector as the actual bottleneck. Clearly this is very different from the normal autoencoder, where the input directly yields a latent vector.

First, an encoder network turns the input sample x into two parameters in a latent space, which we will call z_mean and z_log_sigma. Then, the encoder network randomly samples similar points z from the latent normal distribution that is assumed to generate the data, via z = z_mean + exp(z_log_sigma) * epsilon, where epsilon is a random normal tensor. Finally, a decoder network maps these latent space points back to the original input data. Figure 6-58 depicts the variational encoder neural network.

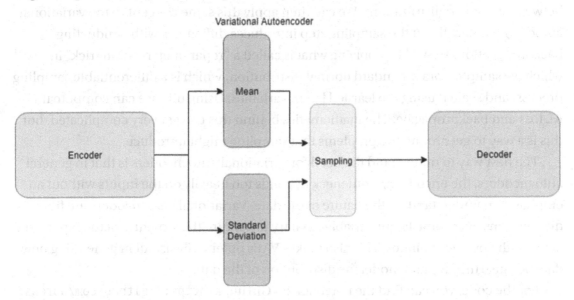

Figure 6-58. *Depiction of the variational encoder neural network*

The parameters of the model are trained via two loss functions: a reconstruction loss forcing the decoded samples to match the initial inputs (just like in the previously covered autoencoders), and the KL divergence between the learned latent distribution and the prior distribution, acting as a regularization term. You could actually get rid of this latter term entirely, although it does help in learning well-formed latent spaces and reducing overfitting to the training data.

The distribution being learned by this latent space is not too different from a Gaussian distribution. However, before we train the variational autoencoder, we have to address the issue with sampling. We sample some vector from the latent space and pass that into the decoder. However, this breaks the flow of data from the input to the output because the sampling process is random and not differentiable, so we can't backpropagate in a straightforward manner like in a usual neural network.

A variational autoencoder is kind of a mix of a neural network and a graphical model. The first paper that came up on variational autoencoders tried to create a graphical model and then turn the graphical model into a neural network. The variational autoencoder is based on variational inference.

In variational inference, assume that there are two different probabilitstic distributions p(x) and q(x). We can then use KL divergence to measure the dissimilarity between these two distributions. We can then apply this same concept to the variational autoencoder. Recall that the sampling step introduces difficulties with conducting backpropagation. We will be applying what is called a "reparameterization trick" in which we sample from a standard normal distribution, which is a differentiable sampling process, and scale it using the learned latent variables. Using this, we can compute the KL loss and backpropagate. The mathematics behind this can get very complicated, but this is a way to get around the problems that sampling might introduce.

The best way to understand the need for variational autoencoders is that in general autoencoders, the encoding / bottleneck depends too heavily on the inputs without an emphasis on understanding the nature of the data. Variational autoencoders are fitting distributions over some latent variables that try to capture the structure of the input data along with any uncertainties. This also makes VAEs potentially useful in generating new data because they try and model the distribution of the data.

For the code, you can find the notebook on GitHub at `https://github.com/apress/beginning-anomaly-detection-python-deep-learning-2e/blob/master/Chapter%20 6%20Autoencoders/chapter6_variationalautoencoder.ipynb`.

Figure 6-59 shows the basic code to import all necessary packages. Also note the versions of various packages we are using.

```
import tensorflow.keras as keras
from tensorflow.keras import optimizers
from tensorflow.keras import losses
from tensorflow.keras import backend as K
from tensorflow.keras.models import Sequential, Model
from tensorflow.keras.layers import Lambda, Dense, Input, Dropout,
Embedding, LSTM
from tensorflow.keras.optimizers import RMSprop, Adam, Nadam
from tensorflow.keras.preprocessing import sequence
from tensorflow.keras.callbacks import TensorBoard
from tensorflow.keras.losses import mse, binary_crossentropy

import sklearn
from sklearn.preprocessing import StandardScaler
from sklearn.model_selection import train_test_split
from sklearn.metrics import confusion_matrix, roc_auc_score
from sklearn.preprocessing import MinMaxScaler

import seaborn as sns
import pandas as pd
import numpy as np
import matplotlib

import matplotlib.pyplot as plt
import matplotlib.gridspec as gridspec
%matplotlib inline

import tensorflow
import sys
print("Python: ", sys.version)

print("pandas: ", pd.__version__)
print("numpy: ", np.__version__)
print("seaborn: ", sns.__version__)
print("matplotlib: ", matplotlib.__version__)
print("sklearn: ", sklearn.__version__)
print("Tensorflow: ", tensorflow.__version__)
```

Figure 6-59. *Importing packages in Jupyter Notebook*

Shown below is code to visualize the results via a confusion matrix, chart for the anomalies and the chart for the errors (difference between predicted and truth) while training. Shown below (Figure 6-60) is code to visualize the results.

```python
class Visualization:
    labels = ["Normal", "Anomaly"]

    def draw_confusion_matrix(self, y, ypred):
        matrix = confusion_matrix(y, ypred)

        plt.figure(figsize=(10, 8))
        colors=[ "orange","green"]
        sns.heatmap(matrix, xticklabels=self.labels,
yticklabels=self.labels, cmap=colors, annot=True, fmt="d")
        plt.title("Confusion Matrix")
        plt.ylabel('Actual')
        plt.xlabel('Predicted')
        plt.show()


    def draw_anomaly(self, y, error, threshold):
        groupsDF = pd.DataFrame({'error': error,
                                 'true': y}).groupby('true')

        figure, axes = plt.subplots(figsize=(12, 8))

        for name, group in groupsDF:
            axes.plot(group.index, group.error, marker='x' if name == 1
else 'o', linestyle='',
                      color='r' if name == 1 else 'g', label="Anomaly" if
name == 1 else "Normal")

        axes.hlines(threshold, axes.get_xlim()[0], axes.get_xlim()[1],
colors="b", zorder=100, label='Threshold')
        axes.legend()

        plt.title("Anomalies")
        plt.ylabel("Error")
        plt.xlabel("Data")
        plt.show()

    def draw_error(self, error, threshold):
            plt.plot(error, marker='o', ms=3.5, linestyle='',
                     label='Point')

            plt.hlines(threshold, xmin=0, xmax=len(error)-1, colors="b",
zorder=100, label='Threshold')
            plt.legend()
            plt.title("Reconstruction error")
            plt.ylabel("Error")
            plt.xlabel("Data")
            plt.show()
```

Figure 6-60. *Code to visualize the results*

We will use the example of credit card data to detect whether a transaction is normal/expected or abnormal/anomalous. Figure 6-61 shows the data being loaded into a Pandas Dataframe.

```
filePath = '../data/creditcard.csv'
df = pd.read_csv(filepath_or_buffer=filePath, header=0, sep=',')
print(df.shape[0])
df.head()
```

Figure 6-61. *Code to load the dataset using pandas*

We will collect 20,000 normal records and 400 abnormal records. You can pick different ratios, but in general, using more normal data examples than abnormal data examples is better, as you want to teach your autoencoder what normal data looks like. Too much abnormal data in training will train the autoencoder to learn that the anomalies are actually normal, which goes against our goal. Figure 6-62 shows the code to take a majority of normal data records with fewer abnormal records.

```
df['Amount'] =
StandardScaler().fit_transform(df['Amount'].values.reshape(-1, 1))
df0 = df.query('Class == 0').sample(20000)
df1 = df.query('Class == 1').sample(400)
df = pd.concat([df0, df1])
```

Figure 6-62. *Code to take a majority of normal data records with fewer abnormal records*

Split the dataframe into training and testing data (80-20 split), as shown in Figure 6-63.

```
x_train, x_test, y_train, y_test =
train_test_split(df.drop(labels=['Time', 'Class'], axis = 1) ,
                                       df['Class'],
test_size=0.2, random_state=42)
print(x_train.shape, 'train samples')
print(x_test.shape, 'test samples')
```

Figure 6-63. *Code to split data into train and test subsets*

The biggest difference between standard autoencoders and variational autoencoders is that here we do not just take the inputs as is; rather, we take the distribution of the input data and then sample the distribution. Figure 6-64 shows the code to implement such a sampling strategy.

```
# reparameterization trick
# instead of sampling from Q(z|X), sample epsilon = N(0,I)
# z = z_mean + sqrt(var) * epsilon
def sampling(args):
    """Reparameterization trick by sampling from an isotropic unit
Gaussian.
    # Arguments
        args (tensor): mean and log of variance of Q(z|X)
    # Returns
        z (tensor): sampled latent vector
    """

    z_mean, z_log_var = args
    batch = K.shape(z_mean)[0]
    dim = K.int_shape(z_mean)[1]
    # by default, random_normal has mean = 0 and std = 1.0
    epsilon = K.random_normal(shape=(batch, dim))
    return z_mean + K.exp(0.5 * z_log_var) * epsilon
```

Figure 6-64. *Code to sample the distributions*

Now it's time to create a simple neural network model with an encoder phase and a decoder phase. We will encode the 29 columns of the input credit card data set into 12 features using the encoder. The encoder uses the special distribution sampling logic to generate two parallel layers and then wraps the sampling output (Figure 6-64) as a Layer object.

The decoder phase uses this latent vector and reconstructs the input. While doing this, it also measures the error of reconstruction in order to minimize it. Figure 6-65 shows the code to create the neural network.

```
#variational autoencoder
logfilename = "variationalautoencoder"


original_dim  = x_train.shape[1]

print(original_dim)

input_shape = (original_dim,)
intermediate_dim = 12
batch_size = 32
latent_dim = 2
epochs = 20

# VAE model = encoder + decoder
# build encoder model
inputs = Input(shape=input_shape, name='encoder_input')
x = Dense(intermediate_dim, activation='relu')(inputs)
z_mean = Dense(latent_dim, name='z_mean')(x)
z_log_var = Dense(latent_dim, name='z_log_var')(x)

# use reparameterization trick to push the sampling out as input
# note that "output_shape" isn't necessary with the TensorFlow backend
z = Lambda(sampling, output_shape=(latent_dim,), name='z')([z_mean,
z_log_var])

# instantiate encoder model
encoder = Model(inputs, [z_mean, z_log_var, z], name='encoder')
encoder.summary()

# build decoder model
latent_inputs = Input(shape=(latent_dim,), name='z_sampling')
x = Dense(intermediate_dim, activation='relu')(latent_inputs)
outputs = Dense(original_dim, activation='sigmoid')(x)

# instantiate decoder model
decoder = Model(latent_inputs, outputs, name='decoder')
decoder.summary()

# instantiate VAE model
outputs = decoder(encoder(inputs)[2])
vae = Model(inputs, outputs, name='vae_mlp')


# VAE loss = mse_loss or xent_loss + kl_loss
reconstruction_loss = mse(inputs, outputs)

reconstruction_loss *= original_dim
kl_loss = 1 + z_log_var - K.square(z_mean) - K.exp(z_log_var)
kl_loss = K.sum(kl_loss, axis=-1)
kl_loss *= -0.5
vae_loss = K.mean(reconstruction_loss + kl_loss)
vae.add_loss(vae_loss)
```

Figure 6-65. *Code to create the neural network*

Figure 6-66 show the encoder and decoder architecture displayed by Figure 6-65.

```
29
Model: "encoder"


 Layer (type)                    Output Shape            Param #          Connected
 to
==========================================================================================
 encoder_input (InputLayer)      [(None, 29)]            0                []

 dense (Dense)                   (None, 12)              360
['encoder_input[0][0]']

 z_mean (Dense)                  (None, 2)               26
['dense[0][0]']

 z_log_var (Dense)               (None, 2)               26
['dense[0][0]']

 z (Lambda)                      (None, 2)               0
['z_mean[0][0]',

'z_log_var[0][0]']


==========================================================================================
Total params: 412
Trainable params: 412
Non-trainable params: 0


Model: "decoder"

 Layer (type)                    Output Shape            Param #
================================================================================
 z_sampling (InputLayer)         [(None, 2)]             0

 dense_1 (Dense)                 (None, 12)              36

 dense_2 (Dense)                 (None, 29)              377

================================================================================
Total params: 413
Trainable params: 413
Non-trainable params: 0
```

Figure 6-66. *Code to show the neural network*

Compile the model using Adam as the optimizer and mean-squared error for the loss computation. As discussed in Chapter 5, Adam is an optimization algorithm that can be used instead of the classical stochastic gradient descent procedure to update network weights iteratively based on training data. Figure 6-67 shows the code to compile the model.

```
vae.compile(optimizer='adam',
                    loss='mean_squared_error',
                    metrics=['accuracy'])
vae.summary()
```

Figure 6-67. *Code to compile the model*

Now, we can start training the model using the training dataset using the testing dataset to validate the model at every step. We chose 32 as the batch size and 20 epochs. As you see the training process outputs the loss, accuracy as well as the validation loss and validation accuracy at each epoch. Shown below (Figure 6-68) is code to train the model.

```
history = vae.fit(x_train, x_train,
                    batch_size=batch_size,
                    epochs=epochs,
                    verbose=1,
                    shuffle=True,
                    validation_data=(x_test, x_test),

callbacks=[TensorBoard(log_dir='./logs/{0}'.format(logfilename))])
```

Figure 6-68. *Code to train the model*

Now that the training process is complete, let's evaluate the model for loss and accuracy, as shown in Figure 6-69.

```
score = vae.evaluate(x_test, x_test, verbose=1)
print('Test loss:', score[0])
print('Test accuracy:', score[1])
```

Figure 6-69. *Code to evaluate the model*

You should see output that looks similar to this, with accuracy of 0.35:

```
128/128 [==============================] - 0s 3ms/step - loss: 46.3963 -
accuracy: 0.3449
Test loss: 46.39627456665039
Test accuracy: 0.3448529541492462
```

The next step is to calculate the errors, detect anomalies, then plot the anomalies and errors. Figure 6-70 shows the code to predict anomalies based on a threshold value of 10.00.

```
threshold=10.00
y_pred = vae.predict(x_test)
y_dist = np.linalg.norm(x_test - y_pred, axis=-1)
z = zip(y_dist >= threshold, y_dist)
y_label=[]
error = []
for idx, (is_anomaly, y_dist) in enumerate(z):
    if is_anomaly:
        y_label.append(1)
    else:
        y_label.append(0)
    error.append(y_dist)
```

Figure 6-70. *Code to predict the anomalies based on threshold*

Figure 6-71 shows the code to compute the AUC score (0.0 to 1.0) and the output of 0.92, which is very high.

```
roc_auc_score(y_test, y_label)
```

0.9159364709499422

Figure 6-71. *Code to calculate AUC*

We can now visualize the confusion matrix to see how well we did with the model, as shown in Figure 6-72.

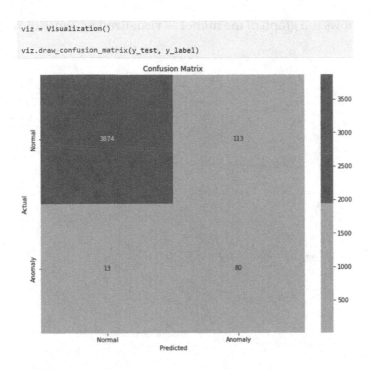

Figure 6-72. *Confusion matrix*

Now, using the predictions of the labels (normal or anomaly), we can plot the anomalies in comparison to the normal data points. Figure 6-73 shows the anomalies relative to the threshold.

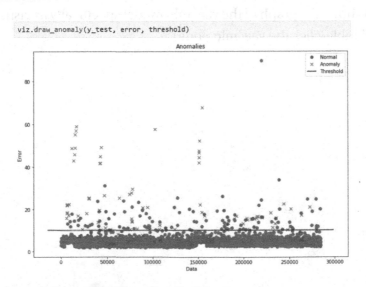

Figure 6-73. *Anomalies relative to threshold*

Figure 6-74 shows the graph of the model as visualized by TensorBoard.

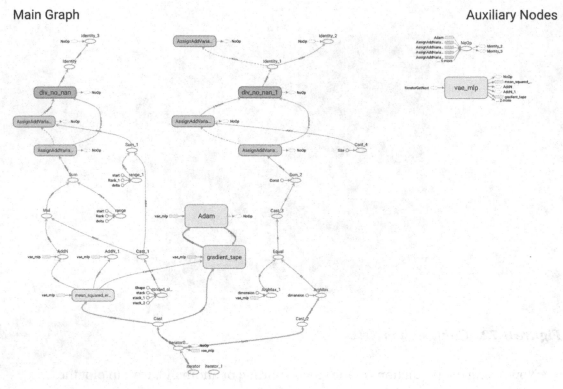

Figure 6-74. *Model graph shown in TensorBoard*

Figure 6-75 shows the graph of the vae_mlp model specifically as visualized by TensorBoard (double-click the vae_mlp node).

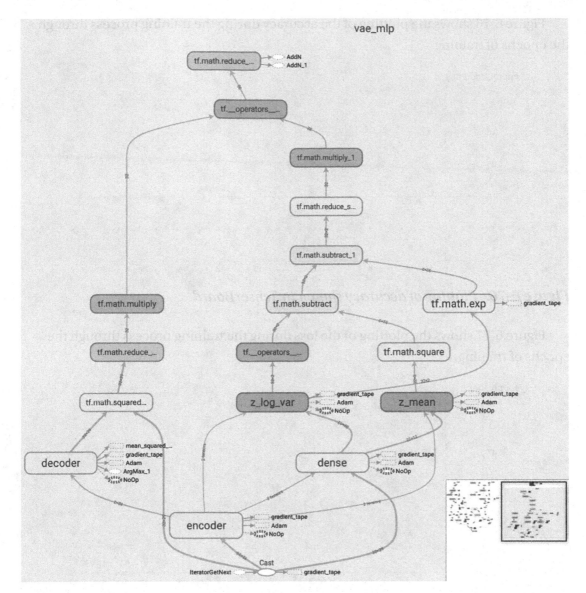

Figure 6-75. *Model graph shown in TensorBoard*

Figure 6-76 shows the plotting of the accuracy during the training process through the epochs of training.

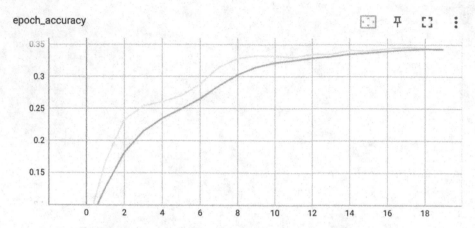

Figure 6-76. *Plotting of accuracy shown in TensorBoard*

Figure 6-77 shows the plotting of the loss during the training process through the epochs of training.

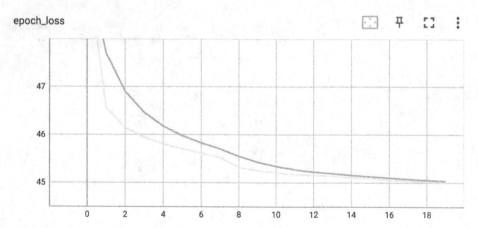

Figure 6-77. *Plotting of loss shown in TensorBoard*

Figure 6-78 shows the plotting of the accuracy of validation during the training process through the epochs of training.

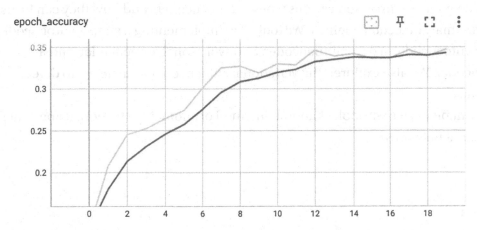

Figure 6-78. *Plotting of validation accuracy shown in TensorBoard*

Figure 6-79 shows the plotting of the loss of validation during the training process through the epochs of training.

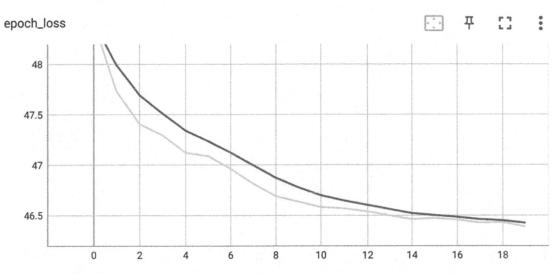

Figure 6-79. *Plotting of validation loss shown in TensorBoard*

Summary

In this chapter, we discussed various types of autoencoders and how they can be used to build anomaly detection engines. We looked at implementing a simple autoencoder, a sparse autoencoder, a deep autoencoder, a convolutional autoencoder, and a denoising autoencoder. We also explored the variational autoencoder as a means to detect anomalies.

In Chapter 7, we will look at another method of anomaly detection, using generative adversarial networks.

CHAPTER 7

Generative Adversarial Networks

In this chapter, you will learn about generative adversarial networks as well as how you can implement anomaly detection using them.

In a nutshell, this chapter covers the following topics:

- What is a generative adversarial network?
- Anomaly detection with generative adversarial networks

Note Code examples are provided in Python 3.8. The code repository for this book is available at `https://github.com/apress/beginning-anomaly-detection-python-deep-learning-2e/tree/master`.

The repository also includes a requirements.txt file to check your packages and their versions.

The notebooks for this chapter can be found at `https://github.com/apress/beginning-anomaly-detection-python-deep-learning-2e/blob/master/Chapter%207%20Generative%20Adversarial%20Networks/chapter7_gan.ipynb`.

Navigate to "Chapter 7 Generative Adversarial Networks" and then click the notebook. The code is provided as .py files as well, though it is the exported version of the notebook.

We will be using JupyterLab to present all of the code examples.

What Is a Generative Adversarial Network?

A generative adversarial network (GAN) is a class of neural network in which two models, a **generator** and a **discriminator**, are trained in tandem. In a typical GAN setup, the role of the **generator** is to produce data points that resemble some real dataset. It then passes these generated data points to the **discriminator**, whose job is to classify these generated points as either real or fake.

The ideal for the generator is to generate data that the discriminator predicts as all real. The ideal for the discriminator, then, is to predict the generated data as all fake. To teach the discriminator what real data looks like, the discriminator is also shown real data throughout the training process. And so, the generator and discriminator are locked in an adversarial cat-and-mouse game where both models are continuously trying to improve and beat the other.

It's a simple concept, yet incredibly powerful. GANs are able to create realistic examples of images, speech, and art, among other things. They have gotten so good at these tasks, in fact, that GANs have been the subject of controversy recently because they have been able to create images, videos, and voices that convincingly portray a real person. You can imagine the sheer danger of this technology through the potential for misuse by malicious entities.

That being said, there are plenty of good uses for GANs:

- **AI-generated art**: Users can bring any type of concept to life by typing in a prompt. The result is high-quality artwork that takes mere seconds to create. However, some of these models have recently become the subject of controversy because of how their training data was sourced. For example, some versions of these models use scraped data that was uploaded by their artists online but taken without permission for training. In some cases, models are trained specifically to replicate a specific artist's style without the knowledge or permission of the original artist.

  Is the art that is generated by the GAN its own creation, or is it plagiarism? These are tough questions that policies do not yet address. In cases in which a model is trained to replicate a particular artist's style, labeling it as plagiarism is much easier. But if the model was trained across many styles, and outputs something that resembles bits and pieces of multiple people's art styles together, is that directly plagiarism? Is the art generated by the GAN its own, novel creation? Neural networks are trained to generalize their training data, not to copy and paste it, and they learn much like a human does to create art of a specific style. So, in this case, is it plagiarism of the work of human artists?

- **Photo modification**: In a similar vein to AI-generated art, GANs can be used to remove objects from photos, restore damaged photos, fill in an area with something else, and so on.

- **Dubbing**: In movies, actors can be dubbed in the native language in their own voice and look as if they're actually saying the words with their lips, enhancing the immersion tremendously for audiences. Different GAN models can be used to regenerate the parts of the frames containing the actor's mouths as well as generate the voices in the native languages.

- **Super resolution**: It's as if the old "enhance" trope from spy movies has come to life. AI can now upscale images and fill in the detail in a realistic manner. This is useful for medical imagery, surveillance, or satellites, but also to upscale old media into 4K quality.

- **Drug discovery**: GANs can generate new molecular structures to discover new drugs.

This list is not comprehensive; GANs can do much more. Nearly any generative task that a human can do can be done by a GAN. Songwriting, poetry, art, and other creative endeavors have long been thought to be an exclusively human domain computers were incapable of entering. However, we were able to automate this with AI before we could properly automate a chorebot.

It turns out that GANs are also good at anomaly detection. The process of training GANs teaches the generator and discriminator what the distribution of normal data looks like. Anomalies fall outside of this distribution, allowing us to pinpoint anomalous data from normal data in a straightforward manner.

More specifically, we can use the GAN setup to have a generator create realistic samples that belong to the normal data distribution. The job of the discriminator, then, is to identify whether these points are real or fake. Once the generator and discriminator have been properly trained, the discriminator can be used to identify anomalies since it will classify them as "fake."

The use of GANs for anomaly detection has the following advantages:

- **Flexibility**: The exact architecture of the generator and the discriminator is totally abstract. You could have an LSTM, autoencoder, transformer, convolutional neural network, or any other type of model as the generator and as the discriminator.

  Furthermore, GANs can be applied to any data types, including time-series data, geospatial data, image data, text data, etc.

- **Generative modeling**: The generator creates new data points that fit within the normal data distribution. This can be useful when you want to synthesize additional data samples to be used in another training task, for example, or to better understand the underlying distribution of the original data.

  - This also means that with the proper setup, you can control what kind of samples you want to generate. In the image-generation domain, an example would be controlling the hair color, complexion, hair style, etc. of a generated person.

  In the task of anomaly detection, you can train the GAN to generate specific types of anomalies that a preexisting anomaly detector is bad at detecting. Then, you can use this newly synthesized dataset of anomalies to supplement your training set and boost your anomaly detector's performance against a specific type of anomaly.

- **Unsupervised learning**: Generators can train in a totally unsupervised fashion without the use of any labels. The discriminator only needs to see the data itself and be told that it's real—there is no actual labeling going on.

On the flip side, GANs do have some serious disadvantages:

- **Training sensitivity**: GANs are notoriously difficult to train because the training is all a balancing act. You do not want a discriminator that constantly overpowers the generator, nor a generator that overpowers the discriminator and starts to produce garbage. It's an adversarial competition, not a one-sided stomp, but this can be tricky to achieve.

Balancing the GAN can involve tuning a number of hyperparameters, redefining the architectures of the generator and the discriminator as well as the number of parameters, and so on.

- **Mode collapse**: This is when the generator produces only limited outputs that still fool the discriminator but fail to properly represent the original data distribution.

- **Computational power**: When you're training a GAN, you're training two or more models at once. This can require more memory and processing power than a typical training setup involving one model.

- **Data hungry**: Properly and accurately modeling the data distribution requires providing the GAN with plenty of real data samples.

- **Long training**: Because GANs are so sensitive, they're usually trained with lower training rates than usual. This means that convergence can take many iterations. Rushing this process with higher learning rates can lead to instability, oscillation, and collapse of the training setup.

Despite the possible disadvantages to using GANs, they are still a worthwhile class of model to explore anomaly detection with. But before we get started, we need to check out what the architecture of a GAN looks like.

Generative Adversarial Network Architecture

Figure 7-1 depicts the architecture of a GAN.

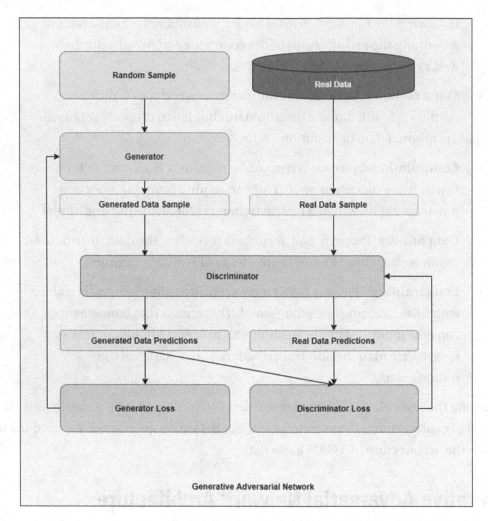

Figure 7-1. *The high-level architecture of a generative adversarial network*

The generator takes a random sample from a space called the *latent space*. This could be a random set of numbers from a normal or uniform distribution, depending on its setup. Using this sample, the generator produces output that, ideally, should look like our real data. This output is then evaluated by the discriminator to determine if it's genuine or synthesized.

The discriminator is also shown a sample of real data, which it must predict as 1 (real) or fake (0). Using these labels, we can compute a loss for the discriminator predictions on the synthesized outputs, and a loss for the predictions on the real outputs.

The generator, on the other hand, wants the discriminator to predict 1 for all the synthesized samples. So, in the loss function, we take the discriminator predictions on the synthesized data and use 1 as the true label.

Recall that the optimizer is trying to minimize the loss, but in this case, we have two competing loss functions: one loss for the generator, and one loss for the discriminator, both at odds with each other in terms of what the true label ought to be for the synthesized outputs.

A simple loss function we can use is binary cross-entropy, and we can assume that the last layer of the discriminator uses sigmoid as the activation function.

However, there are the aforementioned issues with GAN setups, like **mode collapse**, and **training instability/divergence**. To address these issues and stabilize the GAN training further, we will incorporate what is called the **Wasserstein loss** to transform our GAN setup into a WGAN.

Wasserstein GAN

Wasserstein GAN (WGAN) is a type of GAN that uses Wasserstein loss as the loss function, along with a couple other modifications to the training loop. In essence, a WGAN seeks to

- **Stabilize gradients**: The Wasserstein loss function allows for the calculation of more meaningful and stable gradients. In traditional GAN setups, gradients can vanish or explode if the training process becomes imbalanced.

- **Mitigate mode collapse:** Because of the increased stability in gradients, generators can get more consistent feedback from the discriminators, leading to reduced chances of mode collapse where one or two consistent outputs keep getting generated.

- **Enforce Lipschitz constraint:** The Wasserstein loss is structured in a way such that the discriminator is Lipschitz continuous, usually enforced by weight clipping or adding a gradient penalty to ensure that the discriminator after one gradient update does not adjust its weights too much and thus change its behavior too drastically. These changes are to keep the discriminator modifying its behavior at a stable rate and not potentially overpower the generator.

The **Lipschitz constraint** is that the critic (WGAN term for discriminator) must be a 1-Lipschitz function, which effectively imposes a bound on the gradient of the critic and thus avoids rapid rate of change of the critic and stabilizes training.

The discriminator in WGAN setups is referred to as the **critic** because instead of discriminating and assigning a label of real or fake, it gives a score of the realness or fakeness of a sample. Whereas a discriminator has sigmoid as the activation function, a critic has no activiation function applied (linear activation). This is because, instead of a probability, we want the discriminator to output a score whose sign determines the prediction of real or fake and whose magnitude determines the confidence.

With that background in place, let's get into what this loss function looks like.

The loss for the critic is computed as follows:

Critic loss = Average score for real samples – Average score for generated samples

The generator loss is as follows:

Generator loss = -1 * Average Score for Generated Samples

The true label here is 1 and the fake label is –1. As far as the intuition behind these loss function definitions, the optimizer's job is once again to minimize the loss. For the generator, the optimizer wants to make that value as negative as possible, so the scores for the generated samples must be large, positive outputs because the negative sign will make them large-magnitude negative outputs.

The optimizer also wants to minimize the critic loss. Here, the generated sample scores are also incentivized to be large-magnitude positive outputs since that subtraction will turn it into a negative, much like in the generator loss. The only way for the discriminator to improve and further minimize this loss is to produce scores for real samples that are negative, effectively producing two negative terms for the critic loss, minimizing it.

And so, by incentivizing the score for generated samples to be negative and large in magnitude, the generator loss definition works in unison with the critic loss to guide the discriminator into optimally learning how to distinguish between real and fake outputs. If the score is negative, the model is predicting that the data is real. If the score is positive, the model is predicting the data is fake.

The WGAN paper[1] also suggests to train the critic more frequently than the generator, but in our setup, which is a more straightforward extension of a base GAN with Wasserstein loss and gradient penalty principles applied, we don't incorporate this.

---

[1]Arjovsky, Martín et al. "Wasserstein Generative Adversarial Networks." International Conference on Machine Learning (2017).

Another step to perform with a WGAN is weight clipping. The weight of the discriminator can be clipped between some +- range of a parameter **clip_value**. However, we implement the **gradient penalty (GP)** method instead, making our implementation an example of **WGAN-GP**.

WGAN-GP

The gradient penalty method is one way to enforce the Lipschitz constraint and stabilize training. Here's how it works:

1. Compute a random interpolation between a pair of random real (x_{real}) and fake (x_{fake}) data:

$$x' = a * x_{real} + (1-a) * x_{fake}$$

where a is some uniformly sampled value in the range [0, 1].

2. Compute the gradient of the critic score using the interpolated sample as input.

3. Compute the 2 norm of the gradient, subtract 1, and square the difference:

$$GP = \left(\left\| \nabla_{x'} D(x') \right\| - 1 \right)^2$$

Here, x' is the interpolation between real and fake data, and D is the discriminator/critic. D(x'), then, is the output of the discriminator using x' as the input data.

4. Apply the penalty to the critic loss like so:

$$\text{Critic loss} = \text{mean}\left(D\left(x_{real}\right)\right) - \text{mean}\left(D\left(x_{fake}\right)\right) + \lambda * GP$$

where GP is as defined above, and λ is a hyperparameter (commonly set to 10).

With this, we can stabilize the learning and enforce the 1-Lipschitz constraint. Now that we have the definition and some theory out of the way, let's get started with implementing a WGAN-GP to perform anomaly detection.

Anomaly Detection with a GAN

We will be applying a GAN to the KDDCUP 1999 dataset to perform **semi-supervised anomaly detection**. Although GANs can be completely unsupervised, ours will be semi-supervised because we want to teach the generator to generate strictly normal data and have the discriminator/critic predict real or fake data. The idea is that the discriminator will be able to pick out the anomalies as fake and classify the real normal points as real. Since we are basing our implementation off of WGAN-GP, we will designate a critic score above 0 as an anomaly and a critic score below 0 as a real data point.

This task is semi-supervised because we still need the anomaly labels to perform the final evaluation while training on data that we know for sure is labeled normal. This task can be modified to be supervised or unsupervised by having the discriminator receive label feedback in the former case. In the latter case, we'd pass all of the data in with the assumption that anomalies look different from normal data and that they're rare. If this is the case, the generator may likely learn to model just the normal data, because the discriminator would more easily be able to identify the anomaly as fake since it stands apart from the normal data, regardless of our assigning this anomaly to be real.

The path of least resistance for the generator that lets it minimize loss is to just produce normal data. However, since there are no guarantees, it's still possible that the generator will learn to produce authentic-looking anomalies and the discriminator will learn that these are typical of the real data distribution even though they are anomalies.

This is why the semi-supervised approach might be more preferable—at the very least, we know that the GAN is being trained on normal data only and that the generator and discriminator will understand what normal data looks like.

Let's get into the code. The notebook is provided at https://github.com/apress/beginning-anomaly-detection-python-deep-learning-2e/blob/master/Chapter%20 7%20Generative%20Adversarial%20Networks/chapter7_gan.ipynb.

The data is provided on GitHub as well at https://github.com/apress/beginning-anomaly-detection-python-deep-learning-2e/blob/master/data/kddcup.data.gz. Simply download the file and extract it to see kddcup.data.corrected, which is the file we will be loading.

Because of the complicated nature of the loss functions necessitating a custom training loop, training a GAN is not as simple as calling model.fit(). TensorFlow offers a guide for a convolutional GAN that shows how to train a GAN and is a good resource if you are curious about GANs in general: https://www.tensorflow.org/tutorials/generative/dcgan.

We have adapted minor portions of the code to create the training loop, though we have modified it to use Wasserstein loss and gradient penalty.

First, we must import all of the necessary packages, as shown in Figure 7-2.

```
import numpy as np
import pandas as pd
import matplotlib.pyplot as plt
from sklearn.model_selection import train_test_split
from sklearn.preprocessing import LabelEncoder, MinMaxScaler
from sklearn.metrics import precision_score, recall_score, f1_score,
confusion_matrix, roc_auc_score
import tensorflow as tf
print(f'TensorFlow: ', tf.__version__)
```

Figure 7-2. *Importing some of the necessary packages to start off our code*

You should see outputs like the following:

Seaborn: 0.12.2

TensorFlow: 2.7.0

Now, let's define all of the columns of the KDDCUP dataset and import it (Figure 7-3).

```
columns = ["duration", "protocol_type", "service", "flag",
"src_bytes", "dst_bytes", "land", "wrong_fragment", "urgent",
        "hot", "num_failed_logins", "logged_in", "num_compromised",
"root_shell", "su_attempted", "num_root",
        "num_file_creations", "num_shells", "num_access_files",
"num_outbound_cmds", "is_host_login",
        "is_guest_login", "count", "srv_count", "serror_rate",
"srv_serror_rate", "rerror_rate", "srv_rerror_rate",
        "same_srv_rate", "diff_srv_rate", "srv_diff_host_rate",
"dst_host_count", "dst_host_srv_count",
        "dst_host_same_srv_rate", "dst_host_diff_srv_rate",
"dst_host_same_src_port_rate", "dst_host_srv_diff_host_rate",
        "dst_host_serror_rate", "dst_host_srv_serror_rate",
"dst_host_rerror_rate", "dst_host_srv_rerror_rate", "label"]

df = pd.read_csv("../data/kddcup.data.corrected", sep=",",
names=columns, index_col=None)
```

Figure 7-3. *Defining the code to read the dataset*

We want to filter the data to only include HTTP attacks, as shown in Figure 7-4.

```
# Filter to only 'http' attacks
df = df[df["service"] == "http"]
df = df.drop("service", axis=1)
```

Figure 7-4. *Only using the HTTP attack data*

Let's also transform the label column into 0 for normal points and 1 for anything that doesn't have a label of normal as shown in Figure 7-5.

```
df['label'] = df['label'].apply(lambda x: 0 if x=='normal.' else 1)
df['label'].value_counts()
```

Figure 7-5. *Label encoding our label column and condensing the labels into 'normal' and 'anomaly'*

The output should look something like this:

```
label
0    619046
1      4045
Name: count, dtype: int64
```

We need to numerically encode all categorical columns. Refer to Figure 7-6.

```
datatypes = dict(zip(df.dtypes.index, df.dtypes))

encoder_map = {}
for col, datatype in datatypes.items():
    if datatype == 'object':
        encoder = LabelEncoder()
        df[col] = encoder.fit_transform(df[col])
        encoder_map[col] = encoder
```

Figure 7-6. *Label encoding the categorical columns and saving them in a dictionary in case we want to use them later*

Next, we want to select columns that are more than weakly correlated with the label column. Refer to Figure 7-7.

```
# Check the variables with highest correlation with 'label'
df2 = df.copy()
label_corr = df2.corr()['label']

# Filter out anything that has null entry or is not weakly correlated
train_cols = label_corr[(~label_corr.isna()) & (np.abs(label_corr) >
0.2)]
train_cols = list(train_cols[:-1].index)
train_cols
```

Figure 7-7. *Code to select only the columns not weakly correlated with the label column*

You should see output like so:

```
['src_bytes',
 'hot',
 'num_compromised',
 'count',
 'serror_rate',
 'srv_serror_rate',
 'same_srv_rate',
 'diff_srv_rate',
 'dst_host_srv_count',
 'dst_host_same_srv_rate',
 'dst_host_diff_srv_rate',
 'dst_host_serror_rate',
 'dst_host_srv_serror_rate']
```

However, to make the training slightly easier, we will remove both 'hot' and 'num_compromised' because they're highly correlated with 'src_bytes.' We want to make the generation task easier for the generator, so removing any unnecessary columns will help. Refer to Figure 7-8.

```
# Removing these two columns as their correlation with 'src_bytes' is
high
remove_cols = ['hot', 'num_compromised']
for r in remove_cols:
    train_cols.remove(r)
```

Figure 7-8. *Removing the unnecessary columns*

We will now rescale the numerical columns to be in the range [0, 1], as shown
in Figure 7-9. This is once again to improve the data quality to help the generator's
learning task.

```
scaler_map = {}
for col in train_cols:
    scaler = MinMaxScaler()
    df2[col] = scaler.fit_transform(df2[col].values.reshape(-1, 1))
    scaler_map[col] = scaler
```

Figure 7-9. *Code to min-max scale the numerical columns and save them to a
dictionary*

Let's now define the training, validation, and testing splits, as shown in Figure 7-10.
We want to differentiate the normal data and the including anomaly data in the
validation and testing splits will both contain some anomaly data. The reason for
including anomaly data in the validation and testing splits is that we can use the
validation data for seed searching or other hyperparameter tuning, because at the end of
each epoch, we can evaluate the discriminator/critic's performance. If needed, we can
also perform early stopping to maximize the validation performance.

```
normal_df = df2[df2['label'] == 0]
anomaly_df = df2[df2['label'] == 1]

labels_norm = normal_df['label']
x_train_norm, x_test_norm, y_train_norm, y_test_norm =
train_test_split(normal_df[train_cols].values, labels_norm.values,
test_size=0.15, random_state=42)

# Additional split of training dataset to create validation split
x_test_anom, x_val_anom, y_test_anom, y_val_anom =
train_test_split(anomaly_df[train_cols].values,
anomaly_df['label'].values, test_size=0.2, random_state=42)

# Additional split of training dataset to create validation split
x_train, x_val_norm, y_train, y_val_norm =
train_test_split(x_train_norm, y_train_norm, test_size=0.1,
random_state=42)
```

Figure 7-10. *Defining the training, testing, and validation data splits*

Let's now create the final test and validation sets by concatenating the corresponding splits for the normal and anomaly data. Refer to Figure 7-11.

```
x_test = np.concatenate((x_test_norm, x_test_anom))
y_test = np.concatenate((y_test_norm, y_test_anom))
x_val = np.concatenate((x_val_norm, x_val_anom))
y_val = np.concatenate((y_val_norm, y_val_anom))
print("Shapes")
print(f"x_train:{x_train.shape}\ny_train:{y_train.shape}")
print(f"\nx_val:{x_val.shape}\ny_val:{y_val.shape}")
print(f"\nx_test:{x_test.shape}\ny_test:{y_test.shape}")
```

Figure 7-11. *Creating the final test and validation splits containing both normal and anomaly data*

You should see output like this:

```
Shapes
x_train:(473570, 11)
y_train:(473570,)
x_val:(53428, 11)
y_val:(53428,)
x_test:(96093, 11)
y_test:(96093,)
```

With our data preparation out of the way, let's start defining the model architecture. We are setting a random seed, meaning we want the output of the model training to be deterministic across architectures as well as initialize the weights in a specific way. Different seeds can lead to different performances in some cases. In the case of GANs, seed search can be a viable method of finding the best weight initialization setting to lead to optimal performance. Changing this can lead to no convergence and poor performance or perhaps an even better convergence.

The seed was found through a seed searching algorithm, the code for which you can access in the notebook referenced earlier in this section. Over 100 different seeds, seed=10, as shown in Figure 7-12, was found to be the best.

Let's first set the imports and the seed = 10, as shown in Figure 7-12.

```
from tensorflow.keras.models import Model
from tensorflow.keras.layers import Input, Dense, LeakyReLU,
BatchNormalization
from tensorflow.keras.regularizers import L2
from tensorflow.keras.callbacks import EarlyStopping
from tensorflow.keras.losses import BinaryCrossentropy
from tensorflow.keras.optimizers import Adam, RMSprop


seed = 10
tf.random.set_seed(seed)
np.random.seed(seed)
```

Figure 7-12. *Defining model-specific imports and setting the random seed for consistency*

Now, we can define the generator, as shown in Figure 7-13.

```
# The input layer requires you to specify the dimensionality of the x-
features (and not the number of samples)
noise_dimension = 50
data_dim = x_train.shape[-1]
## GENERATOR
g_in = Input(shape=(noise_dimension))
g_h1 = Dense(4*data_dim, activation=LeakyReLU(alpha=0.01),
kernel_initializer = 'he_normal')(g_in)
g_bn1 = BatchNormalization()(g_h1)
g_h2 = Dense(4*data_dim, activation=LeakyReLU(alpha=0.01),
kernel_initializer = 'he_normal')(g_bn1)
g_bn2 = BatchNormalization()(g_h2)
g_h3 = Dense(4*data_dim, activation=LeakyReLU(alpha=0.01),
kernel_initializer = 'he_normal')(g_bn2)
g_bn3 = BatchNormalization()(g_h3)
g_h4 = Dense(data_dim, activation=LeakyReLU(alpha=0.01),
kernel_initializer = 'he_normal')(g_bn3)
g_bn4 = BatchNormalization()(g_h4)
g_out = Dense(data_dim, activation='relu', )(g_bn4)


# Creating a model by specifying the input layer and output layer
generator = Model(g_in, g_out)
```

Figure 7-13. *Defining the architecture of the generator*

For the generator, we are sampling a 50-dimensional noise vector. It then passes through several linear layers with incorporated BatchNormalization layers (to normalize the outputs of the layer and further stabilize the network training).

Next, let's define the discriminator, as shown in Figure 7-14.

```
## DISCRIMINATOR
d_in = Input(shape=(data_dim))
d_h1 = Dense(4*data_dim, activation=LeakyReLU(alpha=0.01),
kernel_initializer = 'he_normal', )(d_in)
d_bn1 = BatchNormalization()(d_h1)
d_h2 = Dense(4*data_dim, activation=LeakyReLU(alpha=0.01),
kernel_initializer = 'he_normal', )(d_bn1)
d_bn2 = BatchNormalization()(d_h2)
d_h3 = Dense(2*data_dim, activation=LeakyReLU(alpha=0.01),
kernel_initializer = 'he_normal', )(d_bn2)
d_bn3 = BatchNormalization()(d_h3)
d_h4 = Dense(data_dim, activation=LeakyReLU(alpha=0.01),
kernel_initializer = 'he_normal', )(d_bn3)

d_out = Dense(1, activation='linear',)(d_h4)

# Creating a model by specifying the input layer and output layer
discriminator = Model(d_in, d_out)
```

Figure 7-14. *Defining the discriminator. The output layer has the linear activation function because this discriminator is actually the critic in the WGAN setup and will be outputting unbounded score predictions*

We must now define the optimizers for the generator and the discriminator. We use RMSProp with a generator learning rate of 5e-5 and a discriminator learning rate of 1e-5. We also define the batch size to be 4096 (to speed up training). Changing this might also change the performance of the model. Refer to Figure 7-15.

```
g_optim = RMSprop(5e-5)
d_optim = RMSprop(1e-5)

batch_size = 4096
train_dataset =
tf.data.Dataset.from_tensor_slices(x_train[:]).shuffle(len(x_train[:])
).batch(batch_size)
```

Figure 7-15. *Defining the optimizers as well as the batched training dataset*

Following this, we will define the gradient penalty function. Refer to Figure 7-16.

```
def gradient_penalty(critic, real_data, generated_data):
    alpha = tf.random.uniform([real_data.shape[0], 1], 0., 1.)
    interpolated_data = alpha * tf.cast(real_data, 'float32') + (1. -
alpha) * generated_data
    with tf.GradientTape() as tape:
        tape.watch(interpolated_data)
        critic_interpolated = critic(interpolated_data)

    gradients = tape.gradient(critic_interpolated,
[interpolated_data])[0]
    gradients_norm = tf.sqrt(tf.reduce_sum(tf.square(gradients),
axis=[1]))
    gradient_penalty = tf.reduce_mean((gradients_norm - 1.)**2)

    return gradient_penalty
```

Figure 7-16. *Defining the gradient penalty function*

Finally, we have the training loop, where it all comes together. Refer to Figure 7-17 for some hyperparameter settings.

```
epochs = 10
lambda_gp = 10

g_loss = 0
d_loss = 0
d_loss_real = 0
d_loss_fake = 0
```

Figure 7-17. *Setting some training hyperparameters*

Figure 7-18 shows the code to define the training loop.

```
for e in range(epochs):
    print(f'\nEpoch {e}')
    for i, batch_x in enumerate(train_dataset):
        print(f'\r  {i}/{len(train_dataset)}: g_loss {g_loss} |
d_loss: {d_loss} real {d_loss_real} fake {d_loss_fake}'.ljust(100, '
'), end='')

        random_noise = tf.random.normal([len(batch_x),
noise_dimension])

        with tf.GradientTape() as gen_tape, tf.GradientTape() as
disc_tape:
            g_output = generator(random_noise)

            real_outputs = discriminator(batch_x)
            fake_outputs = discriminator(g_output)

            g_loss = -
1*tf.keras.backend.mean(tf.ones_like(fake_outputs) * fake_outputs)

            d_loss_real =
tf.keras.backend.mean(tf.ones_like(real_outputs) * real_outputs)
            d_loss_fake =
tf.keras.backend.mean(tf.ones_like(fake_outputs) * fake_outputs)

            # # d_loss = d_loss_real - d_loss_fake
            gp = gradient_penalty(discriminator, batch_x, g_output)

            # Combine losses
            d_loss = d_loss_real - d_loss_fake + lambda_gp * gp

        d_grads = disc_tape.gradient(d_loss,
discriminator.trainable_variables)
        d_optim.apply_gradients(zip(d_grads,
discriminator.trainable_variables))

        g_grads = gen_tape.gradient(g_loss,
generator.trainable_variables)
        g_optim.apply_gradients(zip(g_grads,
generator.trainable_variables))
    preds = discriminator.predict(x_val)
    preds = np.where(preds.reshape(-1) < 0, 0, 1)
    print('\n', precision_score(y_val, preds), recall_score(y_val,
```

Figure 7-18. *Defining the training loop for the GAN. At the end of each epoch, we predict on the validation data and print the precision, recall, and F1-scores*

Your training outputs should look something like Figure 7-19.

```
Epoch 0
  115/116: g_loss -0.23272466659545898 | d_loss: 4.443209171295166
real -0.3765527606010437 fake 0.23272466659545898226
 0.9933422103861518 0.9221260815822002 0.9564102564102565

Epoch 1
  115/116: g_loss -0.8314025402069092 | d_loss: 1.3354390859603882
real -0.44502338767051697 fake 0.83140254020690926
 0.9973404255319149 0.927070457354759 0.9609224855861628

...

Epoch 8
  115/116: g_loss -28.799457550048828 | d_loss: -27.888029098510742
real -0.5944007635116577 fake 28.799457550048828
 0.9912060301507538 0.9752781211372065 0.983177570093458

Epoch 9
  115/116: g_loss -40.05744171142578 | d_loss: -39.108402252197266
real -0.6087197065353394 fake 40.0574417114257844
 0.9899623588456713 0.9752781211372065 0.9825653798256538
```

Figure 7-19. *The training outputs. The three outputs at the end of each epoch are, in order, the precision, recall, and F1-measure. Over the course of training, the F1-score tends to improve*

Once the training finishes, let's make the predictions on the test data. Refer to Figure 7-20.

```
preds = discriminator.predict(x_test)
y_pred = np.where(preds.reshape(-1) < 0, 0, 1)

precision = precision_score(y_test, y_pred)
recall = recall_score(y_test, y_pred)
f1 = f1_score(y_test, y_pred)

print(f"Precision: {precision}")
print(f"Recall: {recall}")
print(f"F1-Measure: {f1}")
```

Figure 7-20. *Using the discriminator to predict on the test set and print out the computed precision, recall, and F1-scores*

If all goes well during training, you should see some outputs like this:

Precision: 0.995598868280415
Recall: 0.9786773794808405
F1-Measure: 0.9870656069814555

Finally, let's plot the confusion matrix, as shown in Figure 7-21.

```
cm = confusion_matrix(y_test, y_pred)
plt.title("Confusion Matrix")
ax = sns.heatmap(cm, annot=True, fmt='0.0f')
ax.invert_yaxis()
ax.invert_xaxis()
ax.set_xlabel('Predicted Label')
ax.set_ylabel('True Label')
```

Figure 7-21. *The code to plot the confusion matrix*

You should see something like Figure 7-22.

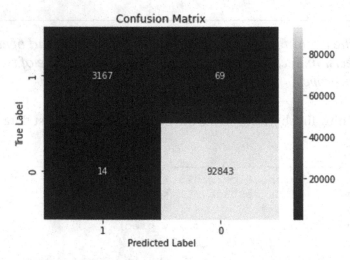

Figure 7-22. *The confusion matrix. There are few misclassifications—the discriminator/critic misclassified 69 anomalies as being normal, while it predicted 14 normal points as being anomalies*

As we can see in Figure 7-22, the GAN did a great job at predicting anomalies despite only being trained on normal data. The precision is 0.9955, meaning that it is very accurate at specifically predicting anomalies. Its recall is 0.9786, which is still a strong number, and indicates that it may capture the vast majority of true anomalies if more new data were predicted on.

There are plenty of experiments you can conduct with different hyperparameter settings, different loss functions, neural network architectures, data splits, and so on.

Overall, this was an example of semi-supervised anomaly detection, since we trained the model only on normal data and used them to successfully predict anomalies at a high rate.

Summary

In this chapter, we discussed how generative adversarial networks can be applied to the task of anomaly detection. In Chapter 8, we will look at anomaly detection in time-series settings.

CHAPTER 8

Long Short-Term Memory Models

In this chapter, you will learn about recurrent neural networks (RNNs) and long short-term memory (LSTM) models. You will also learn how LSTMs work, how they can be used to detect anomalies, and how you can implement anomaly detection using LSTM. You will work through several datasets depicting time Series of different types of data, such as CPU utilization, taxi demand, etc., to illustrate how to detect anomalies. This chapter introduces you to many concepts using LSTM so as to enable you to explore further using the Jupyter notebooks provided as part of the book material.

In a nutshell, this chapter covers the following topics:

- Sequences and time Series analysis

- What is a RNN?

- What is an LSTM?

- LSTM applications

Note Code examples are provided in Python 3.8. The code repository for this book is available at `https://github.com/apress/beginning-anomaly-detection-python-deep-learning-2e/tree/master`.

The repository also includes a requirements.txt file to check your packages and their versions.

The notebook for this chapter can be found at `https://github.com/apress/beginning-anomaly-detection-python-deep-learning-2e/blob/master/Chapter%208%20Long-Short%20Term%20Memory%20Models/chapter8_lstm.ipynb`.

345

Navigate to "Chapter 8 Long Short-Term Memory Models" and then click the notebook. The code is provided as .py files as well, though it is the exported version of the notebook.

We will be using JupyterLab to present all of the code examples.

Sequences and Time Series Analysis

A **time Series** is a Series of data points indexed in time order. Most commonly, a time Series is a sequence taken at successive equally spaced points in time. Thus, it is a sequence of discrete-time data. Examples of time Series are ECG data, weather sensors, and stock prices.

Figure 8-1 shows an example of a time Series.

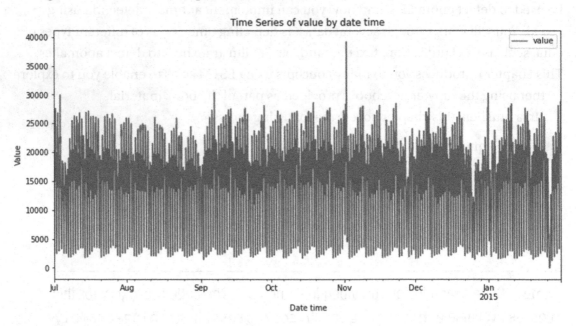

Figure 8-1. *A time-series dataset depicting total taxi pickup count over all taxis in New York City across time*

Figure 8-2 shows the monthly values for the Atlantic Multi-decadal Oscillation (AMO) index for a 157-year time period (AMO measures the variability of sea surface temperatures in the North Atlantic Ocean).

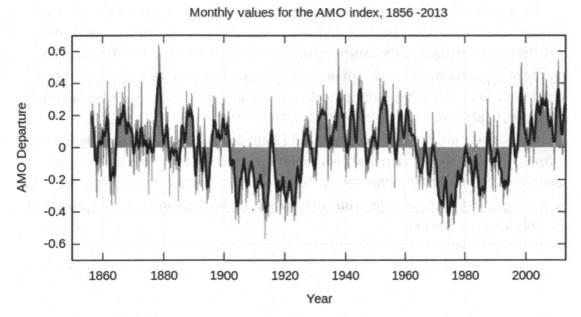

Figure 8-2. *Monthly values of the AMO index across time*

Figure 8-3 shows a chart of the BP stock price for a 20-year time period.

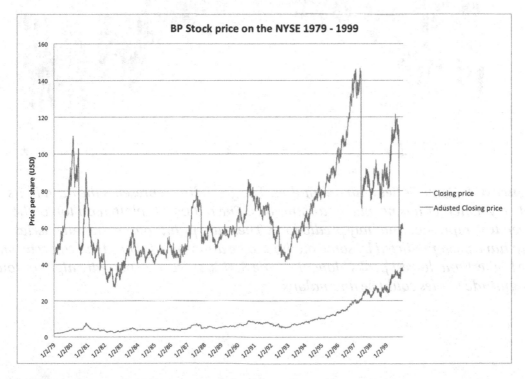

Figure 8-3. *BP stock closing price per share across 20 years*

Time Series analysis, then, refers to the analysis of change in trends of data over a period of time. Time Series analysis comprises methods for analyzing time-series data in order to extract meaningful statistics and other characteristics of the data and has a variety of applications. One such application is the prediction of the future value of an item based on its past values. Future stock price prediction is probably the best example of such an application.

Another very important use case is the ability to detect anomalies. By analyzing and learning the time Series in terms of being able to understand the trends and changes seen from historical data, we can detect abnormal or anomalous data points in the time Series. Figure 8-4 is a time Series with anomalies. It shows the normal data in green and possible anomalies in red.

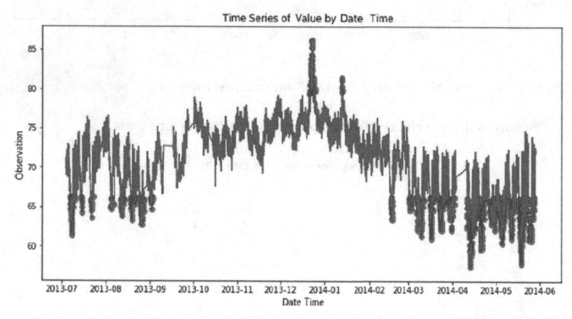

Figure 8-4. *Time Series with anomalies. The green line represents the data in its entirety, whether it is normal or anomalous. The red points plotted on top of the green line represent anomaly predictions. These are points along the time Series that have been predicted by some method to be anomalous. They tend to occur on the highest and lowest peaks along the time Series, since unexpectedly high- or low-magnitude values could be anomalous*

What Is an RNN?

You have seen several types of neural networks throughout the book, so you know that the high-level representation of neural networks looks like Figure 8-5.

***Figure 8-5.** A high-level representation of neural networks*

Clearly, the neural network processes input and produces output, and this works on many types of input data with varying features. However, a critical piece to notice is that this neural network has no notion of the time of the occurrence of the event (input), only that input has come in. So, what happens with events (input) that come in as a stream over long periods of time? How can the neural network shown in Figure 8-5 handle trending in events, seasonality in events, and so on? How can it learn from the past and apply it to the present and future? Recurrent neural networks try to address this by incrementally building neural networks, taking signals from a previous timestamp into the current network. Figure 8-6 depicts an RNN.

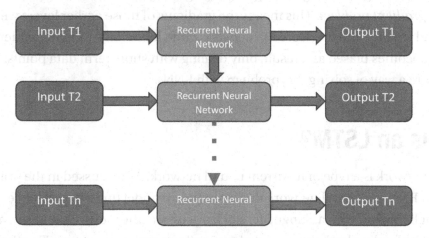

***Figure 8-6.** A recurrent neural network*

You can see that an RNN is a neural network with multiple layers or stages. Each stage represents a time, T; the RNN at T+1 will consider the RNN at time T as one of the signals. Each stage passes its output to the next stage. The hidden state, which is passed from one stage to the next, is the key for the RNN to work so well, and this hidden state is analogous to some sort of memory retention. An RNN layer (or stage) acts as an encoder as it processes the input sequence and returns its own internal state. This state serves as the input of the decoder in the next stage, which is trained to predict the next point of the target sequence, given previous points of the target sequence. Specifically, it is trained to turn the target sequences into the same sequences but offset by one timestep in the future.

Backpropagation is used when training an RNN, as in other neural networks, but in RNNs there is also a time dimension. In backpropagation, we take the derivative (gradient) of the loss with respect to each of the parameters. Using this information (loss), we can then shift the parameters in the opposite direction with a goal to minimize the loss. We have a loss at each timestep since we are moving through time and we can sum the losses across time to get the loss at each timestep. This is the same as summation of gradients across time.

The problem with preceding RNNs constructed from regular neural network nodes is that as we try to model dependencies between sequence values that are separated by a significant number of other values, the gradients of timestep T depend on gradients at T-1's gradients then depend on T-2's gradients, and so on. This leads to the earliest gradient's contribution getting smaller and smaller as we move along the timesteps where the chain of gradients gets longer and longer. This is what is known as the *vanishing gradient problem*. This means the gradients of those earlier layers will become smaller and smaller and therefore the network won't learn long-term dependencies. The RNN becomes biased as a result, only dealing with short-term data points. LSTM networks are a way of solving this problem with RNNs.

What Is an LSTM?

An LSTM network is a type of recurrent neural network. As discussed in the previous section, an RNN is a neural network that attempts to model time- or sequence-dependent behavior, such as language, stock prices, weather sensors, and so on. This is performed by feeding back the output of a neural network layer at time T to the input of the same network layer at time T+1. LSTM builds on top of the RNN, adding a memory component meant to help propagate the information learned at a time T to the future T+1, T+2, and so on. The main idea is that LSTM can forget irrelevant parts of previous state while selectively updating state and then outputting certain parts of the state that are relevant to the future. How does this solve the vanishing gradient problem in RNNs?

Well, now we are discarding some part of the state, updating some part of the state, and propagating forward some part of the state so that we no longer have a long chain of backpropagation as seen in RNNs. Thus, LSTMs are much more efficient than typical RNNs.

Figure 8-7 depicts an RNN with tanh activation.

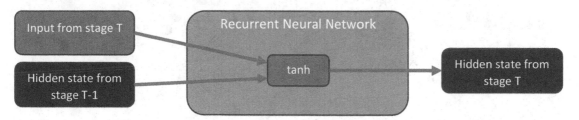

Figure 8-7. *An RNN with tanh activation, a commonly used activation function for RNNs*

The tanh function is called an **activation function**. (For a primer on activation functions, refer to Chapter 5.) There are several types of activation functions that help in applying nonlinear transformations on the inputs at every node in the neural network. Figure 8-8 shows common activation functions.

Common Activation Functions

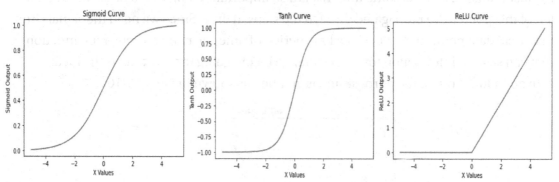

Figure 8-8. *Common activation functions*

The key idea behind activation functions is to add nonlinearity to the data to align better with real-world problems and real-world data. In Figure 8-9, the left graph shows linearity and the right graph shows nonlinearity.

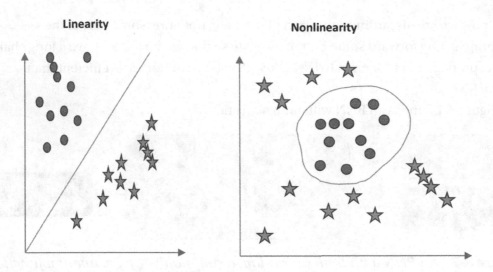

Figure 8-9. *Linear and nonlinear data plots*

Clearly, there is no linear equation to handle the nonlinearity, so we need an activation function to deal with this property. The different activation functions are listed at `https://www.tensorflow.org/api_docs/python/tf/keras/activations`.

In time-series data, the data is spread over a period of time, not some instantaneous set such as seen in Chapter 4 autoencoders, for example. So, not only it is important to look at the instantaneous data at some time T, it is also important for older historical data to the left of this point to be propagated through the steps in time. Since we need the signals from historical data points to survive for a long period of time, we need an activation function that can sustain information for a longer range before going to zero. tanh is the ideal activation function for the purpose and is graphed as shown in Figure 8-10.

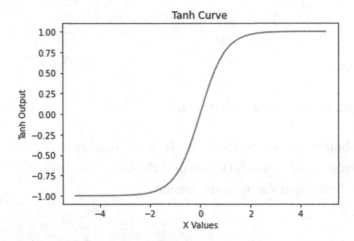

Figure 8-10. *tanh activation*

We also need sigmoid (another activation function) as a way to either remember or forget the information. A sigmoid activation function is shown in Figure 8-11.

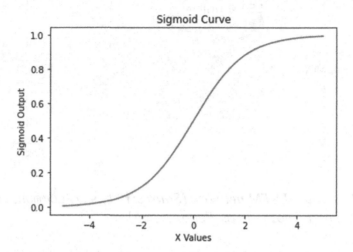

Figure 8-11. *A sigmoid activation function*

Conventional RNNs have a tendency to remember everything, including unnecessary inputs, which results in an inability to learn from long sequences. By contrast, LSTMs selectively remember important inputs, which allows them to handle both short-term and long-term dependencies. LSTMs do this by releasing information between the hidden state and the cell state using three important gates. A common LSTM unit is composed of a cell, an input gate, an output gate, and a forget gate. The cell remembers values over arbitrary time intervals, and the three gates regulate the flow of information into and out of the cell.

A more detailed LSTM architecture is shown in Figure 8-12. The key functions used are the tanh and sigmoid activation functions. F_t is the forget gate, I_t is the input gate, and O_t is the output gate.

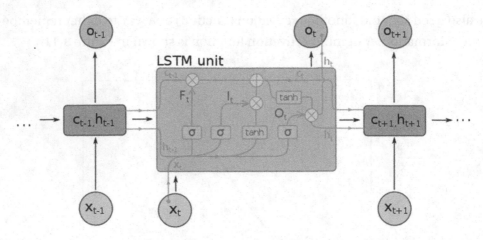

Figure 8-12. *A detailed LSTM network (Source: https://commons.wikimedia. org/wiki/File:Long_Short-Term_Memory.svg)*

A forget gate is the first part of the LSTM stage and it decides how much information from a prior stage should be remembered or forgotten. This is accomplished by passing the previous hidden state hT-1 and current input xT through a sigmoid function.

The input gate helps decide how much information to pass to the current stage by using the sigmoid function and also a tanh function.

The output gate controls how much information will be retained by the hidden state of this stage and passed on to the next stage. Again, the current state passes through the tanh function.

Just for information, the compact forms of the equations for the forward pass of an LSTM unit with a forget gate are as follows, where the initial values are c0 = 0 and h0 = 0, and the subscript indexes the time step:

$$F_t = \sigma\left(W_f \cdot \left[h_{t-1}, x_t\right] + b_f\right)$$

$$I_t = \sigma\left(W_i \cdot \left[h_{t-1}, x_t\right] + b_i\right)$$

$$\tilde{C}_t = tanh\left(W_C \cdot \left[h_{t-1}, x_t\right] + b_C\right)$$

$$C_t = F_t * C_{t-1} + I_t * \tilde{C}_t$$

$$O_t = \sigma\left(W_o \cdot \left[h_{t-1}, x_t\right] + b_o\right)$$

$$h_t = O_t * tanh\left(C_t\right)$$

Here, F_t, I_t, O_t, and C_t correspond to their counterparts in Figure 8-12. $\tilde{C}_t$ is the output of the tanh gate.

The following list summarizes all the variables:

- X_t: Input vector to the LSTM unit.

- F_t: The forget gate's activation vector.

- I_t: The input/update gate's activation vector.

- O_t: The output gate's activation vector.

- h_t: The hidden state vector, also known as the output vector of the entire LSTM unit.

- c_t: The cell state/context vector.

- W, b: The weight matrix and bias vector. Each W matrix and b vector combo with their unique subscript represents a set of learnable parameters that must be learned during training.

As for σ, this is the sigmoid activation function.

LSTM for Anomaly Detection

This section presents LSTM implementations for some use cases using time-series data as examples. A few different time-series datasets are available to use to try to detect anomalies using LSTM. All of them have a timestamp and a value that can easily be plotted in Python.

The basic idea is to train an LSTM on the normal sequence of time-series data. In this case, the first half of the sequence will be our training set, and the second half of the sequence will be the testing set. We will process the dataset into a rolling window of T time steps, where the first T-1 time steps will be passed into the LSTM so it predicts the Tth time step's value.

Then, we will use the LSTM to predict the sequence over the entire time-series sequence, including both the train and testing splits. When predicting each point, we will calculate the loss between the LSTM's predicted output and the actual, true output using mean absolute error. To predict anomalies, we will set a threshold for prediction error that, if crossed, means an automatic anomaly label.

This threshold will be set using the top nth percent training prediction error. This boundary will then be applied to the rest of the testing data, and this time, any prediction error higher than this boundary will be counted as an anomaly. We use the 95th percentile training error, or top 5% error.

The reasoning for training only on the first half of the sequence is to teach the model what normal time-series values look like before we predict on the latter half with anomalies. The idea is that the predicted value will not match the anomaly, producing a higher error and being flagged as an anomaly. Since no labels are required for anomaly detection, this is an unsupervised learning task. This general formula is what we will apply over several different time-series datasets to perform anomaly detection in an unsupervised manner.

Figure 8-13 shows the basic code to import all necessary packages.

```
import tensorflow.keras as keras
from tensorflow.keras import optimizers
from tensorflow.keras import losses
from tensorflow.keras.models import Sequential, Model
from tensorflow.keras.layers import Dense, Input, Dropout, Embedding, LSTM
from tensorflow.keras.optimizers import RMSprop, Adam, Nadam
from tensorflow.keras.preprocessing import sequence
from tensorflow.keras.callbacks import TensorBoard

import sklearn
from sklearn.preprocessing import StandardScaler
from sklearn.model_selection import train_test_split
from sklearn.metrics import confusion_matrix, roc_auc_score
from sklearn.preprocessing import MinMaxScaler

import seaborn as sns
import pandas as pd
import numpy as np
import matplotlib

import matplotlib.pyplot as plt
import matplotlib.gridspec as gridspec
%matplotlib inline

import tensorflow
import sys
print("Python: ", sys.version)

print("pandas: ", pd.__version__)
print("numpy: ", np.__version__)
print("seaborn: ", sns.__version__)
print("matplotlib: ", matplotlib.__version__)
print("sklearn: ", sklearn.__version__)
print("Tensorflow: " tensorflow   version  )
```

Figure 8-13. *Code to import packages*

The versions of the various necessary packages are as follows:

```
Python:  3.8.12 (default, Oct 12 2021, 03:01:40) [MSC v.1916 64 bit (AMD64)]
pandas:  2.0.0
numpy:  1.22.2
seaborn:  0.12.2
matplotlib:  3.7.1
sklearn:  1.2.2
Tensorflow:  2.7.0
```

Figure 8-14 shows the code to visualize the results via a chart for the anomalies and a chart for the errors (the difference between predicted and truth) while training.

```python
class Visualization:
    labels = ["Normal", "Anomaly"]

    def draw_anomaly(self, y, error, threshold):
        groupsDF = pd.DataFrame({'error': error,
                                 'true': y}).groupby('true')

        figure, axes = plt.subplots(figsize=(12, 8))

        for name, group in groupsDF:
            axes.plot(group.index, group.error, marker='x' if name == 1
else 'o', linestyle='',
                      color='r' if name == 1 else 'g', label="Anomaly" if
name == 1 else "Normal")

        axes.hlines(threshold, axes.get_xlim()[0], axes.get_xlim()[1],
colors="b", zorder=100, label='Threshold')
        axes.legend()

        plt.title("Anomalies")
        plt.ylabel("Error")
        plt.xlabel("Data")
        plt.show()

    def draw_error(self, error, threshold):
        plt.figure(figsize=(10, 8))
        plt.plot(error, marker='o', ms=3.5, linestyle='',
                 label='Point')

        plt.hlines(threshold, xmin=0, xmax=len(error)-1, colors="r",
zorder=100, label='Threshold')
        plt.legend()
        plt.title("Reconstruction error")
        plt.ylabel("Error")
        plt.xlabel("Data")
        plt.show()
```

Figure 8-14. *Code to visualize errors and anomalies*

You will use different examples of time-series data to detect whether a point is normal/expected or abnormal/anomalous. Figure 8-15 shows the data being loaded into a Pandas dataframe. It also shows a list of paths to datasets.

```
tensorlogs = ["art_daily_no_noise", #0
              "art_daily_nojump", #1
              "art_daily_jumpsdown",#2
              "art_daily_perfect_square_wave", #3
              "art_increase_spike_density",  #4
              "art_load_balancer_spikes",  #5
              "ambient_temperature_system_failure", #6
              "nyc_taxi",  #7
              "ec2_cpu_utilization", #8
              "rds_cpu_utilization"] #9

dataFilePaths = ['../data/art_daily_no_noise.csv',
                 '../data/art_daily_nojump.csv',
                 '../data/art_daily_jumpsdown.csv',
                 '../data/art_daily_perfect_square_wave.csv',
                 '../data/art_increase_spike_density.csv',
                 '../data/art_load_balancer_spikes.csv',
                 '../data/ambient_temperature_system_failure.csv',
                 '../data/nyc_taxi.csv',
                 '../data/ec2_cpu_utilization.csv',
                 '../data/rds_cpu_utilization.csv']
```

Figure 8-15. *A list of paths to datasets*

You can find all the data at `https://github.com/apress/beginning-anomaly-detection-python-deep-learning-2e/tree/master/data`.

You will work with one of the datasets in more detail now. The dataset is nyc_taxi, which basically consists of timestamps and demand for taxis. This dataset shows the NYC taxi demand from 2014–07–01 to 2015–01–31 with an observation every half hour. There are few detectable anomalies in this dataset: Thanksgiving, Christmas, New Year's Day, a snowstorm, etc. Figure 8-16 shows the code to select the dataset.

```
i = 7

tensorlog = tensorlogs[i]
dataFilePath = dataFilePaths[i]
print("tensorlog: ", tensorlog)
print("dataFilePath: ", dataFilePath)
```

Figure 8-16. *Code to select the nyc_taxi dataset*

You can load the data from the dataFilePath as a csv file using Pandas. Figure 8-17 shows the code to read the csv datafile into Pandas.

```
df = pd.read_csv(filepath_or_buffer=dataFilePath, header=0, sep=',')
print('Shape:' , df.shape[0])
print('Head:')
print(df.head(5))
```

Figure 8-17. *Code to read a csv datafile into Pandas*

The output of Figure 8-17 should look like this:

Shape: 10320
Head:
```
            timestamp  value
0  2014-07-01 00:00:00  10844
1  2014-07-01 00:30:00   8127
2  2014-07-01 01:00:00   6210
3  2014-07-01 01:30:00   4656
4  2014-07-01 02:00:00   3820
```

Figure 8-18a shows the plotting of the time Series, with the months on the x axis and the values on the y axis. It also shows the code to generate a graph of the time Series.

```
df['Datetime'] = pd.to_datetime(df['timestamp'])
print(df.head(3))
df.shape
df.plot(x='Datetime', y='value', figsize=(12,6))
plt.xlabel('Date time')
plt.ylabel('Value')
plt.title('Time Series of value by date time')
```

Figure 8-18a. *Plotting the time Series*

Figure 8-18b shows the resulting plot.

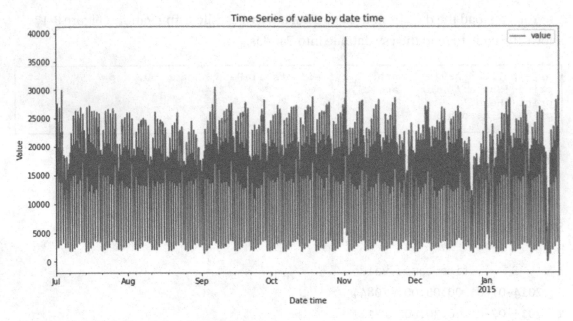

Figure 8-18b. *The plotted time Series*

Let's delve into the data further. You can run the describe() command to look at the value column. Figure 8-19 shows the code to describe the value column.

```
df['Datetime'] = pd.to_datetime(df['timestamp'])
print(df.head(3))
df.shape
df.plot(x='Datetime', y='value', figsize=(12,6))
plt.xlabel('Date time')
plt.ylabel('Value')
plt.title('Time Series of value by date time')
```

Figure 8-19. *Describing the value column*

You should see something like this:

count   10320.000000
mean   15137.569380
std   6939.495808
min   8.000000
25%   10262.000000
50%   16778.000000

```
75%        19838.750000
max        39197.000000
Name: value, dtype: float64
```

You can also plot the data using a seaborn kernel density estimate (KDE) plot, as shown in Figure 8-20a and Figure 8-20b.

```
fig, (ax1) = plt.subplots(ncols=1, figsize=(8, 5))
ax1.set_title('Before Scaling')
sns.kdeplot(df['value'], ax=ax1)
```

Figure 8-20a. *Code to display a KDE plot of the data*

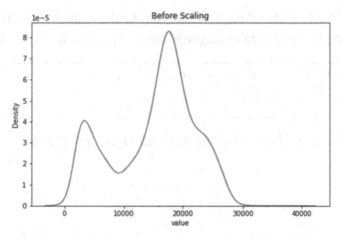

Figure 8-20b. *Using KDE to plot the value column*

The data points have a minimum of 8 and maximum of 39,197, which is a wide range. You can use scaling to normalize the data. The formula for scaling is (x-Min) / (Max-Min). Figure 8-21 shows the code to scale the data.

```
from sklearn.preprocessing import MinMaxScaler
scaler = MinMaxScaler(feature_range = (0, 1))
df['scaled_value'] =
pd.DataFrame(scaler.fit_transform(pd.DataFrame(df['value'])),columns=['valu
e'])
print('Shape:' , df.shape[0])
df.head(5)
```

Figure 8-21. *Code to scale the data*

Your output table should look like Table 8-1.

Table 8-1. *Displaying the Normalized Version of the Original Dataframe*

	timestamp	value	Datetime	scaled_value
0	2014-07-01 00:00:00	10844	2014-07-01 00:00:00	0.276506
1	2014-07-01 00:30:00	8127	2014-07-01 00:30:00	0.207175
2	2014-07-01 01:00:00	6210	2014-07-01 01:00:00	0.158259
3	2014-07-01 01:30:00	4656	2014-07-01 01:30:00	0.118605
4	2014-07-01 02:00:00	3820	2014-07-01 02:00:00	0.097272

Now that you've scaled the data, you can plot the data again. You can plot the data using seaborn kde plot, as shown in Figure 8-22a and the graph output, Figure 8-22b.

```
fig, (ax1) = plt.subplots(ncols=1, figsize=(8, 5))
ax1.set_title('After Scaling')
sns.kdeplot(df['scaled_value'], ax=ax1)
```

Figure 8-22a. *Code to plot the kde for the scaled_value column*

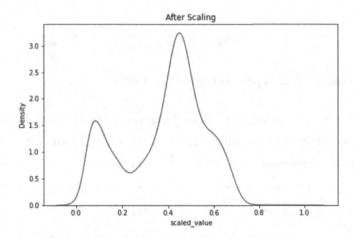

Figure 8-22b. *Using kde to plot the scaled_value column*

You can take a look at the dataframe now that you have scaled the scaled_value column. Figure 8-23 shows the dataframe (from Table 8-1) with the timestamp and value as well as scaled_value and the datetime.

```
df.head(5)
```

	timestamp	value	Datetime	scaled_value
0	2014-07-01 00:00:00	10844	2014-07-01 00:00:00	0.276506
1	2014-07-01 00:30:00	8127	2014-07-01 00:30:00	0.207175
2	2014-07-01 01:00:00	6210	2014-07-01 01:00:00	0.158259
3	2014-07-01 01:30:00	4656	2014-07-01 01:30:00	0.118605
4	2014-07-01 02:00:00	3820	2014-07-01 02:00:00	0.097272

Figure 8-23. *The modified dataframe*

There are 10,320 data points in the sequence, and your goal is to find anomalies. This means you are trying to find out when data points are abnormal. If you can predict a data point at time T based on the historical data until T-1, then you have a way of looking at an expected value compared to an actual value to see if you are within the expected range of values for time T. If you predicted that y_pred number of taxis are in demand on January 1, 2015, then you can compare this y_pred with the actual y_actual. The difference between y_pred and y_actual gives the error, and when you get the errors of all the points in the sequence, you end up with a distribution of just errors.

To accomplish this, you will use a sequential model using Keras. The model consists of two LSTM layers, with the final LSTM layer being the output layer. The first LSTM layer takes as input the time-series data and learns how to learn the values with respect to time. The final LSTM layer is the output layer and will take as input the previous LSTM layer's output and transform it into a single prediction for the next time step. Then, you apply a sigmoid activation on this LSTM layer so that the final output is between 0 and 1.

You also use the Adam optimizer and the mean absolute error as the loss function. Figure 8-24a shows the code to build an LSTM model.

```
time_steps = 16
metric = 'mean_absolute_error'

model = Sequential()
model.add(LSTM(units=64, activation='tanh', input_shape=(time_steps-1, 1),
return_sequences=True))
model.add(LSTM(units=1, activation='sigmoid'))

model.compile(optimizer='adam', loss='mean_absolute_error',
metrics=[metric])
print(model.summary())
```

Figure 8-24a. *Code to build an LSTM model*

The model summary looks like Figure 8-24b.

```
Model: "sequential"

Layer (type)                    Output Shape               Param #
=================================================================
lstm (LSTM)                     (None, 15, 64)             16896

lstm_1 (LSTM)                   (None, 1)                  264


=================================================================
Total params: 17,160
Trainable params: 17,160
Non-trainable params: 0
```

Figure 8-24b. *A text summary of the model created in Figure 8-24a*

As shown above, you used an LSTM layer. Let's look at the details of the LSTM layer function with all the possible parameters (source: https://www.tensorflow.org/api_docs/python/tf/keras/layers/LSTM):

tf.keras.layers.LSTM(
```
        units,
        activation='tanh',
        recurrent_activation='sigmoid',
        use_bias=True,
        kernel_initializer='glorot_uniform',
        recurrent_initializer='orthogonal',
        bias_initializer='zeros',
        unit_forget_bias=True,
```

```
kernel_regularizer=None,
recurrent_regularizer=None,
bias_regularizer=None,
activity_regularizer=None,
kernel_constraint=None,
recurrent_constraint=None,
bias_constraint=None,
dropout=0.0,
recurrent_dropout=0.0,
return_sequences=False,
return_state=False,
go_backwards=False,
stateful=False,
time_major=False,
unroll=False,
**kwargs
)
```

The following is a summary of the arguments:

- units: Positive integer, dimensionality of the output space.

- activation: Activation function to use. Default: hyperbolic tangent (tanh). If you pass None, no activation is applied (i.e., "linear" activation: $a(x) = x$).

- recurrent_activation: Activation function to use for the recurrent step. Default: sigmoid (sigmoid). If you pass None, no activation is applied (i.e., "linear" activation: $a(x) = x$).

- use_bias: Boolean (default True), whether the layer uses a bias vector.

- kernel_initializer: Initializer for the kernel weights matrix, used for the linear transformation of the inputs. Default: glorot_uniform.

- recurrent_initializer: Initializer for the recurrent_kernel weights matrix, used for the linear transformation of the recurrent state. Default: orthogonal.

- bias_initializer: Initializer for the bias vector. Default: zeros.

- `unit_forget_bias`: Boolean (default `True`). If `True`, add 1 to the bias of the forget gate at initialization. Setting it to `True` also forces `bias_initializer="zeros"`. This is recommended in Jozefowicz et al.

- `kernel_regularizer`: Regularizer function applied to the `kernel` weights matrix. Default: `None`.

- `recurrent_regularizer`: Regularizer function applied to the `recurrent_kernel` weights matrix. Default: `None`.

- `bias_regularizer`: Regularizer function applied to the bias vector. Default: `None`.

- `activity_regularizer`: Regularizer function applied to the output of the layer (its "activation"). Default: `None`.

- `kernel_constraint`: Constraint function applied to the `kernel` weights matrix. Default: `None`.

- `recurrent_constraint`: Constraint function applied to the `recurrent_kernel` weights matrix. Default: `None`.

- `bias_constraint`: Constraint function applied to the bias vector. Default: `None`.

- `dropout`: Float between 0 and 1. Fraction of the units to drop for the linear transformation of the inputs. Default: 0.

- `recurrent_dropout`: Float between 0 and 1. Fraction of the units to drop for the linear transformation of the recurrent state. Default: 0.

- `return_sequences`: Boolean. Whether to return the last output in the output sequence, or the full sequence. Default: `False`.

- `return_state`: Boolean. Whether to return the last state in addition to the output. Default: `False`.

- `go_backwards`: Boolean (default `False`). If `True`, process the input sequence backwards and return the reversed sequence.

- `stateful`: Boolean (default `False`). If `True`, the last state for each sample at index i in a batch will be used as initial state for the sample of index i in the following batch.

- `time_major`: The shape format of the `inputs` and `outputs` tensors. If `True`, the inputs and outputs will be in shape [`timesteps`, `batch`, `feature`], whereas in the `False` case, it will be [`batch`, `timesteps`, feature]. Using `time_major` = `True` is a bit more efficient because it avoids transposes at the beginning and end of the RNN calculation. However, most TensorFlow data is batch-major, so by default this function accepts input and emits output in batch-major form.

- `unroll`: Boolean (default `False`). If `True`, the network will be unrolled, else a symbolic loop will be used. Unrolling can speed-up an RNN, although it tends to be more memory-intensive. Unrolling is only suitable for short sequences.

Note that the LSTM call in the code shown in Figure 8-24a includes a parameter `time_steps` = `16`. This is the number of steps in the sequence that is used in training LSTM. Here, 16 means 8 hours because the data points are 30 minutes apart. If it were 48, that would be 24 hours. You can try changing this to 64 or 128 and see what happens to the output.

Figure 8-25 shows the code to split the sequence into a tumbling window of subsequences of length 15, with the 16th point serving as the ground truth prediction. Note the output shapes: there are 10,305 contiguous 16-length windows, and 10,305 corresponding y_actual values that are the 16th element of the aforementioned windows. For each window, the first 15 values are taken as the x points, with the last 16th value taken as the y point.

```
sequence = np.array(df['scaled_value'])
print(sequence)

# Create rolling window sequences as determined by time_steps
x_sequences = []
y_sequences = []

# Number of windows to iterate through
n_iter = len(sequence) - time_steps + 1
for f in range(n_iter):
    window = sequence[f:f+time_steps]
    x_sequences.append(window[:-1])
    y_sequences.append(window[-1:])
x_sequences = np.array(x_sequences)
y_sequences = np.array(y_sequences)
print(x_sequences.shape, y_sequences.shape)
```

Figure 8-25. *Code to create subsequences via rolling window*

You should see some output like so:

```
[0.27650616 0.20717548 0.1582587  ... 0.69664957 0.6783281  0.67059634]
(10305, 15) (10305, 1)
```

Now, let's train the model for 5 epochs (though you may train for longer), using the training set as the validation data. Figure 8-26a shows the code to train the model.

```
sequences_x = x_sequences.reshape(len(x_sequences), time_steps-1, 1)
print("sequences_x: ", sequences_x.shape)
sequences_y = y_sequences.reshape(len(y_sequences), 1)
print("sequences_y: ", sequences_y.shape)


# Training on first half of data only, predicting on whole thing
stop_point = int(0.5 * len(df))
training_x = sequences_x[:stop_point]
print("training_x: ", training_x.shape)
training_y = sequences_y[:stop_point]
print("training_y: ", training_y.shape)


batch_size=32
epochs=5

model.fit(x=training_x, y=training_y,
                    batch_size=batch_size, epochs=epochs,
                    verbose=1, validation_data=(training_x, training_y),

callbacks=[TensorBoard(log_dir='./logs/{0}'.format(tensorlog))])
```

Figure 8-26a. *Code to train the model*

You'll see some training output that looks similar to Figure 8-26b.

```
training_dataset:  (645, 16, 1)
sequences_x:   (10305, 15, 1)
sequences_y:   (10305, 1)
training_x:   (5160, 15, 1)
training_y:   (5160, 1)
Epoch 1/5
162/162 [==============================] - 11s 53ms/step - loss: 0.1047 -
mean_absolute_error: 0.1047 - val_loss: 0.0666 - val_mean_absolute_error:
0.0666

...


Epoch 5/5
162/162 [==============================] - 8s 47ms/step - loss: 0.0346 -
mean_absolute_error: 0.0346 - val_loss: 0.0314 - val_mean_absolute_error:
0.0314
```

Figure 8-26b. *An example of the training output. Much of the output is omitted for ease of display*

Figure 8-27 shows the plotting of the loss during the training process through the epochs of training.

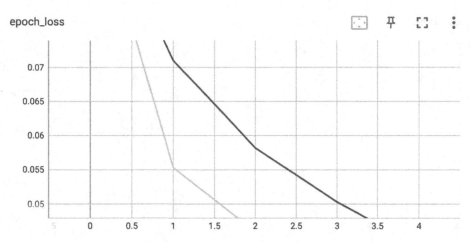

Figure 8-27. *Graph of loss in TensorBoard*

Figure 8-28 shows the plotting of the mean absolute error during the training process through the epochs of training.

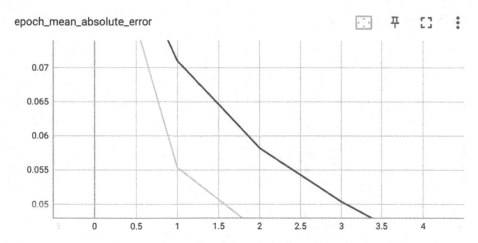

Figure 8-28. *Graph of mean absolute error in TensorBoard*

Figure 8-29 shows the plotting of the loss of validation during the training process through the epochs of training.

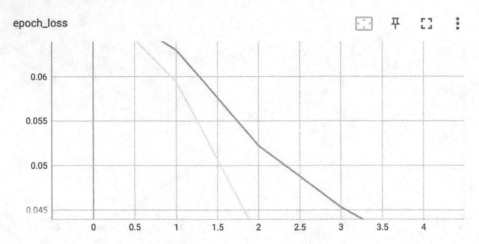

Figure 8-29. *Graph of loss of validation in TensorBoard*

Figure 8-30 shows the plotting of the mean absolute error of validation during the training process through the epochs of training.

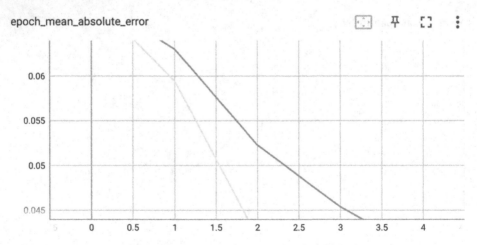

Figure 8-30. *Graph of mean absolute error of validation in TensorBoard*

Figure 8-31 shows the graph of the model as visualized by TensorBoard.

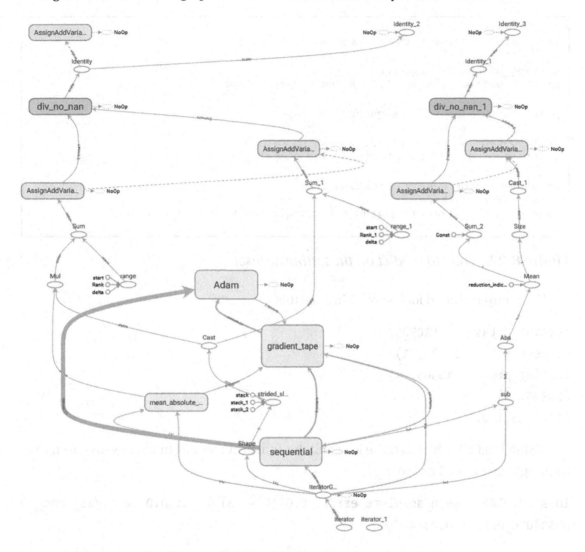

Figure 8-31. *Graph of the model as visualized by TensorBoard*

Once the model is trained, you can predict a test dataset that is split into subsequences of the same length (`time_steps`) as the training datasets. Once this is done, you can then compute the root mean square error (RMSE).

Figure 8-32 shows the code to predict on the testing dataset.

```
import math
from sklearn.metrics import mean_squared_error

testing_dataset = sequences_x
print("testing_dataset: ", testing_dataset.shape)

print("sequences_y: ", sequences_y.shape)

testing_pred = model.predict(x=testing_dataset)
print("testing_pred: ", testing_pred.shape)

errorsDF = sequences_y - testing_pred
print(errorsDF.shape)
rmse = math.sqrt(mean_squared_error(sequences_y, testing_pred))
print('Test RMSE: %.3f' % rmse)
```

Figure 8-32. *Code to predict on the testing dataset*

Your output should look something like this:

```
testing_dataset:  (10305, 15, 1)
sequences_y:  (10305, 1)
testing_pred:  (10305, 1)
(10305, 1)
Test RMSE: 0.050
```

RMSE is 0.050, which is quite low, and this is also evident from the low loss from the training phase after 5 epochs:

```
loss: 0.0346 - mean_absolute_error: 0.0346 - val_loss: 0.0314 - val_mean_
absolute_error: 0.0314
```

Now you can use the predicted dataset and the test dataset to compute the difference as diff, which is then passed through vector norms. Calculating the length or magnitude of vectors is often required directly as a regularization method in machine learning. Then you can sort the scores/diffs and use a cutoff value to pick the threshold. This obviously can change as per the parameters you choose, particularly the cutoff value, which is 0.999 in Figure 8-33. The figure also shows the code to compute the threshold.

```
#based on cutoff after sorting errors
# Calculate threshold using training error
dist = np.linalg.norm(sequences_y[:len(training_y)] -
testing_pred[:len(training_y)], axis=-1)

scores =dist.copy()
print(scores.shape)
scores.sort()
cutoff = int(0.999 * len(scores))
print(cutoff)
threshold= scores[cutoff]

# Calculate total error over full sequence
dist = np.linalg.norm(sequences_y[:] - testing_pred, axis=-1)

print(threshold)
```

Figure 8-33. *Code to compute the threshold*

You should see output like this:

```
(5160,)
5154
0.205043275900064
```

You got 0.205 as the threshold; anything above is considered an anomaly.
Figure 8-34a shows the code to plot the testing dataset (green) and the corresponding predicted dataset (red), and Figure 8-34b shows the resulting plot.

```
plt.figure(figsize=(24,16))
plt.plot(sequences_y, color='green')
plt.plot(training_y, color='blue')
plt.plot(testing_pred, color='red')
```

Figure 8-34a. *Code to plot the training sequence (blue) the model learned from, the total sequence including test data (green), and the predicted values (red)*

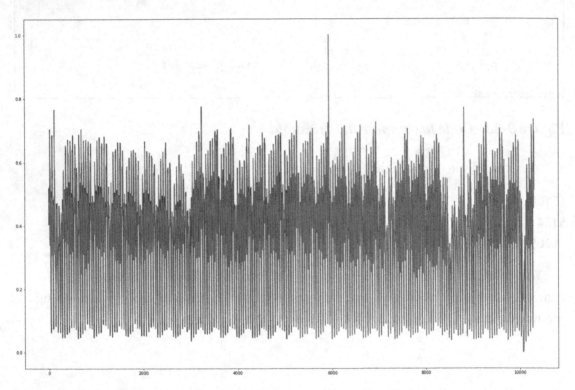

Figure 8-34b. *Plotting the training, testing, and predicted datasets*

Figure 8-35 shows the code to classify a datapoint as anomaly or normal.

```
#label the records anomalies or not based on threshold
z = zip(dist >= threshold, dist)

y_label=[]
error = []
for idx, (is_anomaly, dist) in enumerate(z):
    if is_anomaly:
        y_label.append(1)
    else:
        y_label.append(0)
    error.append(dist)
```

Figure 8-35. *Code to classify a datapoint as anomaly or normal*

Figure 8-36a shows the code to plot the data points with respect to the threshold and Figure 8-36b shows the resulting plot.

```
viz = Visualization()
viz.draw_anomaly(y_label, error, threshold)
```

Figure 8-36a. *Code to plot the data points with respect to the threshold*

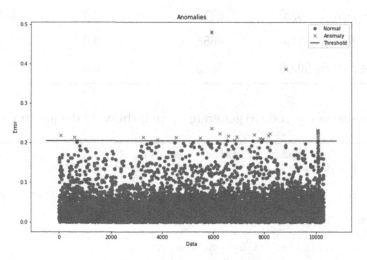

Figure 8-36b. *The plot of the data points with respect to the threshold*

Figure 8-37 shows the code to append the anomaly flag to the dataframe.

```
adf = pd.DataFrame({'Datetime': df['Datetime'], 'observation': df['value'],
                    'error': [0 for f in range((time_steps-1))] + error,
    'anomaly': [0 for f in range((time_steps-1))]+ y_label})
adf.head(5)
```

Figure 8-37. *Code to append the anomaly flag to the dataframe. Since the model must use the first window to actually start predicting, we cannot use the first* time_steps-1 *number of datapoints, thus we need to pad the error and anomaly portions of the dataframe*

Your dataframe output should look like Table 8-2. The first time_steps-1 rows have 0 error due to how we padded it.

Table 8-2. *Dataframe with the Anomaly Flag Appended*

	Datetime	observation	error	anomaly
0	2014-07-01 00:00:00	10844	0.0	0
1	2014-07-01 00:30:00	8127	0.0	0
2	2014-07-01 01:00:00	6210	0.0	0
3	2014-07-01 01:30:00	4656	0.0	0
4	2014-07-01 02:00:00	3820	0.0	0

Figure 8-38a shows the code to generate a graph showing the anomalies.

```
figure, axes = plt.subplots(figsize=(12, 6))
axes.plot(adf['Datetime'], adf['observation'], color='g')
anomaliesDF = adf.query('anomaly == 1')
axes.scatter(anomaliesDF['Datetime'].values, anomaliesDF['observation'],
color='r')
plt.xlabel('Date time')
plt.ylabel('observation')
plt.title('Time Series of value by date time')
```

Figure 8-38a. *Code to plot the anomalies*

In the graph shown in Figure 8-38b, you can spot an anomaly around Thanksgiving Day, one around New Year's Eve, and another one possibly on a snow storm day in January.

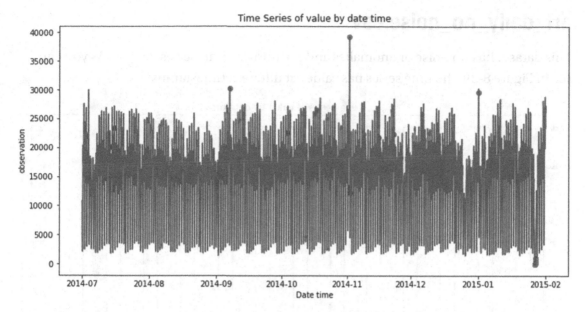

Figure 8-38b. *A graph showing anomalies. Your graph will not exactly match this, but you should be seeing red dots where the time Series seems to deviate from the normal patterns*

If you play around with some of the parameters you used, such as number of time_steps, threshold cutoffs, epochs of the neural network, batch size, and hidden layer, you will see different results.

A good way to improve the detection is to curate good normal data, use identified anomalies, and put it in the mix to have a way to tune the parameters until you get good matches on the identified anomalies.

Examples of Time Series

In all of the examples presented in this section, we use the same notebook but switch which dataset we load by using the "i=0" parameter. All hyperparameters are kept the same, though you are free to experiment. For each dataset, we will build an LSTM model and check whether or not there are anomalies.

art_daily_no_noise.csv

This dataset has no noise or anomalies and is a normal time-series dataset. As you can see in Figure 8-39, the time series has values at different timestamps.

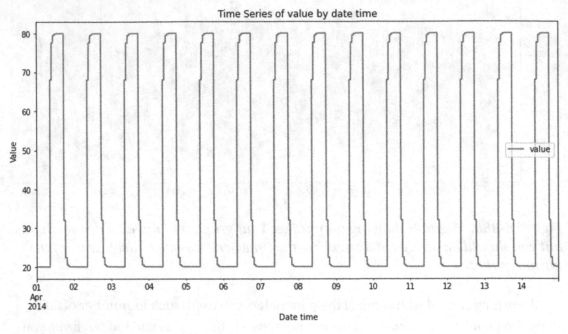

Figure 8-39. *A graph showing the time Series*

Choose i=0 when loading this dataset.

Using the visualization class, you can plot the new time series. As shown below in Figure 8-40, the time series shows the datetime vs. the value column.

The model with the same parameters should fit quite well to this dataset. Feel free to play around with the hyperparameters to achieve a better-fitting model.

Run the code as it appears in the section "LSTM for Anomaly Detection" (or with custom-tuned hyperparameters and model architectures) but with this dataset loaded to run the anomaly detection. The graph in Figure 8-40 shows the anomalies after running the full training script all the way through.

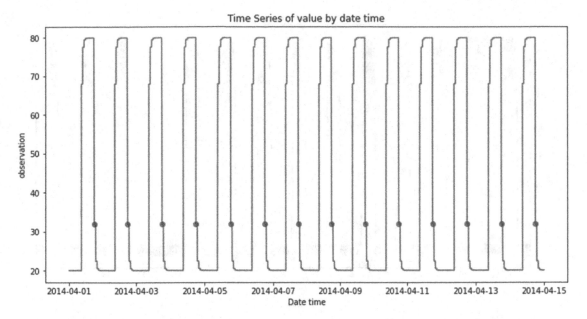

Figure 8-40. *A graph showing anomalies*

Ordinarily, there should not be any anomalies in this dataset. The red "anomalies" you see in Figure 8-40 are the result of the way we compute what an anomaly is. Since we find the 0.999 score and set that as the maximum threshold for a normal point, naturally we will have 0.001% of the data being defined as anomalies. In a sense, we are forcing these red anomalies to be anomalies.

art_daily_nojump.csv

This dataset has no noise or anomalies and is a normal time-series dataset. As you can see in Figure 8-41, the time Series has values at different timestamps.

Using visualization, you can plot the time Series now. You will convert the timestamp to datetime for this work and also drop the timestamp column. As shown below, the time Series shows the datetime vs. the value column.

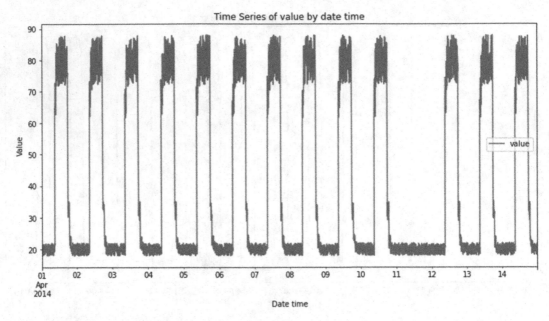

Figure 8-41. *A graph showing the time Series*

Let's add the anomaly column to the original dataframe and prepare a new dataframe. Using visualization, you can plot the new time Series now. As shown below, the time Series shows the datetime vs. the value column. Figure 8-42 shows the generated graph showing predicted anomalies.

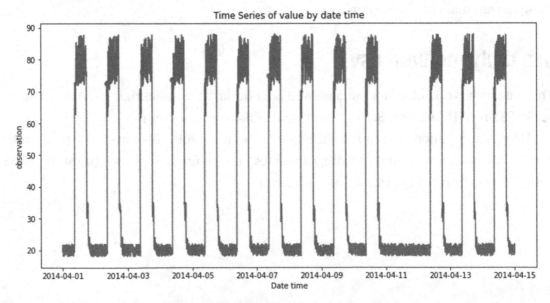

Figure 8-42. *A graph showing anomalies*

Since this dataset has no noise or anomalies and is a normal time-series dataset, there are no real anomalies see the following explanation of the datapoints in red. You will also find that the model has fit quite well, as the predictions align closely with the actual data.

You will notice that there are displayed red points in Figure 8-42. As in the previous example, this is a result of the thresholding, which assumes that the highest reconstruction errors are likely anomalies. If the dataset is normal, then it'll be forced to pick these highest points as anomalies even though they are not.

art_daily_jumpsdown.csv

This dataset has a mixture of normal data and anomalies. As you can see in Figure 8-43, the time Series has values at different timestamps.

Using visualization, you can plot the time Series now. You will convert the timestamp to datetime for this work and also drop the timestamp column. As shown below, the time Series shows the datetime vs. the value column.

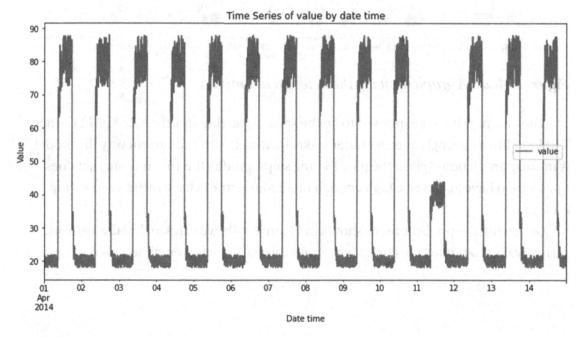

Figure 8-43. *A graph showing the time Series*

Let's add the anomaly column to the original dataframe and prepare a new dataframe. Using visualization, you can plot the new time Series now. As shown below, the time Series shows the datetime vs. the value column. Normal data points are shown in green and anomalies are shown in red. Figure 8-44a shows the generated graph depicting anomalies.

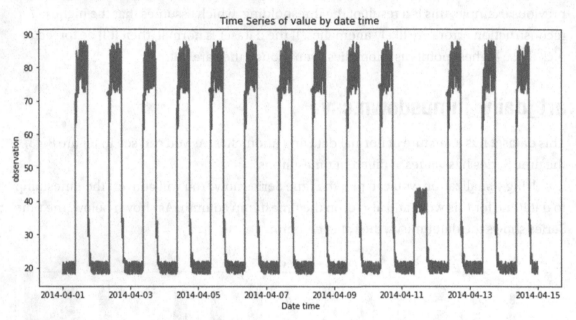

Figure 8-44a. *A graph showing the detected anomalies*

The anomaly here is supposed to be the spike occurring shortly after April 11, since it did not fire as strongly as every other spike. However, with the current way the model is making predictions (given the past 15 time steps, predict the 16th time step), it does not seem to have achieved a high enough prediction error to have registered as a top 0.1% error.

Observing the predictions, as shown in Figure 8-44b, we can see that the network certainly fails to predict the correct values for this particular anomalous spike.

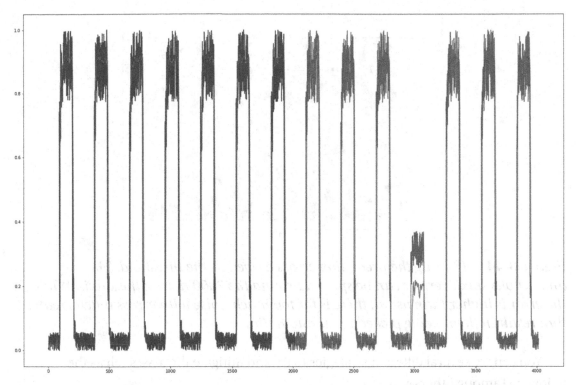

Figure 8-44b. *The model's training data (blue), total sequence (green), and predicted sequence (red). The prediction for the spike is not good, so we should expect that the error here is higher than usual*

Looking at Figure 8-44b, we would expect to see a higher than usual prediction error. However, looking at the actual prediction errors for each point (Figure 8-44c), we can see that there indeed was a higher than usual reconstruction loss for this anomalous spike. However, there is a lot of noise stemming from the network's conservative predictions, as it does not seem to model the individual spikes but rather the general trend of the spikes.

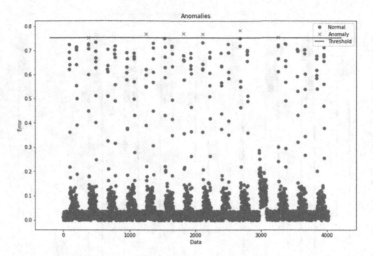

Figure 8-44c. *Plotting the prediction errors as well as the threshold. The anomalous spike's errors can be spotted around the 3000 tick on the x axis. While the error is higher than usual, there is far too much noise with errors much greater that result in them being picked up as anomalies*

You can try several different strategies to try and mitigate the loss, such as the following (among others):

- Increasing the number of training epochs

- Increasing the number of time steps

- Increasing the model capacity by having more neurons or more layers

These strategies can also be used in conjunction with one another to produce a more powerful LSTM network. Assuming we can properly fit the model closely, we can expect to see the anomalous spike itself become the top 0.01% reconstruction error.

art_daily_perfect_square_wave.csv

This dataset has no noise or anomalies and is a normal time-series dataset. As you can see in Figure 8-45, the time Series has values at different timestamps. Using visualization, you can plot the time Series now. You will convert the timestamp to datetime for this work and also drop the timestamp column. As shown below, the time Series shows the datetime vs. the value column.

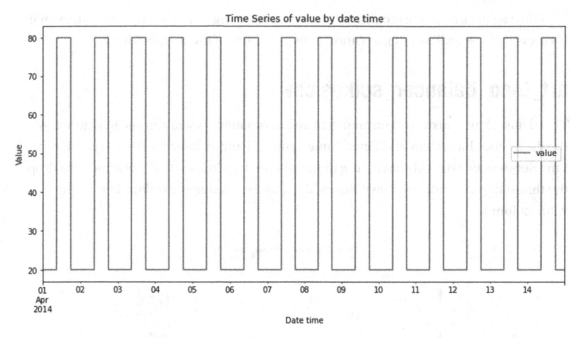

Figure 8-45. *A graph showing the time Series*

Let's add the anomaly column to the original dataframe and prepare a new dataframe. Using visualization, you can plot the new time Series now. As shown below, the time Series shows the datetime vs. value column. Figure 8-46 shows the anomalies.

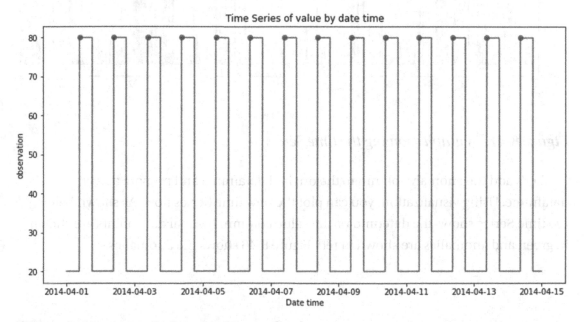

Figure 8-46. *A graph showing anomalies*

Once again, due to the reconstruction errors resulting from the imperfect fit, even if it's a strong fit, we will see flagged anomalies as a result of the thresholding.

art_load_balancer_spikes.csv

This dataset has a mixture of normal data and anomalies. As you can see in Figure 8-47, the time Series has values at different timestamps. Using visualization, you can plot the time Series now. You will convert the timestamp to datetime for this work and also drop the timestamp column. As shown below, the time Series shows the datetime vs. the value column.

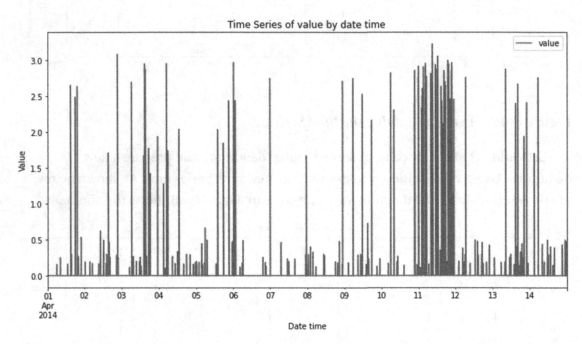

Figure 8-47. *A graph showing the time Series*

Let's add the anomaly column to the original dataframe and prepare a new dataframe. Using visualization, you can plot the new time Series now. As shown below, the time Series shows the datetime vs. the value column. Normal data points are shown in green and anomalies are shown in red. Figure 8-48 shows the anomalies.

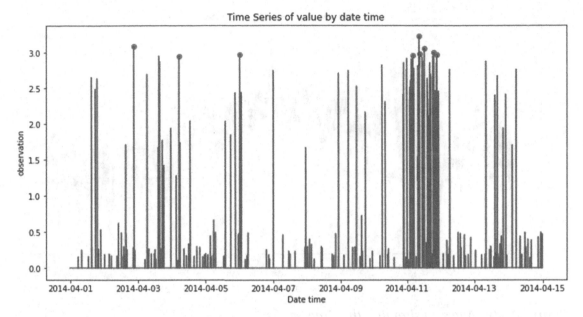

Figure 8-48. *A graph showing anomalies*

Since this dataset has some noise or anomalies, there are anomalies see the following explanation of the datapoints in red shown, and everything else that is normal is green.

ambient_temperature_system_failure.csv

This dataset has a mixture of normal data and anomalies. As you can see in Figure 8-49, the time Series has values at different timestamps. Using visualization, you can plot the time Series now. You will convert the timestamp to datetime for this work and also drop the timestamp column. As shown below, the time Series shows the datetime vs. the value column.

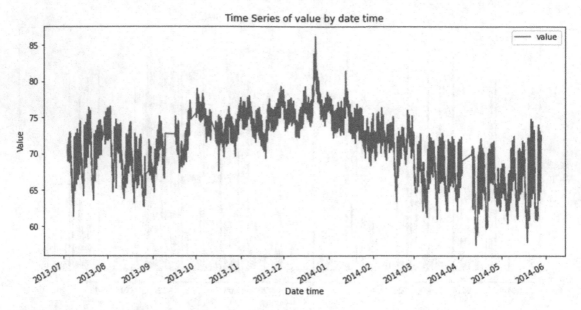

Figure 8-49. *A graph showing the time Series*

Let's add the anomaly column to the original dataframe and prepare a new dataframe. Using visualization, you can plot the new time Series now. As shown below, the time Series shows the datetime vs. the value column. Normal data points are shown in green and anomalies are shown in red. Figure 8-50 shows the anomalies.

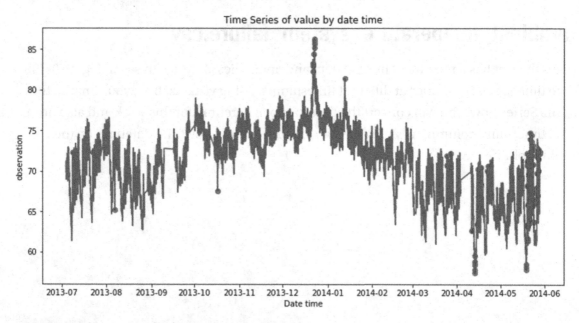

Figure 8-50. *A graph showing anomalies*

Since this dataset has some noise or anomalies, there are anomalies see the following explanation of the datapoints in red shown, and everything else that is normal is green.

ec2_cpu_utilization.csv

This dataset has a mixture of normal data and anomalies. As you can see in Figure 8-51, the time Series has values at different timestamps. Using visualization, you can plot the time Series now. You will convert the timestamp to datetime for this work and also drop the timestamp column. As shown below, the time Series shows the datetime vs. the value column.

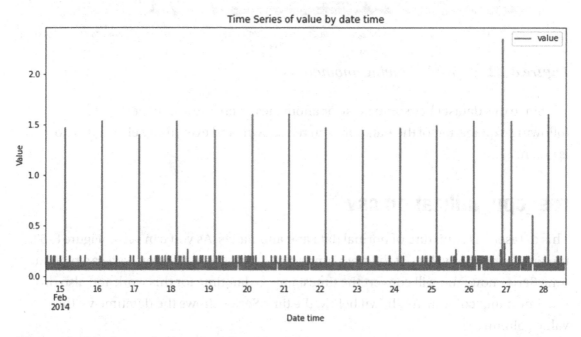

Figure 8-51. *A graph showing the time Series*

Let's add the anomaly column to the original dataframe and prepare a new dataframe. Using visualization, you can plot the new time Series now. As shown below, the time Series shows the datetime vs. the value column. Normal data points are shown in green and anomalies are shown in red. Figure 8-52 shows the anomalies.

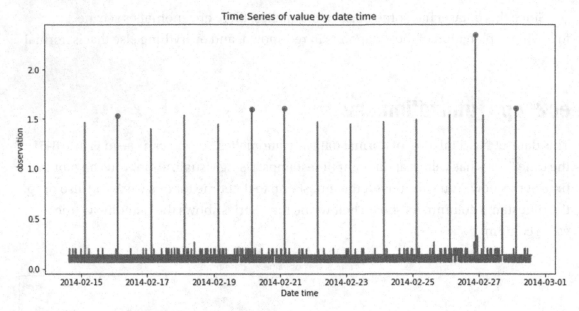

Figure 8-52. *A graph showing anomalies*

Since this dataset has some noise or anomalies, there are anomalies see the following explanation of the datapoints in red shown, and everything else that is normal is green.

rds_cpu_utilization.csv

This dataset has a mixture of normal data and anomalies. As you can see in Figure 8-53, the time Series has values at different timestamps. Using visualization, you can plot the time Series now. You will convert the timestamp to datetime for this work and also drop the timestamp column. As shown below, the time Series shows the datetime vs. the value column.

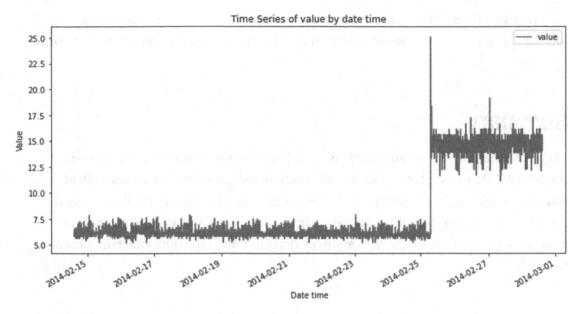

Figure 8-53. *A graph showing the time Series*

Let's add the anomaly column to the original dataframe and prepare a new dataframe. Using visualization, you can plot the new time Series now. As shown below, the time Series shows the datetime vs. the value column. Normal data points are shown in green and anomalies are shown in red. Figure 8-54 depicts a graph showing the anomalies.

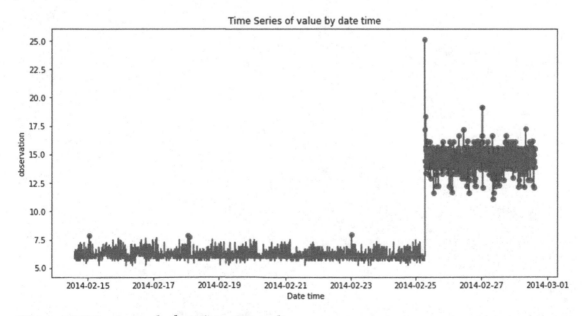

Figure 8-54. *A graph showing anomalies*

Since this dataset has some noise or anomalies, there are anomalies see the following explanation of the datapoints in red shown, and everything else that is normal is green.

Summary

In this chapter, we discussed recurrent neural networks and long short-term memory models. We also looked at LSTMs as a means to detect anomalies. We also walked through several different examples of time-series data with different anomalies and showed how to start detecting anomalies. In the Chapter 9, we will look at another method of performing time-series anomaly detection with the temporal convolutional network.

CHAPTER 9

Temporal Convolutional Networks

In this chapter, you will learn about temporal convolutional networks (TCNs). You will learn how TCNs work, how they can be used to detect anomalies, and how you can implement anomaly detection using a TCN.

In a nutshell, this chapter covers the following topics:

- What is a temporal convolutional network?

- Dilated temporal convolutional network

- Encoder-decoder temporal convolutional network

- TCN application

Note Code examples are provided in Python 3.8. The code repository for this book is available at `https://github.com/apress/beginning-anomaly-detection-python-deep-learning-2e/tree/master`.

The repository also includes a requirements.txt file to check your packages and their versions.

The two notebooks for this chapter are as follows:

- **Dilated TCN**: `https://github.com/apress/beginning-anomaly-detection-python-deep-learning-2e/blob/master/Chapter%209%20Temporal%20Convolutional%20Networks/chapter9_tcn.ipynb`

- **ED-TCN**: `https://github.com/apress/beginning-anomaly-detection-python-deep-learning-2e/blob/master/Chapter%209%20Temporal%20Convolutional%20Networks/chapter9_ed_tcn.ipynb`

Navigate to "Chapter 9 Temporal Convolutional Networks" and then click the notebook. The code is provided as .py files as well, though it is the exported version of the notebook.

We will be using JupyterLab to present all of the code examples.

What Is a Temporal Convolutional Network?

Temporal convolutional networks are a family of architectures that incorporate one-dimensional convolutional layers. More specifically, these convolutions are **causal**, meaning no information from the future is leaked into the past. In other words, the model only processes information going forward in time. One of the problems with recurrent neural networks (covered in Chapter 8) in the context of language translation is that they read sentences from left to right in time, leading them to mistranslate in some cases where the order of the sentence is switched around to create emphasis. To solve this, RNNs use bi-directional encoders, but this means future information is considered in the present. TCNs don't have this problem because, unlike RNNs, they don't rely on information from previous time steps, thanks to their causality. Additionally, TCNs can map an input sequence of any length to an output sequence with the same length, just as RNNs can do.

Basically, TCNs seem to be a great alternative to RNNs. These are the advantages of TCNs, specifically considering RNNs in general:

- **Parallel computations**: Convolutional networks pair well with GPU training, particularly because the matrix-heavy calculations of the convolutional layers are well suited to the structure of GPUs, which are configured to carry out matrix calculations that are part of graphics processing. Because of that, TCNs can train much faster than RNNs.

- **Flexibility**: TCNs can change input size, change filter size, increase dilation factors, stack more layers, etc. in order to easily be applied to various domains.

- **Consistent gradients**: Because TCNs are comprised of convolutional layers, they backpropagate differently than RNNs do, and thus all of the gradients are saved. RNNs have a problem called "exploding," or "vanishing gradients," where sometimes the calculated gradient is either extremely large or extremely small, leading to the readjusted weight to be either too extreme of a change or a relatively nonexistent change. To combat this, types of RNNs such as the LSTM, GRU, and HF-RNN were developed.

- **Lighter on memory**: LSTMs store information in their cell gates, so if the input sequence is long, much more memory is used by the LSTM network. Comparatively, TCNs are relatively straightforward, as they are comprised of several layers that all share their own respective filters. Compared to LSTMs, TCNs are much lighter to run in regard to their memory usage during training.

However, TCNs do have some disadvantages:

- **Memory usage during evaluation mode**: RNNs only need to know some input x_t to generate a prediction, since they maintain a summary of everything they learned through their hidden state vectors. In comparison, TCNs need the entire sequence up until the current point again to make an evaluation, leading to potentially higher memory usage than RNNs when making predictions.

- **Problems with transfer learning**: Transfer learning is when a model has been trained for one particular task (e.g., classifying vehicles) and has the last layer(s) taken out and retrained completely so that the model can be used for a new classification task (e.g., classifying animals).

The of transfer learning is a powerful and useful concept. In computer vision, there are some really powerful models, such as the ResNet152 model, that have been trained on powerful GPUs for quite some time to achieve the performances that they do. Instead of training our own CNN from the ground up (and most of us don't have the GPU hardware or the time to spend that long training an extremely deep model like ResNet152), we can simply take ResNet152, for example, which is really good at extracting features out of images, and train it to associate the features that it extracts with a completely new set of classes. This process takes a lot less time since the weights in the entire network are already well optimized, so we're only concerned with finding the optimal weights for the layers we are retraining.

That's why transfer learning is such a valuable process; it allows us to take a pretrained, high-performance model and simply retrain the last layer(s) with our hardware and teach the model a new classification task (for CNNs).

Going back to TCNs, the model might be required to remember varying levels of sequence history to make predictions. If the old model required some arbitrary length of history to make predictions, but in the new task, it suddenly requires a significantly different length of history to make predictions, that could cause issues and lead to poor performance. Transfer learning doesn't make much sense anymore - you'd have to retrain the entire network because the only way to perform well on this new task is to have a different history length as input. In contrast, images are much more stable and are usually always at a fixed length, unlike one-dimensional sequences.

In a one-dimensional convolutional layer, we still have a parameter **k** to determine the size of our kernel, or filter.

As an example of what the one-dimensional convolutional operation looks like, assume that we have an input vector defined as shown in Figure 9-1 and a filter initialized as shown in Figure 9-2.

$$x = \begin{bmatrix} 10 & 5 & 15 & 20 & 10 & 20 \end{bmatrix}$$

Figure 9-1. *An input vector x defined with corresponding values*

$$\begin{array}{c} \text{Filter weights} \\ \begin{bmatrix} 1 & 0.2 & 0.1 \end{bmatrix} \end{array}$$

Figure 9-2. *The filter weights associated with this one-dimensional convolutional layer*

The output of the convolutional layer is calculated as shown in Figure 9-3, Figure 9-4, Figure 9-5, and Figure 9-6.

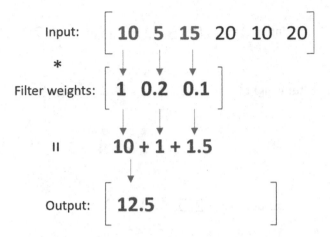

Figure 9-3. *How the first entry of the output vector is calculated using the filter weights. The filter weights are multiplied element-wise with the first three entries in the input, and the results are summed up to produce the output value*

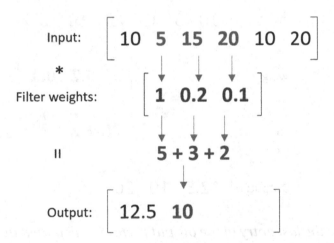

Figure 9-4. *How the second entry of the output vector is calculated using the filter weights. The procedure is the same as in Figure 9-3, but the filter weights are shifted right one*

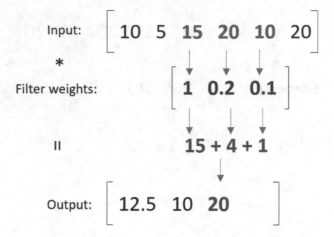

Figure 9-5. *How the third entry of the output vector is calculated using the filter weights*

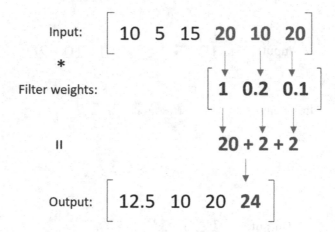

Figure 9-6. *How the last entry of the output vector is calculated using the filter weights*

Now, we have the output of the one-dimensional convolutional layer. These one-dimensional convolutional layers are quite similar to how two-dimensional convolutional layers work, and they comprise nearly the entirety of the two different TCNs we will look at: the **dilated temporal convolutional network**, and the **encoder-decoder temporal convolutional network**. It is important to note that both models involve **supervised anomaly detection**, though the encoder-decoder TCN is capable of semi-supervised anomaly detection since it is an autoencoder.

Dilated Temporal Convolutional Network

In this type of TCN, we deal with a new property known as a **dilation**. Basically, when the dilation factor is greater than 1, we introduce gaps in the output data that correspond to the dilation factor. To understand the concept of dilation better, let's look at how it works for a **two-dimensional convolutional layer**.

This is a standard convolution, equivalent to what we looked at in Chapter 3. You can also think of a standard convolutional layer as having a **dilation factor** of 1 (refer to Figure 9-7).

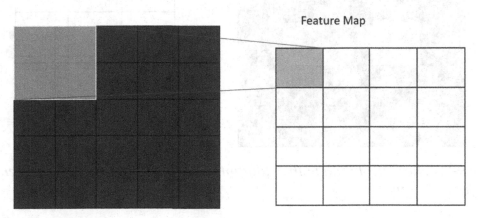

Figure 9-7. *A standard convolution with a dilation factor of 1*

Now, let's look at what happens when we increase the dilation factor to 2. For the first entry in the feature map, the convolution looks like Figure 9-8.

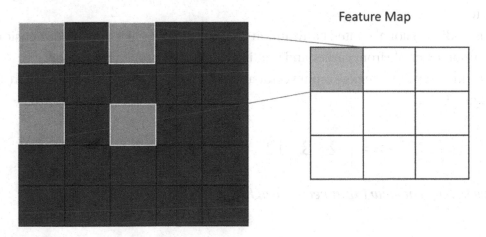

Figure 9-8. *A standard convolution with a dilation factor of 2 defining the first entry in the feature map*

Notice that the spacing between each sampled entry has increased by one across all directions. Vertically, horizontally, and diagonally, the sampled entries are all spaced apart by one entry. Essentially, this spacing is determined by finding what **d – 1** is, where **d** is the **dilation factor**. For a dilation factor of 3, this spacing will be two apart. Now, for the second entry, the convolution process proceeds as normal, as shown in Figure 9-9.

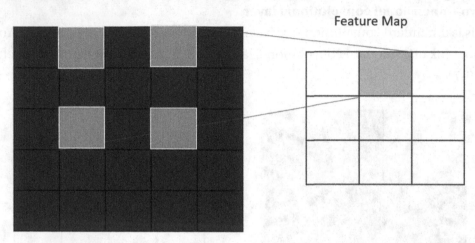

Figure 9-9. *The convolution with a dilation factor of 2 defining the second entry in the feature map*

Once the process terminates, we will have our feature map. Notice the reduction in dimensionality of the feature map, which is a direct result of increasing the dilation factor. In the standard two-dimensional convolutional layer, we had a 4×4 feature map since the dilation factor was 1, but now we have a 3×3 feature map after increasing this factor to 2.

A one-dimensional dilated convolution is similar. Let's revisit the one-dimensional convolution example from earlier and modify it a bit to illustrate this concept.

Assume now that the new input vector and filter weights are as shown in Figure 9-10 and Figure 9-11.

$$x = \begin{bmatrix} 2 & 8 & 12 & 4 & 6 & 4 & 2 & 12 \end{bmatrix}$$

Figure 9-10. *The new input vector weights*

Filter weights

$$\begin{bmatrix} 0.5 & 0.2 & 0.4 \end{bmatrix}$$

Figure 9-11. *The new filter weights*

Let's also assume now that the dilation factor is 2, not 1. The new output vector would be the following, using dilated one-dimensional convolutions with a dilation factor of 2 (Figure 9-12, Figure 9-13, Figure 9-14, and Figure 9-15).

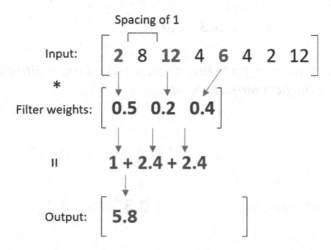

Figure 9-12. *Calculating the first entry in the output factor using dilated one-dimensional convolutions with a dilation factor of 2*

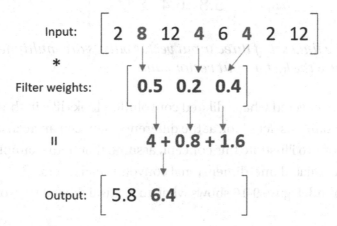

Figure 9-13. *The next set of three input vector values are multiplied with the filter weights to produce the next output vector value*

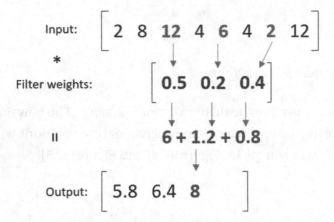

Figure 9-14. *The third set of three input vector values are multiplied with the filter weights to produce the next output vector value*

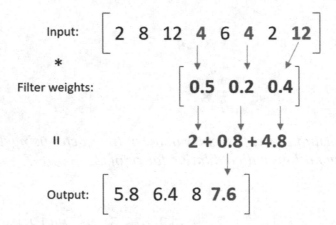

Figure 9-15. *The final set of three input vector values are multiplied with the filter weights to produce the last output vector value*

Now that we've covered what a dilated convolution looks like in the context of one-dimensional convolutions, let's look at the difference between an acausal and a causal dilated convolution. To illustrate this concept, assume that both examples that follow are referring to a set of dilated one-dimensional convolutional layers.

With that in mind, Figure 9-16 shows what an acausal dilated network looks like.

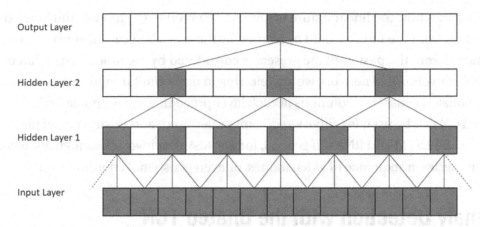

Figure 9-16. *An acausal dilated network. The first hidden layer has a dilation factor of 2, and the second hidden layer has a dilation factor of 4. Notice how inputs "forward in the sequence" contribute to the next layer's node as well*

It might not be that apparent from the way the architecture is structured, but if you think of the input layer as a sequence of some data going forward in time, you might be able to see that information from the future would be accounted for when selecting the output. In a causal network, we only want information that we've learned up until the present, so none of the information from the future will be accounted for in the model's predictions. Figure 9-17 shows what a causal dilated network looks like.

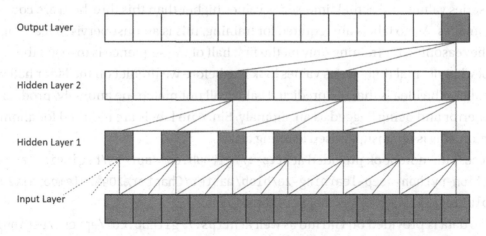

Figure 9-17. *A causal dilated network. The first hidden layer has a dilation factor of 2, and the second hidden layer has a dilation factor of 4. Notice how no inputs forward in the sequence contribute to the next layer's node. This type of structure is ideal if the goal is to preserve some sort of flow within the data set, which is time in our case*

This shows how the linear nature of time is preserved in the model, and how no information from the future would be learned by the model. In causal networks, only information from the past until the present is considered by the model. The dilated temporal convolutional network we are referring to has a similar model architecture, utilizing dilated causal convolutions in each layer preceding the output layer.

The TCN can be used for supervised, semi-supervised, and unsupervised tasks, much like the LSTM and RNNs in general, for any task that involves sequences, whether it's a time series, a sequence of video frames, or even geospatial coordinate data.

Anomaly Detection with the Dilated TCN

Now that you know more about what a TCN is and how it works, let's try applying a dilated TCN to the New York taxi dataset to see how it performs **unsupervised anomaly detection** in a time-series setting. Much like in the LSTM example from Chapter 8, the TCN will be trained to predict the Tth time step's value given the time step values from 1 to T-1, where T is some window size. An example would be using the first 15 values to predict the 16th value. Similarly, we will train it on the first half of the time series on strictly normal data. We will set the error threshold (95th percentile error on the training set, so top 5% largest error) on the training prediction errors as the anomaly threshold. Any points with prediction errors given by the mean absolute error between predicted time-series values and actual time-series values higher than this threshold are counted as anomalies. As no labels are required for training, this is an unsupervised learning task.

The reasoning for training only on the first half of the sequence is to teach the model what normal time-series values look like before we predict on the latter half with anomalies. The idea is that the predicted value will not match the anomaly, producing a higher error and being flagged as an anomaly. Since no labels are required for anomaly detection, this is an unsupervised learning task.

Access the notebook provided at `https://github.com/apress/beginning-anomaly-detection-python-deep-learning-2e/blob/master/Chapter%209%20Temporal%20Convolutional%20Networks/chapter9_tcn.ipynb`.

The data is provided on GitHub as well at `https://github.com/apress/beginning-anomaly-detection-python-deep-learning-2e/tree/master/data`.

First, import all of the necessary packages as shown in Figure 9-18.

```
import tensorflow.keras as keras
from tensorflow.keras import regularizers
from tensorflow.keras.models import Model
from tensorflow.keras.layers import Input, Conv1D, Flatten, Dense
from tensorflow.keras.layers import Activation, SpatialDropout1D
from tensorflow.keras.preprocessing import sequence

from tensorflow.keras.callbacks import ModelCheckpoint, TensorBoard
from tensorflow.keras.utils import to_categorical

import sklearn
from sklearn.preprocessing import StandardScaler, MinMaxScaler
from sklearn.model_selection import train_test_split
from sklearn.metrics import confusion_matrix, roc_auc_score
from sklearn.metrics import classification_report

import seaborn as sns
import pandas as pd
import numpy as np
import matplotlib

import matplotlib.pyplot as plt
import matplotlib.gridspec as gridspec
%matplotlib inline

import tensorflow
import sys
print("Python: ", sys.version)

print("pandas: ", pd.__version__)
print("numpy: ", np.__version__)
print("seaborn: ", sns.__version__)
print("matplotlib: ", matplotlib.__version__)
print("sklearn: ", sklearn. version )
```

Figure 9-18. *Importing all of the necessary packages*

You should see output like the following:

```
Python:  3.8.12 (default, Oct 12 2021, 03:01:40) [MSC v.1916 64 bit
(AMD64)]
pandas:  2.0.0
numpy:  1.22.2
seaborn:  0.12.2
matplotlib:  3.7.1
sklearn:  1.2.2
Tensorflow:  2.7.0
```

After that, we proceed to defining our `Visualization` class, as shown in Figure 9-19.

```
class Visualization:
    labels = ["Normal", "Anomaly"]

    def draw_confusion_matrix(self, y, ypred):
        matrix = confusion_matrix(y, ypred)

        plt.figure(figsize=(10, 8))
        colors=[ "orange","green"]
        sns.heatmap(matrix, xticklabels=self.labels,
yticklabels=self.labels, cmap=colors, annot=True, fmt="d")
        plt.title("Confusion Matrix")
        plt.ylabel('Actual')
        plt.xlabel('Predicted')
        plt.show()


    def draw_anomaly(self, y, error, threshold):
        groupsDF = pd.DataFrame({'error': error,
                                    'true': y}).groupby('true')

        figure, axes = plt.subplots(figsize=(12, 8))

        for name, group in groupsDF:
            axes.plot(group.index, group.error, marker='x' if name ==
1 else 'o', linestyle='',
                        color='r' if name == 1 else 'g', label="Anomaly"
if name == 1 else "Normal")

        axes.hlines(threshold, axes.get_xlim()[0], axes.get_xlim()[1],
colors="b", zorder=100, label='Threshold')
        axes.legend()

        plt.title("Anomalies")
        plt.ylabel("Error")
        plt.xlabel("Data")
        plt.show()

    def draw_error(self, error, threshold):
            plt.plot(error, marker='o', ms=3.5, linestyle='',
                    label='Point')

            plt.hlines(threshold, xmin=0, xmax=len(error)-1,
colors="b", zorder=100, label='Threshold')
            plt.legend()
            plt.title("Reconstruction error")
            plt.ylabel("Error")
            plt.xlabel("Data")
            plt.show()
```

Figure 9-19. *Defining the* `Visualization` *class*

Next, let's define the mechanism to load our data, as shown in Figure 9-20.

```
tensorlogs = ["art_daily_no_noise", #0
              "art_daily_nojump", #1
              "art_daily_jumpsdown",#2
              "art_daily_perfect_square_wave", #3
              "art_increase_spike_density",  #4
              "art_load_balancer_spikes",  #5
              "ambient_temperature_system_failure", #6
              "nyc_taxi",  #7
              "ec2_cpu_utilization", #8
              "rds_cpu_utilization"] #9

dataFilePaths = ['../data/art_daily_no_noise.csv',
                 '../data/art_daily_nojump.csv',
                 '../data/art_daily_jumpsdown.csv',
                 '../data/art_daily_perfect_square_wave.csv',
                 '../data/art_increase_spike_density.csv',
                 '../data/art_load_balancer_spikes.csv',
                 '../data/ambient_temperature_system_failure.csv',
                 '../data/nyc_taxi.csv',
                 '../data/ec2_cpu_utilization.csv',
                 '../data/rds_cpu_utilization.csv']
```

Figure 9-20. *Defining the dataset paths to be able to switch between them easily*

Let's now select our dataset (Figure 9-21).

```
i = 7

tensorlog = tensorlogs[i]
dataFilePath = dataFilePaths[i]
print("tensorlog: ", tensorlog)
print("dataFilePath: ", dataFilePath)
```

Figure 9-21. *Selecting our dataset. We have picked* i = 7, *which should load the NYC taxi dataset*

Your output should look like this:

tensorlog: nyc_taxi
dataFilePath: ../data/nyc_taxi.csv

Let's continue to actually loading our selected dataset, as shown in Figure 9-22.

```
df = pd.read_csv(filepath_or_buffer=dataFilePath, header=0, sep=',')
print('Shape:' , df.shape[0])
print('Head:')
print(df.head(5))
```

Figure 9-22. *Code to load the selected dataset*

Your output should look like this:

```
Shape: 10320
Head:
            timestamp  value
0  2014-07-01 00:00:00  10844
1  2014-07-01 00:30:00   8127
2  2014-07-01 01:00:00   6210
3  2014-07-01 01:30:00   4656
4  2014-07-01 02:00:00   3820
```

Let's plot our time-series data. Refer to Figure 9-23a and Figure 9-23b.

```
df['Datetime'] = pd.to_datetime(df['timestamp'])
print(df.head(3))
df.shape
df.plot(x='Datetime', y='value', figsize=(12,6))
plt.xlabel('Date time')
plt.ylabel('Value')
plt.title('Time Series of value by date time')
```

Figure 9-23a. *Code to plot the time-series dataset*

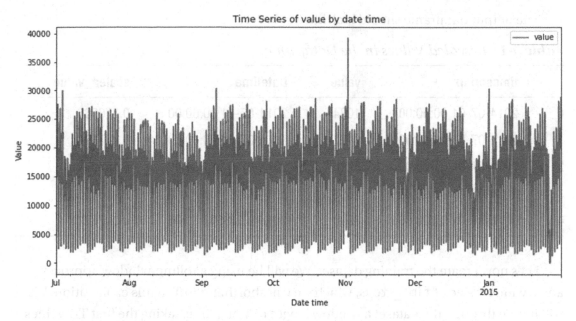

Figure 9-23b. *The plotted time-series dataset. This is the NYC taxi dataset*

We should rescale our dataset next. Refer to Figure 9-24.

```
from sklearn.preprocessing import MinMaxScaler
scaler = MinMaxScaler(feature_range = (0, 1))
df['scaled_value'] =
pd.DataFrame(scaler.fit_transform(pd.DataFrame(df['value'])),columns=[
'value'])
print('Shape:' , df.shape[0])
df.head(5)
```

Figure 9-24. *Code to rescale the dataset to a range of 0 to 1*

The output dataframe should look like Table 9-1.

Table 9-1. *Rescaled Values in the Dataframe*

	timestamp	value	Datetime	scaled_value
0	2014-07-01 00:00:00	10844	2014-07-01 00:00:00	0.276506
1	2014-07-01 00:30:00	8127	2014-07-01 00:30:00	0.207175
2	2014-07-01 01:00:00	6210	2014-07-01 01:00:00	0.158259
3	2014-07-01 01:30:00	4656	2014-07-01 01:30:00	0.118605
4	2014-07-01 02:00:00	3820	2014-07-01 02:00:00	0.097272

Let's now create the training dataset. We will be using a rolling-window approach and a window size of time_steps, which we will shorthand to **T** in this explanation. We will iterate through the dataset a window length of T at a time, taking the first T-1 values as the x and the Tth values as the y. We will teach the model to read an input of T-1 values and predict the Tth. Refer to Figure 9-25.

```
time_steps = 16

sequence = np.array(df['scaled_value'])
print(sequence)

# Create rolling window sequences as determined by time_steps
x_sequences = []
y_sequences = []

# Number of windows to iterate through
n_iter = len(sequence) - time_steps + 1
for f in range(n_iter):
    window = sequence[f:f+time_steps]
    x_sequences.append(window[:-1])
    y_sequences.append(window[-1:])
x_sequences = np.array(x_sequences)
y_sequences = np.array(y_sequences)
print(x_sequences.shape, y_sequences.shape)
```

Figure 9-25. *Creating the rolling-window features to feed into the model*

Now that we have our data created, let's define the model architecture. Refer to Figure 9-26.

```
input_layer = Input(shape=(time_steps-1, 1))

#Series of temporal convolutional layers with dilations increasing by
powers of 2.
conv_1 = Conv1D(filters=128, kernel_size=2, dilation_rate=1,
                padding='causal', strides=1,
                kernel_regularizer=regularizers.l2(0.01),
                kernel_initializer='he_normal',
                activation='relu')(input_layer)

#Dropout layer after each 1D-convolutional layer
drop_1 = SpatialDropout1D(0.05)(conv_1)

conv_2 = Conv1D(filters=64, kernel_size=2, dilation_rate=2,
                padding='causal',strides=1,
kernel_regularizer=regularizers.l2(0.01),
                kernel_initializer='he_normal',
                activation='relu')(drop_1)

drop_2 = SpatialDropout1D(0.05)(conv_2)

conv_3 = Conv1D(filters=32, kernel_size=2, dilation_rate=4,
                padding='causal',
strides=1,kernel_regularizer=regularizers.l2(0.01),
                kernel_initializer='he_normal',
                activation='relu')(drop_2)

drop_3 = SpatialDropout1D(0.05)(conv_3)

conv_4 = Conv1D(filters=4, kernel_size=2, dilation_rate=8,
                padding='causal',
strides=1,kernel_regularizer=regularizers.l2(0.05),
                kernel_initializer='he_normal',
                activation='relu')(drop_3)

flat = Flatten()(conv_4)
output = Dense(1  activation='relu')(flat)
tcn = Model(input_layer, output)
```

Figure 9-26. *Defining the TCN architecture*

There are a lot of new parameters to specify:

- kernel_size: This controls the length of each filter that is being applied over the data.

- dilation_rate: This controls d, the dilation size.

- **padding**: Set to `'causal'`, this specifies that we want a causal layer, not an acausal layer.

- **strides**: The stride length. Set to 1, this means each filter shifts over by one value of the input data once it's done with its computation. It's like the rolling-window code described earlier in that each consecutive window is one stride or value apart.

- **kernel_regularizer**: Specifying the L2 regularization prevents overfitting.

- **kernel_initializer**: Set to `'he_normal'`, we are initializing the weights using the Kaiming/He normal method. This helps with training stability with networks that rely on ReLU activation functions.

We have also added `SpatialDropout1D` layers, which act as dropout layers for one-dimensional CNN settings and helps with the regularization of the model.

The `Flatten` layer simply concatenates all of the outputs of the previous layer to make it one-dimensional into a (batch_size, 1) shape.

Let's now compile the model and print the architecture. Refer to Figure 9-27a.

```
metric = 'mean_absolute_error'
tcn.compile(optimizer='adam', loss='mean_absolute_error',
metrics=[metric])
print(tcn.summary())
```

Figure 9-27a. *Compiling the model to use the Adam optimizer and mean absolute error as the loss function*

You should see output like that shown in Figure 9-27b.

```
Model: "model_1"

_____
Layer (type)                 Output Shape              Param #
=================================================================
input_2 (InputLayer)         [(None, 15, 1)]           0

conv1d_4 (Conv1D)            (None, 15, 128)           384

spatial_dropout1d_3 (Spatia  (None, 15, 128)           0
lDropout1D)

conv1d_5 (Conv1D)            (None, 15, 64)            16448

spatial_dropout1d_4 (Spatia  (None, 15, 64)            0
lDropout1D)

conv1d_6 (Conv1D)            (None, 15, 32)            4128

spatial_dropout1d_5 (Spatia  (None, 15, 32)            0
lDropout1D)

conv1d_7 (Conv1D)            (None, 15, 4)             260

flatten_1 (Flatten)          (None, 60)                0

dense_1 (Dense)              (None, 1)                 61

=================================================================
Total params: 21,281
Trainable params: 21,281
Non-trainable params: 0
_____
```

Figure 9-27b. *The output architecture printed out*

Let's initiate the training process. Refer to Figure 9-28.

```python
sequences_x = x_sequences.reshape(len(x_sequences), time_steps-1, 1)
print("sequences_x: ", sequences_x.shape)
sequences_y = y_sequences.reshape(len(y_sequences), 1)
print("sequences_y: ", sequences_y.shape)

# Training on first half of data only, predicting on whole thing
stop_point = int(0.5 * len(df))
training_x = sequences_x[:stop_point]
print("training_x: ", training_x.shape)
training_y = sequences_y[:stop_point]
print("training_y: ", training_y.shape)

batch_size=32
epochs=10

tcn.fit(x=training_x, y=training_y,
                    batch_size=batch_size, epochs=epochs,
                    verbose=1, validation_data=(training_x,
training_y),

callbacks=[TensorBoard(log dir='./logs/{0}'.format(tensorlog))])
```

Figure 9-28. Specifying the training dataset to be half of the total dataset to teach the model what is normal before predicting on the entire dataset later on. This also specifies the training loop

The output should look somewhat like Figure 9-29.

```
sequences_x:   (10305, 15, 1)
sequences_y:   (10305, 1)
training_x:   (5160, 15, 1)
training_y:   (5160, 1)
Epoch 1/10
162/162 [==============================] - 10s 58ms/step - loss:
3.2908 - mean_absolute_error: 0.1490 - val_loss: 2.2728 -
val_mean_absolute_error: 0.0654
Epoch 2/10
162/162 [==============================] - 8s 50ms/step - loss: 1.8847
- mean_absolute_error: 0.0665 - val_loss: 1.5691 -
val_mean_absolute_error: 0.0572


...


Epoch 9/10
162/162 [==============================] - 2s 10ms/step - loss: 0.3016
- mean_absolute_error: 0.0467 - val_loss: 0.2648 -
val_mean_absolute_error: 0.0421
Epoch 10/10
162/162 [==============================] - 2s 10ms/step - loss: 0.2415
- mean_absolute_error: 0.0451 - val_loss: 0.2172 -
val_mean_absolute_error: 0.0451
```

Figure 9-29. *The output of the training code, with some of the outputs removed for a clearer display*

One thing to note about TCN/one-dimensional CNNs in general is that they tend to be a bit sensitive, meaning that sometimes the training completely fails and the CNN just produces a straight-line output. The He Normal initialization technique should help combat this collapse, but sometimes you'll just need to retrain the model to achieve a convergence.

If your results are not satisfactory or are a failure, feel free to rerun the model training a few times if necessary. Because neural networks in general depend on random weight initializations that are not privy to the geometry of the loss curve, they can get trapped into reaching suboptimal convergences or even a divergence.

With the training completed, let's move on to making the predictions. Execute the code shown in Figure 9-30.

```
import math
from sklearn.metrics import mean_squared_error

testing_dataset = sequences_x
print("testing_dataset: ", testing_dataset.shape)

print("sequences_y: ", sequences_y.shape)

testing_pred = tcn.predict(x=testing_dataset)
print("testing_pred: ", testing_pred.shape)

errorsDF = sequences_y - testing_pred
print(errorsDF.shape)
rmse = math.sqrt(mean_squared_error(sequences_y, testing_pred))
print('Test RMSE: %.3f' % rmse)
```

Figure 9-30. *Making predictions using the trained model*

If all went well during training, the output should look something like the following:

testing_dataset: (10305, 15, 1)

sequences_y: (10305, 1)

testing_pred: (10305, 1)

(10305, 1)

Test RMSE: 0.060

Let's now calculate the distance cutoffs to determine the anomaly threshold. Refer to Figure 9-31.

```
#based on cutoff after sorting errors
# Calculate threshold using training error
dist = np.linalg.norm(sequences_y[:len(training_y)] -
testing_pred[:len(training_y)], axis=-1)

scores =dist.copy()
print(scores.shape)
scores.sort()
cutoff = int(0.999 * len(scores))
print(cutoff)
threshold= scores[cutoff]

# Calculate total error over full sequence
dist = np.linalg.norm(sequences_y[:] - testing_pred, axis=-1)

print(threshold)
```

Figure 9-31. *Calculating the threshold for maximum prediction error that's still normal*

You should see output like this:

(5160,)

5154

0.16645004726088353

Your determined threshold may differ, as it all depends on the model training process.

Let's now plot the model's predictions. Refer to Figure 9-32a and Figure 9-32b.

```
plt.figure(figsize=(24,16))
plt.plot(sequences_y, color='green')
plt.plot(training_y, color='blue')
plt.plot(testing_pred, color='red')
```

Figure 9-32a. *Code to plot the time-series sequences, color-coded so that* green *represents the true value of the entire sequence of values,* blue *only represents the training split, and* red *represents the model's predictions on the whole dataset's time-series values*

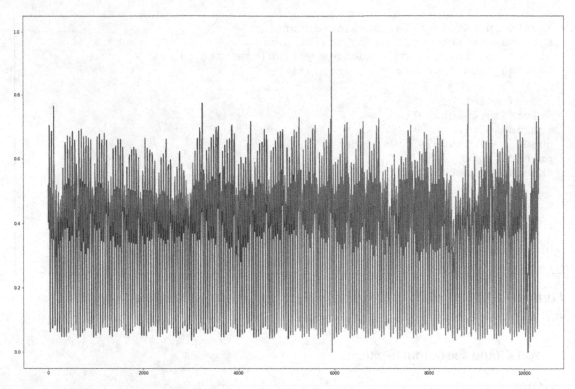

Figure 9-32b. *The resulting plot. The model seems to have learned a pretty good fit for the dataset*

Let's start making predictions for what the anomalies are using our calculated threshold. Refer to Figure 9-33.

```
#label the records anomalies or not based on threshold
z = zip(dist >= threshold, dist)

y_label=[]
error = []
for idx, (is_anomaly, dist) in enumerate(z):
    if is_anomaly:
        y_label.append(1)
    else:
        y_label.append(0)
    error.append(dist)
```

Figure 9-33. *Code to calculate the anomaly label for each data point*

We can visualize this error threshold and find out what points lie beyond it. Refer to Figure 9-34 for the code to plot this threshold graph.

```
viz = Visualization()
viz.draw_anomaly(y_label, error, threshold)
```

Figure 9-34. *Code to plot the error threshold graph*

Figure 9-35 shows the resulting plot.

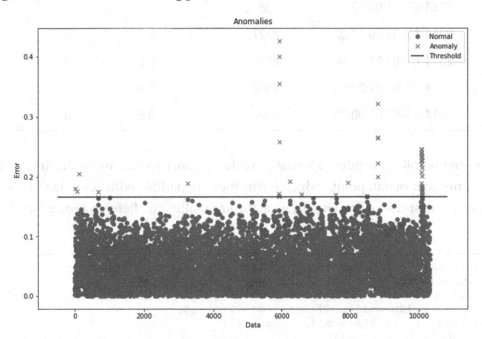

Figure 9-35. *The plot showing the points that lie beyond the error threshold, thus being classified as anomalies*

Let's create an anomaly prediction dataframe to more easily plot which points exactly were determined to be anomalies. Refer to Figure 9-36.

```
adf = pd.DataFrame({'Datetime': df['Datetime'], 'observation':
df['value'],
                    'error': [0 for f in range((time_steps-1))] +
error, 'anomaly': [0 for f in range((time_steps-1))]+ y_label})
adf.head(5)
```

Figure 9-36. *Creating a dataframe of the original values as well as the observed prediction error and anomaly flag*

The output should look like Table 9-2.

Table 9-2. *Resulting Dataframe Table of Code in Figure 9-36*

	Datetime	observation	error	anomaly
0	2014-07-01 00:00:00	10844	0.0	0
1	2014-07-01 00:30:00	8127	0.0	0
2	2014-07-01 01:00:00	6210	0.0	0
3	2014-07-01 01:30:00	4656	0.0	0
4	2014-07-01 02:00:00	3820	0.0	0

Due to the rolling-window approach, we don't have predictions for the first `time_steps-1` number of datapoints, which is why they are padded with zeros. Let's now plot the final graph showing visually what the anomalies are. Refer to Figure 9-37 for this code.

```
figure, axes = plt.subplots(figsize=(12, 6))
axes.plot(adf['Datetime'], adf['observation'], color='g')
anomaliesDF = adf.query('anomaly == 1')
axes.scatter(anomaliesDF['Datetime'].values,
anomaliesDF['observation'], color='r')
plt.xlabel('Date time')
plt.ylabel('observation')
plt.title('Time Series of value by date time')
```

Figure 9-37. *Code to plot the anomalous points on the original time-series sequence*

The result looks something like Figure 9-38.

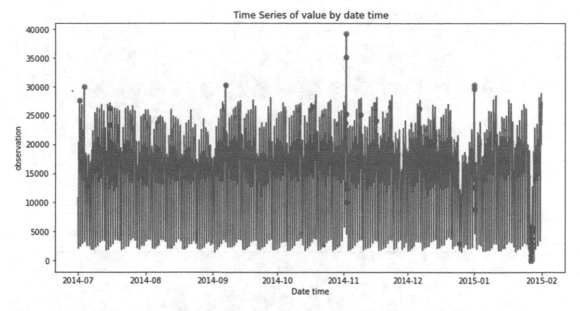

Figure 9-38. *The identified anomalies using the TCN's predictions*

Figure 9-38 shows that quite a few points are labeled as anomalies. Most importantly, the model managed to predict as anomalies Thanksgiving, Christmas, New Year's Eve, and a snowstorm-type weather anomaly, among some other points.

Overall, this was an example of **unsupervised anomaly detection**. We took a subset of the original sequence, trained the model to output a prediction given rolling windows, and calculated a threshold by which we can identify anomalies. None of the steps in this process involved a training label.

For our next example, we will be implementing the encoder-decoder temporal convolutional network, essentially combining the TCN with the concept of an autoencoder to perform unsupervised anomaly detection.

Encoder-Decoder Temporal Convolutional Network

The version of the encoder-decoder TCN (ED-TCN) we will be exploring involves a combination of one-dimensional causal convolutional and pooling layers to encompass the encoding stage, and a series of upsampling and one-dimensional causal convolutional layers to comprise the decoding stage. The convolutional layers in this model aren't dilated, but they still count as layers of a TCN. To better understand the structure of this model, take a look at Figure 9-39.

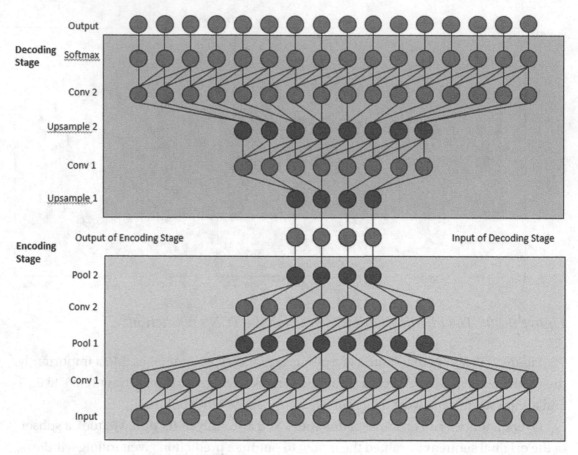

Figure 9-39. *In both the encoding and decoding stages, the model is comprised of causal convolutional layers, and is structured so that the layers are always causal*

The diagram might seem pretty complicated, so let's break it down layer by layer.

First, look at the encoding stage and start with the input layer at the very bottom. From this layer, we perform a **causal convolution** on the input as part of the first convolutional layer. The outputs of the first convolutional layer, which we will call `conv_1`, are now the inputs of the first **max pooling layer**, which we will call `pool_1`.

Recall from Chapter 3 that the pooling layer emphasizes the maximum value in the areas it passes through, effectively generalizing the inputs by choosing the heaviest values. From here, we have another set of causal convolutions and max pooling with layers `conv_2` and `pool_2`. Note the progressive reduction in size of the data as it passes through the encoding stage, a feature characteristic to autoencoders. Finally, we have a dense layer in the middle of the two stages, representing the final, encoded output of the encoding stage, as well as the encoded input of the decoding stage.

The decoding stage is a bit different in this case, since we make use of what is called **upsampling**. Upsampling is a technique in which we repeat the data **n** number of times to scale it up by a factor **n**. In the max pooling layers, the data was reduced by a factor of 2. So, to upsample and increase the data by the same factor of 2, we repeat the data twice. In this case, we are using one-dimensional upsampling, so the layer would repeat each step **n** times with respect to the axis of time. To get a better understanding of what upsampling does, let's apply one-dimensional upsampling as shown in Figure 9-40 and Figure 9-41.

$$x = \begin{bmatrix} 4 & 2 & 6 & 7 & 1 & 6 & 9 \end{bmatrix}$$

Figure 9-40. *A vector x defined with the corresponding values*

n = 2
So data increases by factor of 2 /
repeat each step two times

Figure 9-41. *The upsampling factor n*

Keeping in mind that each individual temporal step is repeated twice, we would see something like procedure shown in Figure 9-42, Figure 9-43, and Figure 9-44.

$$\begin{bmatrix} 4 & 4 & & & & & & & & \end{bmatrix}$$
$$\begin{bmatrix} 4 & 2 & 6 & 7 & 1 & 6 & 9 & 5 \end{bmatrix}$$

Figure 9-42. *The first entry in the input is repeated twice to form the first two entries in the upsampled output vector*

$$\begin{bmatrix} 4 & 4 & 2 & 2 & & & & & & \end{bmatrix}$$
$$\begin{bmatrix} 4 & 2 & 6 & 7 & 1 & 6 & 9 & 5 \end{bmatrix}$$

Figure 9-43. *The next entry is repeated twice to form the next two entries in the output vector of the upsampling operation*

$$\begin{bmatrix} 4 & 4 & 2 & 2 & 6 & 6 & & & & & \end{bmatrix}$$

$$\begin{bmatrix} 4 & 2 & 6 & 7 & 1 & 6 & 9 & 5 \end{bmatrix}$$

Figure 9-44. *This process is repeated with the third entry in the input vector to form the next third pair of entries in the output vector*

This continues until we finally get to the result shown in Figure 9-45.

$$\begin{bmatrix} 4 & 4 & 2 & 2 & 6 & 6 & 7 & 7 & 1 & 1 & 6 & 6 & 9 & 9 & 5 & 5 \end{bmatrix}$$

$$\begin{bmatrix} 4 & 2 & 6 & 7 & 1 & 6 & 9 & 5 \end{bmatrix}$$

Figure 9-45. *The output vector after the upsampling operation compared to the original input vector below it*

Going back to the model, each upsampling layer is then connected to a one-dimensional convolutional layer, and the pair of upsampling layer and one-dimensional convolutional layer repeats again until the final output is passed through a `softmax()` function to result in the output/prediction.

Anomaly Detection with the ED-TCN

Let's put this model to the test by applying it to the credit card dataset. Once again, this example will be another instance of **unsupervised learning**, so we will not have any anomaly or normal data labeled. Follow along with the notebook at `https://github.com/apress/beginning-anomaly-detection-python-deep-learning-2e/blob/master/Chapter%209%20Temporal%20Convolutional%20Networks/chapter9_ed_tcn.ipynb`.

The data is also available on GitHub. Much of the following code is very similar to the earlier dilated TCN example.

First, import all of the necessary packages, as shown in Figure 9-46.

```
import tensorflow.keras as keras
from tensorflow.keras import regularizers
from tensorflow.keras.models import Model
from tensorflow.keras.layers import Input, Conv1D, Dense,\
UpSampling1D, SpatialDropout1D, MaxPooling1D
from tensorflow.keras.callbacks import TensorBoard

import sklearn
from sklearn.preprocessing import StandardScaler, MinMaxScaler
from sklearn.model_selection import train_test_split
from sklearn.metrics import confusion_matrix, roc_auc_score
from sklearn.metrics import classification_report

import seaborn as sns
import pandas as pd
import numpy as np
import matplotlib

import matplotlib.pyplot as plt
import matplotlib.gridspec as gridspec
%matplotlib inline

import tensorflow
import sys
print("Python: ", sys.version)

print("pandas: ", pd.__version__)
print("numpy: ", np.__version__)
print("seaborn: ", sns.__version__)
print("matplotlib: ", matplotlib.__version__)
print("sklearn: ", sklearn.__version__)
print("Tensorflow: ", tensorflow.__version__)
```

Figure 9-46. *Importing all of the necessary packages*

You should see output like this:

```
Python:  3.8.12 (default, Oct 12 2021, 03:01:40) [MSC v.1916 64 bit
(AMD64)]
pandas:  2.0.0
numpy:  1.22.2
seaborn:  0.12.2
matplotlib:  3.7.1
sklearn:  1.2.2
Tensorflow:  2.7.0
```

After that, we proceed to defining our `Visualization` class, as shown in Figure 9-47.

```python
class Visualization:
    labels = ["Normal", "Anomaly"]

    def draw_confusion_matrix(self, y, ypred):
        matrix = confusion_matrix(y, ypred)

        plt.figure(figsize=(10, 8))
        colors=[ "orange","green"]
        sns.heatmap(matrix, xticklabels=self.labels,
yticklabels=self.labels, cmap=colors, annot=True, fmt="d")
        plt.title("Confusion Matrix")
        plt.ylabel('Actual')
        plt.xlabel('Predicted')
        plt.show()


    def draw_anomaly(self, y, error, threshold):
        groupsDF = pd.DataFrame({'error': error,
                                 'true': y}).groupby('true')

        figure, axes = plt.subplots(figsize=(12, 8))

        for name, group in groupsDF:
            axes.plot(group.index, group.error, marker='x' if name ==
1 else 'o', linestyle='',
                      color='r' if name == 1 else 'g', label="Anomaly"
if name == 1 else "Normal")

            axes.hlines(threshold, axes.get_xlim()[0], axes.get_xlim()[1],
colors="b", zorder=100, label='Threshold')
        axes.legend()

        plt.title("Anomalies")
        plt.ylabel("Error")
        plt.xlabel("Data")
        plt.show()

    def draw_error(self, error, threshold):
            plt.plot(error, marker='o', ms=3.5, linestyle='',
                     label='Point')

            plt.hlines(threshold, xmin=0, xmax=len(error)-1,
colors="b", zorder=100, label='Threshold')
            plt.legend()
            plt.title("Reconstruction error")
            plt.ylabel("Error")
            plt.xlabel("Data")
            plt.show()
```

Figure 9-47. *Defining the Visualization class*

Next, let's define the mechanism to load our data, as shown in Figure 9-48.

```
tensorlogs = ["art_daily_no_noise", #0
              "art_daily_nojump", #1
              "art_daily_jumpsdown",#2
              "art_daily_perfect_square_wave", #3
              "art_increase_spike_density",  #4
              "art_load_balancer_spikes",  #5
              "ambient_temperature_system_failure", #6
              "nyc_taxi",  #7
              "ec2_cpu_utilization", #8
              "rds_cpu_utilization"] #9

dataFilePaths = ['../data/art_daily_no_noise.csv',
                 '../data/art_daily_nojump.csv',
                 '../data/art_daily_jumpsdown.csv',
                 '../data/art_daily_perfect_square_wave.csv',
                 '../data/art_increase_spike_density.csv',
                 '../data/art_load_balancer_spikes.csv',
                 '../data/ambient_temperature_system_failure.csv',
                 '../data/nyc_taxi.csv',
                 '../data/ec2_cpu_utilization.csv',
                 '../data/rds_cpu_utilization.csv']
```

Figure 9-48. *Defining the dataset paths to be able to switch between them easily*

Let's now select our dataset (Figure 9-49).

```
i = 7

tensorlog = tensorlogs[i]
dataFilePath = dataFilePaths[i]
print("tensorlog: ", tensorlog)
print("dataFilePath: ", dataFilePath)
```

Figure 9-49. *Selecting our dataset. We have picked i = 7, which should load the NYC taxi dataset*

Your output should look like this:

```
tensorlog:  nyc_taxi
dataFilePath:  ../data/nyc_taxi.csv
```

Let's continue to actually loading our selected dataset, as shown in Figure 9-50.

```
df = pd.read_csv(filepath_or_buffer=dataFilePath, header=0, sep=',')
print('Shape:' , df.shape[0])
print('Head:')
print(df.head(5))
```

Figure 9-50. *Code to load the selected dataset*

Your output should look like this:

Shape: 10320
Head:

	timestamp	value
0	2014-07-01 00:00:00	10844
1	2014-07-01 00:30:00	8127
2	2014-07-01 01:00:00	6210
3	2014-07-01 01:30:00	4656
4	2014-07-01 02:00:00	3820

Let's plot our time-series data. Refer to Figure 9-51a and Figure 9-51b.

```
df['Datetime'] = pd.to_datetime(df['timestamp'])
print(df.head(3))
df.shape
df.plot(x='Datetime', y='value', figsize=(12,6))
plt.xlabel('Date time')
plt.ylabel('Value')
plt.title('Time Series of value by date time')
```

Figure 9-51a. *Code to plot the time-series dataset*

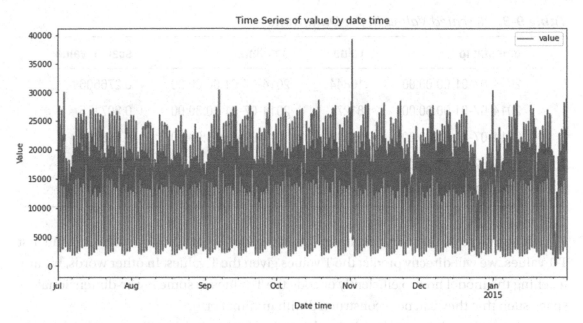

Figure 9-51b. *The plotted time-series dataset. This is the NYC taxi dataset*

We should rescale our dataset next. Refer to Figure 9-52.

```
from sklearn.preprocessing import MinMaxScaler
scaler = MinMaxScaler(feature_range = (0, 1))
df['scaled_value'] =
pd.DataFrame(scaler.fit_transform(pd.DataFrame(df['value'])),columns=[
'value'])
print('Shape:' , df.shape[0])
df.head(5)
```

Figure 9-52. *Code to rescale the dataset to a range of 0 to 1*

The output dataframe should look like Table 9-3.

Table 9-3. *Rescaled Values in the Dataframe*

	timestamp	value	Datetime	scaled_value
0	2014-07-01 00:00:00	10844	2014-07-01 00:00:00	0.276506
1	2014-07-01 00:30:00	8127	2014-07-01 00:30:00	0.207175
2	2014-07-01 01:00:00	6210	2014-07-01 01:00:00	0.158259
3	2014-07-01 01:30:00	4656	2014-07-01 01:30:00	0.118605
4	2014-07-01 02:00:00	3820	2014-07-01 02:00:00	0.097272

Let's now create the training dataset. Instead of predicting the Tth value using the first T-1 values, we will directly predict the T values given the T values. In other words, we are teaching the model how to efficiently encode the T values to some lower-dimensional space such that they can be reconstructed with minimal error.

And so, we won't have a y set, but rather only the x set. Refer to Figure 9-53.

```
time_steps = 16

sequence = np.array(df['scaled_value'])
print(sequence)

# Create rolling window sequences as determined by time_steps
x_sequences = []

# Number of windows to iterate through
n_iter = len(sequence) - time_steps + 1
for f in range(n_iter):
    window = sequence[f:f+time_steps]
    x_sequences.append(window)
x_sequences = np.array(x_sequences)
print(x_sequences.shape)
```

Figure 9-53. *Creating the rolling-window features to feed into the model*

Your output should look somewhat like the following:

```
[0.27650616 0.20717548 0.1582587  ... 0.69664957 0.6783281  0.67059634]
(10305, 16)
```

Now that we have our data created, let's define the model architecture. Refer to Figure 9-54a and Figure 9-54b.

```
input_shape = (time_steps, 1)
input_layer = Input(shape=input_shape)


### ENCODING STAGE
# Pairs of causal 1D convolutional layers and pooling layers
comprising the encoding stage
conv_1 = Conv1D(filters=int(input_shape[0]), kernel_size=2,
dilation_rate=1,
                padding='causal', strides=1,input_shape=input_shape,
                kernel_regularizer=regularizers.l2(0.01),
                kernel_initializer='he_normal',
activation='relu')(input_layer)

pool_1 = MaxPooling1D(pool_size=2, strides=2)(conv_1)

conv_2 = Conv1D(filters=int(input_shape[0] / 2), kernel_size=2,
dilation_rate=1,
                padding='causal',strides=1,
kernel_regularizer=regularizers.l2(0.01),
                kernel_initializer='he_normal',
activation='relu')(pool_1)

pool_2 = MaxPooling1D(pool_size=2, strides=2)(conv_2)

conv_3 = Conv1D(filters=int(input_shape[0] / 4), kernel_size=2,
dilation_rate=1,
                padding='causal',
strides=1,kernel_regularizer=regularizers.l2(0.01),
                kernel_initializer='he_normal',
activation='relu')(pool_2)


### OUTPUT OF ENCODING STAGE
encoder = Dense(int(input_shape[0] / 6), activation='relu')(conv_3)
```

Figure 9-54a. *Defining the encoder portion of the ED-TCN architecture*

```
### DECODING STAGE
# Pairs of upsampling and causal 1D convolutional layers comprising
the decoding stage
upsample_1 = UpSampling1D(size=2)(encoder)

conv_4 = Conv1D(filters=int(input_shape[0]/4), kernel_size=2,
dilation_rate=1,
                padding='causal',strides=1,
kernel_regularizer=regularizers.l2(0.01),
                kernel_initializer='he_normal',
activation='relu')(upsample_1)

upsample_2 = UpSampling1D(size=2)(conv_4)

conv_5 = Conv1D(filters=int(input_shape[0]/2), kernel_size=2,
dilation_rate=1,
                padding='causal',
strides=1,kernel_regularizer=regularizers.l2(0.05),
                kernel_initializer='he_normal',
activation='relu')(upsample_2)

# zero_pad_1 = ZeroPadding1D(padding=(0,1))(conv_5)

conv_6 = Conv1D(filters=int(input_shape[0]), kernel_size=2,
dilation_rate=1,
                padding='causal',
strides=1,kernel_regularizer=regularizers.l2(0.05),
                kernel_initializer='he_normal',
activation='relu')(conv_5)

conv_7 = Conv1D(filters=1, kernel_size=1, dilation_rate=1,
                padding='causal',
strides=1,kernel_regularizer=regularizers.l2(0.05),
                kernel_initializer='he_normal',
activation='relu')(conv_6)
tcn = Model(inputs=input layer, outputs=conv 7)
```

Figure 9-54b. *Defining the decoder portion of the ED-TCN architecture as well as the final model*

The MaxPooling1D layers apply a filter of length pool_size=2, moving through the data two at a time (as specified by strides=2). It picks the maximum value in the filter to pass through to the next layer. It has the effect of halving the data dimension (not the filters), so we are lowering not only the filter dimension space, but also the data dimension space itself.

We have an UpSampling1D layer in the decoder portion functions similarly to how we described UpSampling right before the coding section for the ED-TCN.

Let's now compile the model and print the architecture. Refer to Figure 9-55a.

```
metric = 'mean_absolute_error'
tcn.compile(optimizer='adam', loss='mean_absolute_error',
metrics=[metric])
print(tcn.summary())
```

Figure 9-55a. *Compiling the model to use the Adam optimizer and mean absolute error as the loss function*

You should see output like that shown in Figure 9-55b.

```
Model: "model_6"
_____
Layer (type)                 Output Shape              Param #
=================================================================
input_7 (InputLayer)         [(None, 16, 1)]           0

conv1d_42 (Conv1D)           (None, 16, 16)            48

max_pooling1d_12 (MaxPoolin  (None, 8, 16)             0
g1D)

conv1d_43 (Conv1D)           (None, 8, 8)              264

max_pooling1d_13 (MaxPoolin  (None, 4, 8)              0
g1D)

conv1d_44 (Conv1D)           (None, 4, 4)              68

dense_6 (Dense)              (None, 4, 2)              10

up_sampling1d_12 (UpSamplin  (None, 8, 2)              0
g1D)

conv1d_45 (Conv1D)           (None, 8, 4)              20

up_sampling1d_13 (UpSamplin  (None, 16, 4)             0
g1D)

conv1d_46 (Conv1D)           (None, 16, 8)             72

conv1d_47 (Conv1D)           (None, 16, 16)            272

conv1d_48 (Conv1D)           (None, 16, 1)             17

=================================================================
Total params: 771
Trainable params: 771
Non-trainable params: 0
_____
```

Figure 9-55b. *The output architecture printed out*

As you can see, there is a compression of dimensions occurring throughout the model along with an expanding. We are getting the model to learn the attributes of the time-series model by having it learn an efficient representation of the input to reconstruct the original input back out from. The idea is that it will learn the properties of what makes a normal time-series dataset, and when applying these learned properties to encoding and decoding an anomalous sequence, it will have high reconstruction errors that would flag these sequences as anomalous.

Let's initiate the training process. Refer to Figure 9-56.

```
sequences_x = x_sequences.reshape(len(x_sequences), time_steps, 1)
print("sequences_x: ", sequences_x.shape)


# Training on first half of data only, predicting on whole thing
stop_point = int(0.5 * len(df))
training_x = sequences_x[:stop_point]
print("training_x: ", training_x.shape)



batch_size=32
epochs=10

tcn.fit(x=training_x, y=training_x,
                    batch_size=batch_size, epochs=epochs,
                    verbose=1, validation_data=(training_x,
training_x),

callbacks=[TensorBoard(log dir='./logs/{0}'.format(tensorlog))])
```

Figure 9-56. *Specifying the training dataset to be half of the total dataset to teach the model what is normal before predicting on the entire dataset later on. This also specifies the training loop*

The output should look somewhat like Figure 9-57.

```
sequences_x:   (10305, 16, 1)
training_x:    (5160, 16, 1)
Epoch 1/10
162/162 [==============================] - 4s 15ms/step - loss: 2.5748
- mean_absolute_error: 0.1274 - val_loss: 1.8833 -
val_mean_absolute_error: 0.0694
Epoch 2/10
162/162 [==============================] - 2s 14ms/step - loss: 1.4558
- mean_absolute_error: 0.0489 - val_loss: 1.1054 -
val_mean_absolute_error: 0.0410


...


Epoch 9/10
162/162 [==============================] - 2s 14ms/step - loss: 0.1754
- mean_absolute_error: 0.0348 - val_loss: 0.1655 -
val_mean_absolute_error: 0.0344
Epoch 10/10
162/162 [==============================] - 2s 14ms/step - loss: 0.1581
- mean_absolute_error: 0.0343 - val_loss: 0.1510 -
val_mean_absolute_error: 0.0339
```

Figure 9-57. *The output of the training code, with some of the outputs removed for a clearer display*

As this model is also based on the TCN/1D-CNN, it may also possibly fail training and produce a single-line output. Rerun the model training in this case, though we have initialized the weights with He Normal for stability.

With the training completed, let's move on to making the predictions. Execute the code in Figure 9-58.

```
import math
from sklearn.metrics import mean_squared_error

testing_dataset = sequences_x
print("testing_dataset: ", testing_dataset.shape)

testing_pred = tcn.predict(x=testing_dataset)
print("testing_pred: ", testing_pred.shape)

testing_dataset = testing_dataset.reshape(-1, time_steps)
testing_pred = testing_pred.reshape(-1, time_steps)
errorsDF = testing_dataset - testing_pred
print(errorsDF.shape)
rmse = math.sqrt(mean_squared_error(testing_dataset, testing_pred))
print('Test RMSE: %.3f' % rmse)
```

Figure 9-58. *Making predictions using the trained model*

If all went well during training, the output should look something like the following:

testing_dataset: (10305, 16, 1)

testing_pred: (10305, 16, 1)

(10305, 16)

Test RMSE: 0.052

Let's now calculate the distance cutoffs to determine the anomaly threshold. Refer to Figure 9-59.

```
#based on cutoff after sorting errors
# Calculate threshold using training error
dist = np.linalg.norm(testing_dataset[:len(training_x)] -
testing_pred[:len(training_x)], axis=-1)

scores =dist.copy()
print(scores.shape)
scores.sort()
cutoff = int(0.999 * len(scores))
print(cutoff)
threshold= scores[cutoff]

# Calculate total error over full sequence
dist = np.linalg.norm(testing_dataset - testing_pred, axis=-1)

print(threshold)
```

Figure 9-59. *Calculating the threshold for maximum prediction error that's still normal*

You should see output like this:

```
(5160,)
5154
0.380293597408704
```

Your determined threshold may differ, as it all depends on the model training process.

Before we plot the predictions, we must process the rolling-window sequences to be in terms of the original, continuous sequence of values. For that, run the code in Figure 9-60.

```
plot_x = np.array([])
plot_preds = []
for f in range(0, len(sequence), time_steps):
    plot_x = np.concatenate((plot_x, sequence[f:f+time_steps]))
    plot_preds = np.concatenate((plot_preds, testing_pred[f].reshape(-1)))

plot_training_x = np.array([])
for f in range(0, stop_point, time_steps):
    plot_training_x = np.concatenate((plot_training_x,
training_x[f].reshape(-1)))
```

Figure 9-60. *Processing the predicted sequences to be in terms of the original continuous sequence to plot them*

Let's now plot the model's predictions. Refer to Figure 9-61a and Figure 9-61b.

```
plt.figure(figsize=(24,16))
plt.plot(plot_x, color='green')
plt.plot(plot_training_x, color='blue')
plt.plot(plot_preds, color='red')
```

Figure 9-61a. *Code to plot the time-series sequences color-coded so that 'green' represents the true value of the entire sequence of values, 'blue' only represents the training split, and 'red' represents the model's predictions on the whole dataset's time-series values*

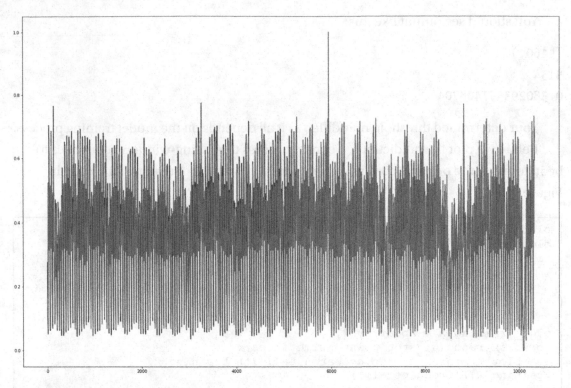

Figure 9-61b. *The resulting plot. The model seems to have learned a pretty good fit for the dataset*

Let's start making predictions for what the anomalies are using our calculated threshold. Refer to Figure 9-62.

```
#label the records anomalies or not based on threshold
z = zip(dist >= threshold, dist)

y_label=[]
error = []
for idx, (is_anomaly, dist) in enumerate(z):
    if is_anomaly:
        y_label.append(1)
    else:
        y_label.append(0)
    error.append(dist)
```

Figure 9-62. *Code to calculate the anomaly label for each data point*

We can visualize this error threshold and find out what points lie beyond it. Refer to Figure 9-63 for the code to plot this threshold graph.

```
viz = Visualization()
viz.draw_anomaly(y_label, error, threshold)
```

Figure 9-63. *Code to plot the error threshold graph*

Refer to Figure 9-64 to see the resulting plot.

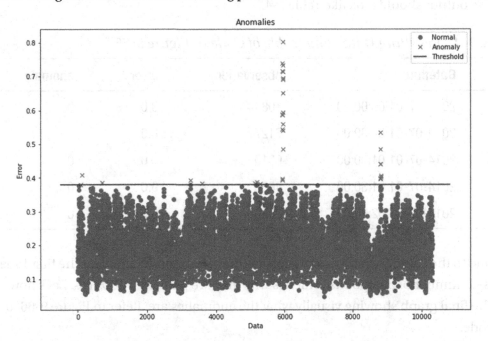

Figure 9-64. *The plot showing the points that lie beyond the error threshold, thus being classified as anomalies*

Let's create an anomaly prediction dataframe to more easily plot which points exactly were determined to be anomalies. Refer to Figure 9-65.

```
adf = pd.DataFrame({'Datetime': df['Datetime'], 'observation':
df['value'],
                    'error': [0 for f in range((time_steps-1))] +
error, 'anomaly': [0 for f in range((time_steps-1))]+ y_label})
adf.head(5)
```

Figure 9-65. *Creating a dataframe of the original values as well as the observed prediction error and anomaly flag*

The output should look like Table 9-4.

Table 9-4. *Resulting Dataframe Table of Code in Figure 9-65*

	Datetime	observation	error	anomaly
0	2014-07-01 00:00:00	10844	0.0	0
1	2014-07-01 00:30:00	8127	0.0	0
2	2014-07-01 01:00:00	6210	0.0	0
3	2014-07-01 01:30:00	4656	0.0	0
4	2014-07-01 02:00:00	3820	0.0	0

Due to the rolling-window approach, we don't have predictions for the first `time_steps-1` number of datapoints, which is why they are padded with zeros. Let's now plot the final graph showing visually what the anomalies are. Refer to Figure 9-66 for this code.

```
figure, axes = plt.subplots(figsize=(12, 6))
axes.plot(adf['Datetime'], adf['observation'], color='g')
anomaliesDF = adf.query('anomaly == 1')
axes.scatter(anomaliesDF['Datetime'].values,
anomaliesDF['observation'], color='r')
plt.xlabel('Date time')
plt.ylabel('observation')
plt.title('Time Series of value by date time')
```

Figure 9-66. *Code to plot the anomalous points on the original time-series sequence*

The result looks something like Figure 9-67.

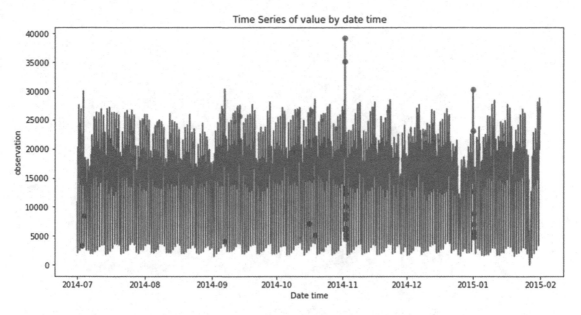

Figure 9-67. *The identified anomalies using the TCN's predictions*

Figure 9-67 shows that quite a few points are labeled as anomalies. There were more errors than in the Dilated TCN plot, but most importantly, the model heavily predicted the spikes around Thanksgiving and New Year's Eve.

It did miss Christmas and the snowstorm event in late January. Try playing around with the model architecture and parameters, as well as the `time_steps` parameter, in order to get better predictions.

Finally, this was another example of **unsupervised anomaly detection**. We took a subset of the original sequence, trained the model to encode and decode each rolling-window sequence, and calculated a threshold by which we can identify anomalies. None of the steps in this process involved a training label.

Summary

In this chapter, we discussed temporal convolutional networks and saw how they fare when applied to anomaly detection.

In Chapter 10, we will look at the practical use case of anomaly detection.

Transformers

In this chapter, you will learn about transformer networks and how you can implement anomaly detection using a transformer.

In a nutshell, this chapter covers the following topics:

- What is a transformer?
- Anomaly detection with transformers

Note Code examples are provided in Python 3.8. The code repository for this book is available at `https://github.com/apress/beginning-anomaly-detection-python-deep-learning-2e/tree/master`.

The repository also includes a requirements.txt file to check your packages and their versions.

The notebooks for this chapter can be found at `https://github.com/apress/beginning-anomaly-detection-python-deep-learning-2e/blob/master/Chapter%20 10%20Transformers/chapter10_transformer.ipynb`.

Navigate to "Chapter 10 Transformers" and then click the notebook. The code is provided as .py files as well, though it is the exported version of the notebook.

We will be using JupyterLab to showcase all of the code examples.

What Is a Transformer?

A transformer is a class of deep learning architecture that has revolutionized the field of natural language processing (NLP). Succeeding recurrent neural network (RNN)-based architectures like the LSTM model, transformer networks are able to scale up to extremely large parameter and dataset sizes thanks to their ability to train and make inferences in parallel. This parallelization capability is possible because the layers of the network rely on the concept of self-attention.

Self-attention, on a high level, is the mechanism by which the model learns to assign importance weights to tokens in a sequence in relation to a specific token being evaluated. For example, take the sentence "The dog ate some food and it became full," and consider that the model is trying to evaluate what "it" means in this context. The word "it" could refer to either "dog" or "food." The model, recognizing language patterns from all the training it's done, learns to assign a higher importance to "dog" than "food" in this case, and thus pays more attention to "dog" instead.

In the numerical terms that a model operates on, the model places a higher weight on the embedding representation of "dog" as it is evaluating the embedding representation of "it" to understand the sequence more accurately. At a very high level, self-attention basically means that the model pays attention to specific details of the input to understanding of context. Since this process can operate on the entire sequence at once, it is able to accurately capture long-range dependencies, avoid vanishing/exploding gradients, and avoid losing context over very long sequences.

Transformer networks have outclassed RNN-based counterparts at many NLP tasks, such as text classification, question answering, sentiment analysis, machine translation, and so on. Perhaps the most recent and exciting type of transformer model is the large language model (LLM). Trained on enormous amounts of text data, an LLM performs tasks like text generation, question answering, summarization, translation, code writing, and even brainstorming. Some LLMs can be tuned to serve as conversational chatbots, able to perform multiple of these tasks in a human-like manner. Perhaps the most famous and successful of these LLMs is ChatGPT, a model so powerful that its foundational model, GPT4, has been described by some researchers to be a key point toward achieving artificial general intelligence.

While we won't be creating any sentient AI in this chapter, we can take advantage of the ability of the transformer model to perform well on sequence-based tasks to perform time-series anomaly detection. In comparison to RNNs, transformer models carry the following advantages:

- **Parallelization**: The ability of the transformer model to process sequences in parallel enables it to make fuller use of GPU computing power, allowing it to train and make inferences in parallel and in less time than an RNN-based network can do so.

- **Large sequences**: Because of the self-attention mechanism, the transformer pays attention to every token in the sequence at once, regardless of distance. This allows for better, more accurate modeling of relationships between tokens anywhere in the sequence.

- **Scalability**: Because of the parallelization of transformers, we can scale up to far larger sizes than is feasible for RNN-based networks.

- **Vanishing gradient**: Though transformers are very deep, they make use of residual skip connections over the blocks, allowing the gradient to propagate deeper through the network as necessary and stabilize training, allowing for ease of scale by making the networks deeper.

- **Rich feature learning**: A transformer can have multiple attention heads, each of which can learn a specific set of features given the input data. It is sort of comparable to how CNN filters can learn different kernels, such as one filter learning horizontal edge detection while another learns vertical edge detection. Having multiple attention heads enables a transformer to gain a deeper and better understanding of the given inputs.

Given all these advantages, it's easy to see why transformers have overtaken RNNs, even though some RNNs have been modified to use attention mechanisms. That being said, transformers do have some disadvantages:

- **Computational demand**: Transformers scale very well, but they are incredibly computationally expensive due to the heavy parameter count. They require large amounts of GPU memory, which is both resource intensive and monetarily expensive.

- **Data hungry**: Transformers also require lots of data to train well, which can be troublesome to provide without the proper computational resources.

- **Overfitting**: Transformers have many parameters, meaning their model capacity is very high. Though this enables them to perform very powerfully, it can also lead to some serious overfitting to smaller datasets.

- **Sequential inference**: Although transformers perform sequence processing in parallel, they generate tokens sequentially—one at a time using the previous tokens. It's just that the sequence processing itself is done in parallel, whereas in an LSTM it's still sequential.

- **Hallucination**: More applicable to text generation tasks, transformers can "hallucinate" plausible-sounding outputs that are actually wrong. This is noticeable in LLMs that tend to confidently make up incorrect information.

Transformer Architecture

Figure 10-1 shows what the architecture of a transformer looks like[1].

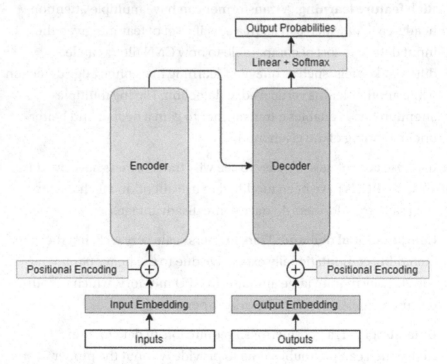

Figure 10-1. *The high-level architecture of a transformer, which is composed of an encoder (the blue block) and a decoder (the red block)*

[1] *Vaswani, Ashish et al. "Attention Is All You Need." NIPS (2017).* https://proceedings.neurips.cc/paper_files/paper/2017/file/3f5ee243547dee91fbd053c1c4a845aa-Paper.pdf

On a high level, the job of the encoder is to learn to encode the input sequence in such a manner that allows for a rich understanding of the original input sequence. The decoder then takes the output embedding and the encoding output of the encoder, processes it, and outputs some type of value. This could be a word token in the case of text generation models, or a real-value output in the case of a regression model. In Figure 10-1, as well as the original paper from which the figure was adapted, the decoder output is processed by a linear layer and a `softmax()` layer to create probabilities for determining the output word token.

The input embedding is a representation of the original input token, meant to capture its meaning. Input embeddings can be pre-initialized to specific, learned embeddings, or they can be learned through the training process by the model. The same goes for the output embedding. Because the model has no recurrence and processes the whole sequence in parallel, the transformer doesn't understand the order of tokens in the sequence. That is why we must include a positional embedding, which helps encode the relative positions of all the tokens in the sequence.

Transformer Encoder

Next, let's look at the encoder in detail. Refer to Figure 10-2.

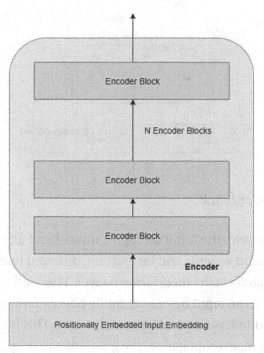

Figure 10-2. *The encoder takes the combined input embedding and positional embedding and passes it through a Series of N encoder blocks*

The encoder is itself comprised of a Series of encoder blocks that the inputs pass through. By the end of the entire encoder, the output sequence should be some kind of encoded embedding that represents the original input sequence.

So, what exactly is contained within this encoder block? Refer to Figure 10-3.

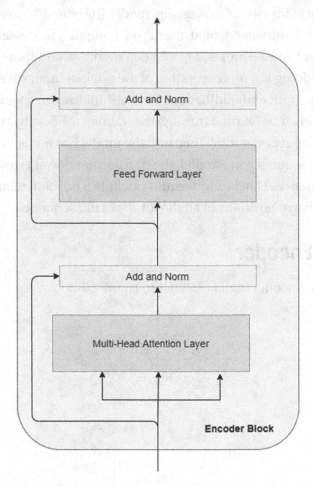

Figure 10-3. *An encoder block*

The encoder block passes the input through a multi-head attention layer, responsible for computing the attention weights and for learning different features per attention head that all get aggregated via the feed-forward layer. The "add and norm" step basically involves adding the input before the multi-head attention layer to the output of the multi-head attention layer and then normalizing it. This is called a *residual skip-connection*, since during backpropagation, the gradient can propagate up the network more directly, skipping over the multi-head attention layer if needed. Because the gradient can skip over the multi-head attention layer, it avoids the issue of vanishing gradient since chain-rule would add more values to be multiplied with the gradient and potentially diminish its magnitude (if these multipliers were also less than 1 in magnitude). This allows the transformer to become much deeper, since the gradient signal goes farther up the network without being diminished in magnitude.

A feed forward layer is a regular neural network layer, so what is a multi-head attention layer? Refer to Figure 10-4.

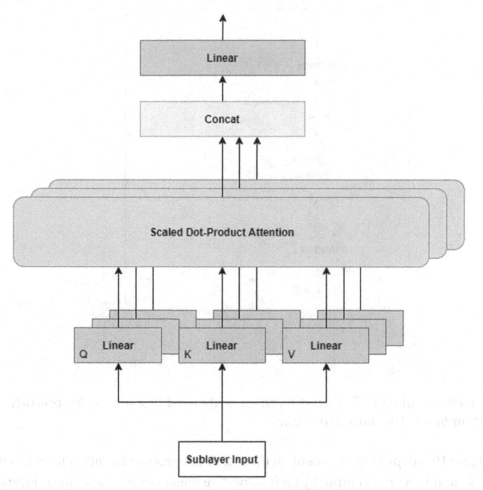

Figure 10-4. *The multi-head attention layer zoomed in*

The sublayer refers to the multi-head attention sublayer. Focusing on one particular head of this multi-head attention layer, the sublayer input is passed through three separate linear layers, each responsible for computing the **Query (Q)**, **Key (K)**, and **Value (V)** matrices. Let's zoom into the scaled dot-product attention block itself and find out how the Q, K, and V matrices are used. Refer to Figure 10-5.

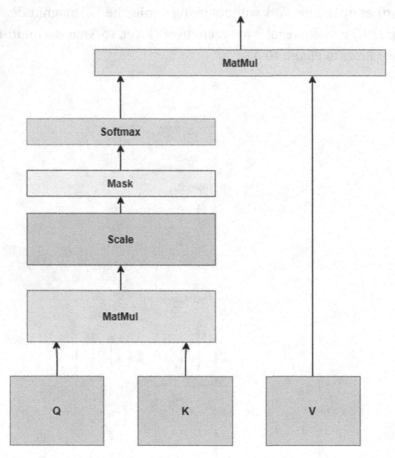

Figure 10-5. *How the Q, K, and V matrices are used in a scaled dot-product attention head. The mask is optional*

Figure 10-5 depicts what goes on in the scaled dot-product attention layer given the Q, K, and V matrices output by their respective linear layers. These linear layers are neural networks layers without any activation functions.

Let's focus on the high-level. Q, K, and V can be thought of as follows:

- **Q (Query)**: The "question" about certain parts of the input. Different attention heads can focus on different parts of the input and thus ask different "questions" about the input. Example: What is the phrase "the animal" most relevant to in the input "the animal is eating food"?

- **K (Key)**: The "reference" material, transforming the input in a particular way. When the scaled dot-product of Q, K followed by `softmax()` is computed, we get attention scores that "answer" the question and give us a weighting of importance of what tokens to pay "attention" to. Example: "eating" might have the highest attention score in the input "the animal is eating food."

- **V (Value)**: The "answer" and a different transformation of the input. When multiplied with the attention scores, we get the final output embedding of this scaled dot-product attention mechanism. Example: Q and K tell us that "eating" is most relevant to "the animal." The generated embedding is then some kind of interpretation of the input sentence focusing on "the animal" and "eating," to be further transformed by subsequent layers.

Keep in mind that because of the multiple attention heads, each will learn to ask different "questions" and in the end produce a richer and more diverse understanding of the original input. When scaled up and fed a lot of diverse data, these models can learn to link concepts and ideas across different fields, allowing LLMs like ChatGPT to produce more compelling, informed, and accurate responses across a wide variety of fields (aside from the tendency of LLMs to sometimes produce confident but factually untrue answers).

Back to the multi-head attention layer, the outputs of each scaled dot-product attention head are concatenated and passed through a linear layer. This should now bring us to the add and norm step of Figure 10-3. With the output of this multi-head attention layer added to its input and normalized, we pass it through a feedforward layer (a neural network layer with an activation function), and then repeat another add and norm step.

With all the preceding discussion, we have explored the inner workings of one encoder block. An encoder layer is simply a sequence of N of these encoder blocks, the final output of which is the final encoding generated.

Transformer Decoder

The decoder architecture is very similar to that of the encoder but with a few key differences. Let's look at the high-level overview of the decoder architecture, shown in Figure 10-6.

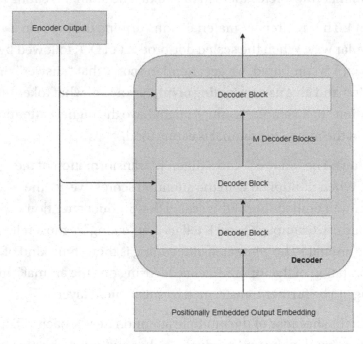

Figure 10-6. *The high-level overview of the decoder in a transformer*

The decoder is comprised of M decoder blocks, and each of them takes the encoder's final output as an additional input to the preceding decoder block's output (or the output embedding with positional encoding if it's the first decoder block).

What does each of these decoder blocks look like? Refer to Figure 10-7a.

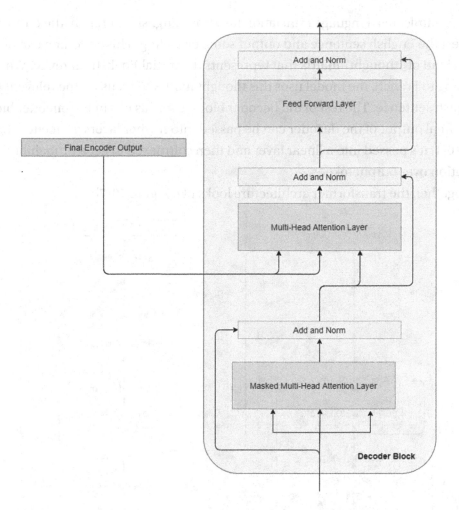

Figure 10-7a. *The decoder has an additional multi-head attention layer that performs cross-attention (self-attention that focuses on the encoder's final output)*

The masked multi-head attention layer performs the same function as the multi-head attention layer from the encoder, except that it applies a causality mask so that attention is computed on previous tokens, not future tokens.

As for the second multi-head attention layer, this one is not masked. The query comes from the output of the masked multi-head attention layer, but the key and value matrices are computed using the output of the final encoder layer. This allows for computing attention scores with the context of the encoder's representation of the original input sequence.

For example, in a language translation task from English to French, the encoder can ingest the English sentence and output some encoding. This encoding can be seen as some kind of "thought" matrix that represents the initial English sentence. While translating to French, the model uses this thought matrix to focus on the relevant parts of the English sentence. The rest of the decoder block is just as it is in the encoder block.

The final output of the decoder can be passed into further layers as needed. In Figure 10-1, it's passed into a linear layer and then softmaxed to get a probability distribution over output tokens.

Altogether, the transformer architecture looks like Figure 10-7b.

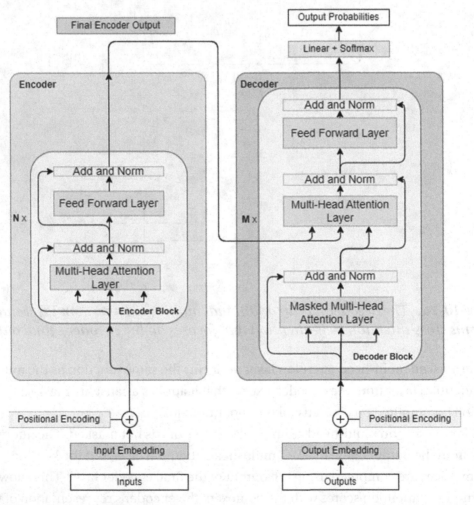

Figure 10-7b. *The full transformer architecture including the encoder blocks and decoder blocks. N and M may be equal, as is the case in the original Attention Is All You Need*

Transformer Inference

Inference with transformers is done in an autoregressive manner. The following is how it works in a machine translation setting:

1. Encode the input sentence using the encoder.

2. Use <sos> as the token to indicate the start of a sentence for the decoder.

3. Using <sos> and the encoding, get the probability distribution for the next word.

   - Use a selection technique to pick a word—a greedy search algorithm takes the highest probability word, but you can also sample the top K tokens or conduct beam search.

4. Take the next word, add it to the output sequence to get <sos> and <first_output>, where <first_output> is meant to be whatever first word was selected.

5. Repeating this for the other words until the cutoff length was reached or the end of sequence <eos> token is reached, indicating the end of the sentence.

With that, you should have a high-level understanding of how transformer networks operate. We can now get started with implementing a transformer and using it to perform time-series anomaly detection.

Anomaly Detection with the Transformer

Let's try applying a transformer to the New York taxi dataset to see how it performs **unsupervised anomaly detection** in a time-series setting. Just like with the LSTM and TCN examples in the prior two chapters, we will be training the transformer to take an input sequence of time steps 1 through T-1 and predict the Tth time-series value. Once again, the first half of the time-series data will be set as the training set, since it shouldn't contain the anomalies seen in the latter half. The 95th percentile of the prediction error (given by mean absolute error between predicted values and actual time-series values) is used as the error threshold for determining anomalies. If a prediction error is higher than this threshold, it is labeled as an anomaly.

The reasoning for training only on the first half of the sequence is to teach the model what normal time-series values look like before we predict on the latter half with anomalies. The idea is that the predicted value will not match the anomaly, producing a higher error and being flagged as an anomaly. Since no labels are required for anomaly detection, this is an unsupervised learning task.

Access the notebook provided at `https://github.com/apress/beginning-anomaly-detection-python-deep-learning-2e/blob/master/Chapter%2010%20Transformers/chapter10_transformer.ipynb`.

The data is provided on GitHub as well at `https://github.com/apress/beginning-anomaly-detection-python-deep-learning-2e/tree/master/data`.

One thing to note is that there isn't a ready implementation of the transformer layers in Keras, so they must be implemented and defined from scratch. TensorFlow offers a guide that implements the basic transformer at `https://www.tensorflow.org/text/tutorials/transformer`. In this section we incorporate that code and make some minor modifications to fit our task, as instead of machine translation (which expects token representations of words), we are operating on direct values.

First, import all of the necessary packages as shown in Figure 10-8.

```
import tensorflow as tf

import sklearn
from sklearn.preprocessing import MinMaxScaler
from sklearn.model_selection import train_test_split
from sklearn.metrics import confusion_matrix, roc_auc_score
from sklearn.metrics import classification_report

import seaborn as sns
import pandas as pd
import numpy as np
import matplotlib

import matplotlib.pyplot as plt
import matplotlib.gridspec as gridspec
%matplotlib inline

import sys
print("Python: ", sys.version)

print("pandas: ", pd.__version__)
print("numpy: ", np.__version__)
print("seaborn: ", sns.__version__)
print("matplotlib: ", matplotlib.__version__)
print("sklearn: ", sklearn.__version__)
print("Tensorflow: ", tf.__version__)
```

Figure 10-8. *Importing all of the necessary packages*

You should see output like the following:

```
Python:  3.8.12 (default, Oct 12 2021, 03:01:40) [MSC v.1916 64 bit
(AMD64)]
pandas:  2.0.0
numpy:  1.22.2
seaborn:  0.12.2
matplotlib:  3.7.1
sklearn:  1.2.2
Tensorflow:  2.7.0
```

After that, we proceed to defining our Visualization class, as shown in Figure 10-9.

```
class Visualization:
    labels = ["Normal", "Anomaly"]

    def draw_confusion_matrix(self, y, ypred):
        matrix = confusion_matrix(y, ypred)

        plt.figure(figsize=(10, 8))
        colors=[ "orange","green"]
        sns.heatmap(matrix, xticklabels=self.labels,
yticklabels=self.labels, cmap=colors, annot=True, fmt="d")
        plt.title("Confusion Matrix")
        plt.ylabel('Actual')
        plt.xlabel('Predicted')
        plt.show()


    def draw_anomaly(self, y, error, threshold):
        groupsDF = pd.DataFrame({'error': error,
                                 'true': y}).groupby('true')

        figure, axes = plt.subplots(figsize=(12, 8))

        for name, group in groupsDF:
            axes.plot(group.index, group.error, marker='x' if name ==
1 else 'o', linestyle='',
                      color='r' if name == 1 else 'g', label="Anomaly"
if name == 1 else "Normal")

        axes.hlines(threshold, axes.get_xlim()[0], axes.get_xlim()[1],
colors="b", zorder=100, label='Threshold')
        axes.legend()

        plt.title("Anomalies")
        plt.ylabel("Error")
        plt.xlabel("Data")
        plt.show()

    def draw_error(self, error, threshold):
            plt.plot(error, marker='o', ms=3.5, linestyle='',
                     label='Point')

            plt.hlines(threshold, xmin=0, xmax=len(error)-1,
colors="b", zorder=100, label='Threshold')
            plt.legend()
            plt.title("Reconstruction error")
            plt.ylabel("Error")
            plt.xlabel("Data")
            plt.show()
```

Figure 10-9. *Defining the Visualization class*

Next, let's define the mechanism to load our data, as shown in Figure 10-10.

```
tensorlogs = ["art_daily_no_noise", #0
              "art_daily_nojump", #1
              "art_daily_jumpsdown",#2
              "art_daily_perfect_square_wave", #3
              "art_increase_spike_density",  #4
              "art_load_balancer_spikes",  #5
              "ambient_temperature_system_failure", #6
              "nyc_taxi",  #7
              "ec2_cpu_utilization", #8
              "rds_cpu_utilization"] #9

dataFilePaths = ['../data/art_daily_no_noise.csv',
                 '../data/art_daily_nojump.csv',
                 '../data/art_daily_jumpsdown.csv',
                 '../data/art_daily_perfect_square_wave.csv',
                 '../data/art_increase_spike_density.csv',
                 '../data/art_load_balancer_spikes.csv',
                 '../data/ambient_temperature_system_failure.csv',
                 '../data/nyc_taxi.csv',
                 '../data/ec2_cpu_utilization.csv',
                 '../data/rds_cpu_utilization.csv']
```

Figure 10-10. *Defining the dataset paths to be able to switch between them easily*

Let's now select our dataset (Figure 10-11).

```
i = 7

tensorlog = tensorlogs[i]
dataFilePath = dataFilePaths[i]
print("tensorlog: ", tensorlog)
print("dataFilePath: ", dataFilePath)
```

Figure 10-11. *Selecting our dataset. We have picked* i *= 7, which should load the NYC taxi dataset*

Your output should look like this:

```
tensorlog:  nyc_taxi
dataFilePath:  ../data/nyc_taxi.csv
```

Let's continue to actually loading our selected dataset (Figure 10-12).

```
df = pd.read_csv(filepath_or_buffer=dataFilePath, header=0, sep=',')
print('Shape:' , df.shape[0])
print('Head:')
print(df.head(5))
```

Figure 10-12. *Code to load the selected dataset*

Your output should look like this:

Shape: 10320
Head:

	timestamp	value
0	2014-07-01 00:00:00	10844
1	2014-07-01 00:30:00	8127
2	2014-07-01 01:00:00	6210
3	2014-07-01 01:30:00	4656
4	2014-07-01 02:00:00	3820

Let's plot our time-series data. Refer to Figure 10-13a and Figure 10-13b.

```
df['Datetime'] = pd.to_datetime(df['timestamp'])
print(df.head(3))
df.shape
df.plot(x='Datetime', y='value', figsize=(12,6))
plt.xlabel('Date time')
plt.ylabel('Value')
plt.title('Time Series of value by date time')
```

Figure 10-13a. *Code to plot the time-series dataset*

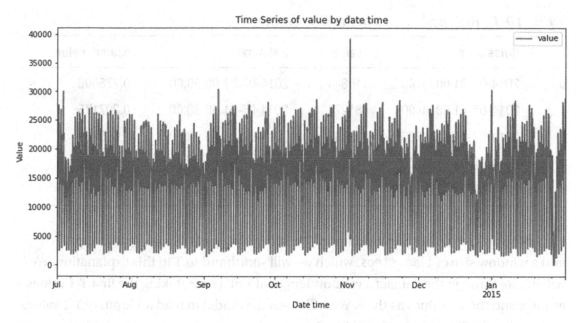

Figure 10-13b. *The plotted time-series dataset. This is the NYC taxi dataset*

We should rescale our dataset next. Refer to Figure 10-14.

```
from sklearn.preprocessing import MinMaxScaler
scaler = MinMaxScaler(feature_range = (0, 1))
df['scaled_value'] =
pd.DataFrame(scaler.fit_transform(pd.DataFrame(df['value'])),columns=[
'value']) - 0.5
print('Shape:' , df.shape[0])
df.head(5)
```

Figure 10-14. *Code to rescale the dataset to a range of 0 to 1. Then, we shift it all down by 0.5 so that the range is between –0.5 to 0.5, making it easier for the model to train since the output is zero-centered*

The output dataframe should look like Table 10-1.

Table 10-1. *Rescaled Values in the Dataframe*

	timestamp	value	Datetime	scaled_value
0	2014-07-01 00:00:00	10844	2014-07-01 00:00:00	0.276506
1	2014-07-01 00:30:00	8127	2014-07-01 00:30:00	0.207175
2	2014-07-01 01:00:00	6210	2014-07-01 01:00:00	0.158259
3	2014-07-01 01:30:00	4656	2014-07-01 01:30:00	0.118605
4	2014-07-01 02:00:00	3820	2014-07-01 02:00:00	0.097272

Let's now create the training dataset. We will be using a rolling-window approach and a window size of `time_steps`, which we will shorthand to **T** in this explanation. We will iterate through the dataset a window length of T at a time, taking the first T-1 values as the x and the Tth values as the y. We will teach the model to read an input of T-1 values and predict the Tth. Refer to Figure 10-15.

```
time_steps = 16

sequence = np.array(df['scaled_value'])
print(sequence)

# Create rolling window sequences as determined by time_steps
x_sequences = []
y_sequences = []

# Number of windows to iterate through
n_iter = len(sequence) - time_steps + 1
for f in range(n_iter):
    window = sequence[f:f+time_steps]
    x_sequences.append(window[:-1])
    y_sequences.append(window[-1:])
x_sequences = np.array(x_sequences)
y_sequences = np.array(y_sequences)
print(x_sequences.shape, y_sequences.shape)
```

Figure 10-15. *Creating the rolling-window features to feed into the model*

You should see something like the following:

[0.27650616 0.20717548 0.1582587   ... 0.69664957 0.6783281   0.67059634]
(10305, 15) (10305, 1)

Let's create our training sequences as well. Refer to Figure 10-16.

```
sequences_x = x_sequences.reshape(len(x_sequences), time_steps-1, 1)
print("sequences_x: ", sequences_x.shape)
sequences_y = y_sequences.reshape(len(y_sequences), 1)
print("sequences_y: ", sequences_y.shape)

# Training on first half of data only, predicting on whole thing
stop_point = int(0.5 * len(df))
training_x = sequences_x[:stop_point]
print("training_x: ", training_x.shape)
training_y = sequences_y[:stop_point]
print("training_y: ", training_y.shape)
```

Figure 10-16. *Defining the training data*

You should see output like this:

sequences_x: (10305, 15, 1)
sequences_y: (10305, 1)
training_x: (5160, 15, 1)
training_y: (5160, 1)

With our data preparation out of the way, let's start defining the model architecture. First, let's define the positional embedding code. Refer to Figure 10-17.

```
def positional_encoding(length, depth):
  depth = depth/2

  positions = np.arange(length)[:, np.newaxis]      # (seq, 1)
  depths = np.arange(depth)[np.newaxis, :]/depth   # (1, depth)

  angle_rates = 1 / (10000**depths)             # (1, depth)
  angle_rads = positions * angle_rates          # (pos, depth)

  pos_encoding = np.concatenate(
      [np.sin(angle_rads), np.cos(angle_rads)],
      axis=-1)

  return tf.cast(pos_encoding, dtype=tf.float32)

class PositionalEmbedding(tf.keras.layers.Layer):
  def __init__(self, vocab_size, d_model):
    super().__init__()
    self.d_model = d_model
    self.pos_encoding = positional_encoding(length=2048,
depth=d_model)

  def call(self, x):
    length = tf.shape(x)[1]

    # This factor sets the relative scale of the embedding and
positonal_encoding.
    x *= tf.math.sqrt(tf.cast(self.d_model, tf.float32))
    x = x + self.pos_encoding[tf.newaxis, :length, :]
    return x
```

Figure 10-17. *The definition for the positional encoding code*

If you are following along with TensorFlow's tutorial, you'll notice that the call to the embedding layer has been removed here. This is because we are passing real-value inputs, and we are not trying to embed them. That being said, we still want to positionally encode these inputs, so we will need the positional encoding to tell the transformer the relative order of each element in the input sequence.

Next, we will define the attention layers. Refer to Figure 10-18.

```
class BaseAttention(tf.keras.layers.Layer):
  def __init__(self, **kwargs):
    super().__init__()
    self.mha = tf.keras.layers.MultiHeadAttention(**kwargs)
    self.layernorm = tf.keras.layers.LayerNormalization()
    self.add = tf.keras.layers.Add()

class CrossAttention(BaseAttention):
  def call(self, x, context):
    attn_output, attn_scores = self.mha(
        query=x,
        key=context,
        value=context,
        return_attention_scores=True)

    # Cache the attention scores for plotting later.
    self.last_attn_scores = attn_scores

    x = self.add([x, attn_output])
    x = self.layernorm(x)

    return x

class GlobalSelfAttention(BaseAttention):
  def call(self, x):
    attn_output = self.mha(
        query=x,
        value=x,
        key=x)
    x = self.add([x, attn_output])
    x = self.layernorm(x)
    return x

class CausalSelfAttention(BaseAttention):
  def call(self, x):
    attn_output = self.mha(
        query=x,
        value=x,
        key=x,)

        # use_causal_mask = True)
    x = self.add([x, attn_output])
    x = self.layernorm(x)
    return x
```

Figure 10-18. *Defining the attention layers*

For TensorFlow 2.7.0 specifically, we had to remove the `use_causal_mask=True` argument from the `CausalSelfAttention` class.

Next, let's define the feedforward block, as shown in Figure 10-19.

```python
class FeedForward(tf.keras.layers.Layer):
  def __init__(self, d_model, dff, dropout_rate=0.1):
    super().__init__()
    self.seq = tf.keras.Sequential([
      tf.keras.layers.Dense(dff, activation='relu'),
      tf.keras.layers.Dense(d_model),
      tf.keras.layers.Dropout(dropout_rate)
    ])
    self.add = tf.keras.layers.Add()
    self.layer_norm = tf.keras.layers.LayerNormalization()

  def call(self, x):
    x = self.add([x, self.seq(x)])
    x = self.layer_norm(x)
    return x
```

Figure 10-19.   The feedforward block

We can now define the encoding layer. Refer to Figure 10-20.

```python
class EncoderLayer(tf.keras.layers.Layer):
  def __init__(self,*, d_model, num_heads, dff, dropout_rate=0.1):
    super().__init__()

    self.self_attention = GlobalSelfAttention(
        num_heads=num_heads,
        key_dim=d_model,
        dropout=dropout_rate)

    self.ffn = FeedForward(d_model, dff)

  def call(self, x):
    x = self.self_attention(x)
    x = self.ffn(x)
    return x
```

Figure 10-20.   The encoding layer definition

With the encoding layer code defined, let's do the same for the entire encoder itself. Refer to Figure 10-21.

```python
class Encoder(tf.keras.layers.Layer):
  def __init__(self, *, num_layers, d_model, num_heads,
               dff, vocab_size, dropout_rate=0.1):
    super().__init__()

    self.d_model = d_model
    self.num_layers = num_layers

    self.pos_embedding = PositionalEmbedding(
        vocab_size=vocab_size, d_model=d_model)

    self.enc_layers = [
        EncoderLayer(d_model=d_model,
                     num_heads=num_heads,
                     dff=dff,
                     dropout_rate=dropout_rate)
        for _ in range(num_layers)]
    self.dropout = tf.keras.layers.Dropout(dropout_rate)

  def call(self, x):
    # `x` is token-IDs shape: (batch, seq_len)
    x = self.pos_embedding(x)  # Shape `(batch_size, seq_len,
d_model)`.

    # Add dropout.
    x = self.dropout(x)

    for i in range(self.num_layers):
      x = self.enc_layers[i](x)

    return x  # Shape `(batch size, seq len, d model)`.
```

Figure 10-21. *The encoder definition*

We must do the same for the decoder. To define the decoder layer, refer to Figure 10-22.

```python
class DecoderLayer(tf.keras.layers.Layer):
  def __init__(self,
               *,
               d_model,
               num_heads,
               dff,
               dropout_rate=0.1):
    super(DecoderLayer, self).__init__()

    self.causal_self_attention = CausalSelfAttention(
        num_heads=num_heads,
        key_dim=d_model,
        dropout=dropout_rate)

    self.cross_attention = CrossAttention(
        num_heads=num_heads,
        key_dim=d_model,
        dropout=dropout_rate)

    self.ffn = FeedForward(d_model, dff)

  def call(self, x, context):
    x = self.causal_self_attention(x=x)
    x = self.cross_attention(x=x, context=context)

    # Cache the last attention scores for plotting later
    self.last_attn_scores = self.cross_attention.last_attn_scores

    x = self.ffn(x)  # Shape `(batch_size, seq_len, d_model)`.
    return x
```

Figure 10-22. *The decoder layer definition*

Figure 10-23 shows the decoder definition in its entirety.

```
class Decoder(tf.keras.layers.Layer):
  def __init__(self, *, num_layers, d_model, num_heads, dff,
vocab_size,
               dropout_rate=0.1):
    super(Decoder, self).__init__()

    self.d_model = d_model
    self.num_layers = num_layers

    self.pos_embedding = PositionalEmbedding(vocab_size=vocab_size,
                                       d_model=d_model)
    self.dropout = tf.keras.layers.Dropout(dropout_rate)
    self.dec_layers = [
        DecoderLayer(d_model=d_model, num_heads=num_heads,
                     dff=dff, dropout_rate=dropout_rate)
        for _ in range(num_layers)]

    self.last_attn_scores = None

  def call(self, x, context):
    # `x` is token-IDs shape (batch, target_seq_len)
    x = self.pos_embedding(x)  # (batch_size, target_seq_len, d_model)

    x = self.dropout(x)

    for i in range(self.num_layers):
      x  = self.dec_layers[i](x, context)

    self.last_attn_scores = self.dec_layers[-1].last_attn_scores

    # The shape of x is (batch_size, target_seq_len, d_model).
    return x
```

Figure 10-23. *The decoder definition*

With all the components of a transformer defined, we can now define the transformer model code itself. Refer to Figure 10-24.

```python
class Transformer(tf.keras.Model):
  def __init__(self, *, num_layers, d_model, num_heads, dff,
               input_vocab_size, target_vocab_size, dropout_rate=0.1):
    super().__init__()
    self.encoder = Encoder(num_layers=num_layers, d_model=d_model,
                           num_heads=num_heads, dff=dff,
                           vocab_size=input_vocab_size,
                           dropout_rate=dropout_rate)

    self.decoder = Decoder(num_layers=num_layers, d_model=d_model,
                           num_heads=num_heads, dff=dff,
                           vocab_size=target_vocab_size,
                           dropout_rate=dropout_rate)
    self.flat = tf.keras.layers.Flatten()
    self.final_layer = tf.keras.layers.Dense(1)

  def call(self, inputs):
    # To use a Keras model with `.fit` you must pass all your inputs
in the
    # first argument.
    context, x  = inputs

    context = self.encoder(context)  # (batch_size, context_len,
d_model)

    x = self.decoder(x, context)  # (batch_size, target_len, d_model)
    x = self.flat(x)
    # Final linear layer output.
    logits = self.final_layer(x)  # (batch_size, target_len,
target_vocab_size)

    try:
      # Drop the keras mask, so it doesn't scale the losses/metrics.
      # b/250038731
      del logits._keras_mask
    except AttributeError:
      pass

    # Return the final output and the attention weights.
    return logits
```

Figure 10-24. *The transformer model definition*

Now we can finally get started with training. Let's set the random seed so that the training is more consistent, and let's instantiate our model with some hyperparameters. Refer to Figure 10-25.

```
# Setting seed
tf.random.set_seed(2)
np.random.seed(2)

num_layers = 6
d_model = 64
dff = 128
num_heads = 4
dropout_rate = 0.1

transformer = Transformer(
    num_layers=num_layers,
    d_model=d_model,
    num_heads=num_heads,
    dff=dff,
    input_vocab_size=1,
    target_vocab_size=1,
    dropout_rate=dropout_rate)

output = transformer((training_x, training_x))

transformer.summary()
```

Figure 10-25. *Setting random seed, instantiating, and building the transformer network. The* output = transformer((training_x, training_x)) *constructs the transformer object so that we can print out the summary and use it in the subsequent steps*

You should see output like that shown in Figure 10-26.

```
Model: "transformer"
_____
 Layer (type)                Output Shape              Param #
=================================================================
 encoder (Encoder)           multiple                  499200

 decoder (Decoder)           multiple                  898176

 flatten (Flatten)           multiple                  0

 dense_24 (Dense)            multiple                  961


=================================================================
Total params: 1,398,337
Trainable params: 1,398,337
Non-trainable params: 0
_____
```

Figure 10-26. *Summary of the transformer network architecture*

The model has over a million parameters! Let's now compile our model. Refer to Figure 10-27.

```
transformer.compile(
    loss='mean_absolute_error',
    optimizer='adam',
    metrics=['mean_absolute_error'])
```

Figure 10-27. *Compiling the model to use the Adam optimizer and mean absolute error as the loss function*

Let's initiate the training process. Refer to Figure 10-28.

```
history = transformer.fit((training_x, training_x), training_y,
                epochs=8,)
```

Figure 10-28. *Starting the training process. The x input is a tuple (training_x, training_x). We are telling the transformer that it should encode the training input, and have the decoder decode it back the same way. Then, based on the output of the decoder, we predict the next timestep*

The training output should look somewhat like Figure 10-29.

```
Epoch 1/8
162/162 [==============================] - 38s 166ms/step - loss:
0.2355 - mean_absolute_error: 0.2355
Epoch 2/8
162/162 [==============================] - 27s 166ms/step - loss:
0.0885 - mean_absolute_error: 0.0885
Epoch 3/8
162/162 [------------------------------] - 26s 164ms/step - loss:
0.0679 - mean_absolute_error: 0.0679
Epoch 4/8
162/162 [==============================] - 27s 164ms/step - loss:
0.0557 - mean_absolute_error: 0.0557
Epoch 5/8
162/162 [==============================] - 26s 162ms/step - loss:
0.0495 - mean_absolute_error: 0.0495
Epoch 6/8
162/162 [==============================] - 26s 162ms/step - loss:
0.0552 - mean_absolute_error: 0.0552
Epoch 7/8
162/162 [==============================] - 27s 164ms/step - loss:
0.0434 - mean_absolute_error: 0.0434
Epoch 8/8
162/162 [==============================] - 27s 164ms/step - loss:
0.0455 - mean_absolute_error: 0.0455
```

Figure 10-29. *The output of the training code*

If you do not see loss values decreasing and matching in magnitude to what Figure 10-29 shows (given by the loss and mean absolute error metrics), and it gets stuck around a loss of 0.15, you should restart the training process. For this particular dataset, if the epochs consistently produce a loss of around 0.15, that indicates that the transformer isn't learning and only produces a straight line or some other faulty output. You can visualize this with the subsequent code blocks if desired.

Given the seed setting and zero-centering of the scaled data, training should be consistent and your model should be able to learn the task.

With the training completed, let's move on to making the predictions. Execute the code in Figure 10-30.

```
import math
from sklearn.metrics import mean_squared_error

testing_dataset = sequences_x
print("testing_dataset: ", testing_dataset.shape)

print("sequences_y: ", sequences_y.shape)

testing_pred = transformer.predict(x=(testing_dataset,
testing_dataset))
print("testing_pred: ", testing_pred.shape)

errorsDF = sequences_y - testing_pred
print(errorsDF.shape)
rmse = math.sqrt(mean_squared_error(sequences_y, testing_pred))
print('Test RMSE: % 3f' % rmse)
```

Figure 10-30. *Making predictions using the trained model*

If all went well during training, the output should look something like this:

testing_dataset: (10305, 15, 1)

sequences_y: (10305, 1)

testing_pred: (10305, 1)

(10305, 1)

Test RMSE: 0.039

Let's now calculate the distance cutoffs to determine the anomaly threshold. Refer to Figure 10-31.

```
#based on cutoff after sorting errors
# Calculate threshold using training error
dist = np.linalg.norm(sequences_y[:len(training_y)] -
testing_pred[:len(training_y)], axis=-1)

scores =dist.copy()
print(scores.shape)
scores.sort()
cutoff = int(0.999 * len(scores))
print(cutoff)
threshold= scores[cutoff]

# Calculate total error over full sequence
dist = np.linalg.norm(sequences_y[:] - testing_pred, axis=-1)

print(threshold)
```

Figure 10-31. *Calculating the threshold for maximum prediction error that's still normal*

You should see output like this:

```
(5160,)
5154
0.13164996056429368
```

Your determined threshold may differ, as it all depends on the model training process.

Let's now plot the model's predictions. Refer to Figure 10-32a and Figure 10-32b.

```
plt.figure(figsize=(24,16))
plt.plot(sequences_y, color='green')
plt.plot(training_y, color='blue')
plt.plot(testing_pred, color='red')
```

Figure 10-32a. *Code to plot the time-series sequences, color-coded so that green represents the true value of the entire sequence of values, blue only represents the training split, and red represents the model's predictions on the whole dataset's time-series values*

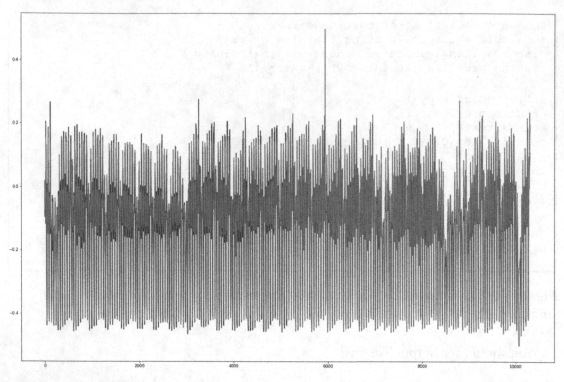

Figure 10-32b. *The resulting plot. The model seems to have learned a pretty good fit for the dataset*

Let's start making predictions for what the anomalies are by using our calculated threshold. Refer to Figure 10-33.

```
#label the records anomalies or not based on threshold
z = zip(dist >= threshold, dist)

y_label=[]
error = []
for idx, (is_anomaly, dist) in enumerate(z):
    if is_anomaly:
        y_label.append(1)
    else:
        y_label.append(0)
    error.append(dist)
```

Figure 10-33. *Code to calculate the anomaly label for each data point*

We can visualize this error threshold and find out what points lie beyond it. Refer to Figure 10-34 for the code to plot this threshold graph.

```
viz = Visualization()
viz.draw_anomaly(y_label, error, threshold)
```

Figure 10-34. *Code to plot the error threshold graph*

Refer to Figure 10-35 to see the resulting plot.

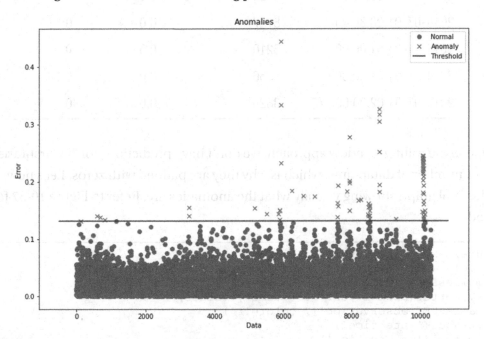

Figure 10-35. *The plot showing the points that lie beyond the error threshold, thus being classified as anomalies*

Let's create an anomaly prediction dataframe to more easily plot which points exactly were determined to be anomalies. Refer to Figure 10-36.

```
adf = pd.DataFrame({'Datetime': df['Datetime'], 'observation':
df['value'],
                    'error': [0 for f in range((time_steps-1))] +
error, 'anomaly': [0 for f in range((time_steps-1))]+ y_label})
adf.head(5)
```

Figure 10-36. *Creating a dataframe of the original values as well as the observed prediction error and anomaly flag*

The output should look like Table 10-2.

Table 10-2. *Resulting Dataframe Table of Code in Figure 10-36*

	Datetime	observation	error	anomaly
0	2014-07-01 00:00:00	10844	0.0	0
1	2014-07-01 00:30:00	8127	0.0	0
2	2014-07-01 01:00:00	6210	0.0	0
3	2014-07-01 01:30:00	4656	0.0	0
4	2014-07-01 02:00:00	3820	0.0	0

Due to the rolling-window approach, we don't have predictions for the first `time_steps-1` number of datapoints, which is why they are padded with zeros. Let's now plot the final graph showing visually what the anomalies are. Refer to Figure 10-37 for this code.

```
figure, axes = plt.subplots(figsize=(12, 6))
axes.plot(adf['Datetime'], adf['observation'], color='g')
anomaliesDF = adf.query('anomaly == 1')
axes.scatter(anomaliesDF['Datetime'].values,
anomaliesDF['observation'], color='r')
plt.xlabel('Date time')
plt.ylabel('observation')
plt.title('Time Series of value by date time')
```

Figure 10-37. *Code to plot the anomalous points on the original time-series sequence*

The result looks something like Figure 10-38.

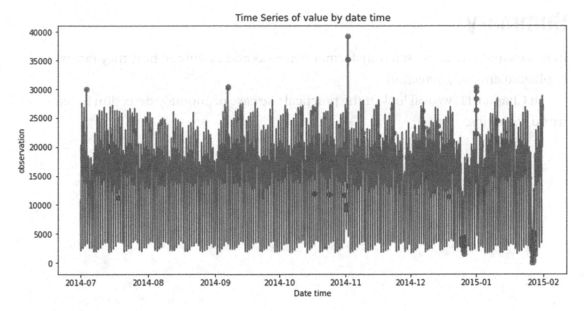

Figure 10-38. *The identified anomalies using the transformer predictions*

Figure 10-38 shows that quite a few points are labeled as anomalies. Most importantly, the model managed to predict as anomalies Thanksgiving, Christmas, New Year's Eve, and a snowstorm-type weather anomaly, among some other points. That being said, there are quite a few points that don't seem like they should be anomalies. One problem with the transformer is that it does tend to overfit due to its high model capacity. However, you can play with the hyperparameters of the transformer model as well as add early stopping or any other mechanism to your training loop to see if you can manage the overfitting.

It's a balancing act, because how well the transformer fits the training data during the training process informs the anomaly detection algorithm later on. This is due to significant deviations from the time Series as determined by the predicted outputs of the transformer will be flagged as anomalies. But if the model underfits the training data, you may not pick up on anomalies that you should.

Overall, this was an example of **unsupervised anomaly detection**. We took a subset of the original sequence, trained the model to output a prediction given rolling windows, and we calculated a threshold by which we can identify anomalies. None of the steps in this process involved a training label.

Summary

In this chapter, we discussed transformer networks and examined how they fare when applied to anomaly detection.

In Chapter 11, we will look at the practical use case of anomaly detection in real-world scenarios.

Practical Use Cases and Future Trends of Anomaly Detection

In this chapter, you will learn about how anomaly detection can be used in several industry verticals. You will explore how anomaly detection techniques can be used to address practical use cases and address real-life problems in the business landscape. Every business and use case is different, and we cannot simply copy and paste code and build a successful model to detect anomalies in any dataset, so this chapter covers many use cases to give you an idea of the possibilities and concepts behind the thought process.

In a nutshell, this chapter covers the following topics:

- What is anomaly detection?

- Real-world use cases of anomaly detection

  - Telecom

  - Banking

  - Environmental

  - Health care

  - Transportation

  - Social media

  - Finance and insurance

  - Cybersecurity

- Video surveillance

- Manufacturing

- Smart homes

- Retail

- Implementation of deep learning–based anomaly detection

- Future trends

Anomaly Detection

Anomaly detection involves finding patterns that do not adhere to what is considered as normal or expected behavior. Businesses could lose millions of dollars due to abnormal events. Consumers can also lose millions of dollars. In fact, there are many situations every day where people's lives are at risk and where their property is at risk. If your bank account gets cleaned out, that's a problem. If your water line breaks, flooding your basement, that's a problem. If all flights at an airport get delayed due to a technical glitch in the traffic control system, that's a problem. If you have a health issue that is misdiagnosed or not diagnosed at all, that's a very big problem that directly impacts your well-being.

Figure 11-1 shows an example of an anomaly, a smiling rainbow-colored fish in a school of frowning blue fish.

The Fish Family portrait

Figure 11-1. *An anomalous rainbow-colored smiling fish among a group of frowning blue fish*

In business use cases, everything is centered on data, and in this context anomaly detection is the identification of abnormal data points, events, or observations that raise suspicion because they differ significantly from the data perceived as normal or typical. Many types of such anomalies can impact the business operations or bottom line significantly, which is why anomaly detection is gaining a lot of traction in the industry and many businesses are investing heavily in technologies that can help them identify abnormal behavior before it is too late. Such proactive anomaly detection is becoming more and more visible, and new technologies developed as part of the AI revolution are helping to detect anomalies faster and in ways never possible before.

Figure 11-2 is an example of the daily number of cars that cross the Golden Gate Bridge in San Francisco.

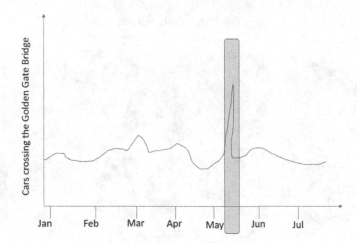

Figure 11-2.   Daily count of cars crossing the Golden Gate Bridge over time. The anomalous spike in the data is highlighted

The kind of anomaly detection that could potentially help some arbitrary business avoid losses due to anomalies depends very much on the type of data that is collected as part of business operations as well as the assortment of techniques and algorithms used as part of the strategy to perform anomaly detection.

There is also a strong importance for real-time anomaly detector models that operate on streams of incoming data. The sooner anomalies are caught, the better, since the potential damage can be lessened if they are dealt with as quickly as possible. It would be great if anomalies could be foreseen before they even occur, but by this point, they no longer become anomalies but rather an expected pattern. By nature, anomalies are unpredictable and unforeseen.

Timeliness in anomaly detection is key in many sectors. For example, in the banking sector, the quicker banking fraud is detected, the sooner the customer's assets can be protected. In the manufacturing sector, consider the example of a high-speed production facility that manufactures screws. Batches are sampled and tested for quality assurance, and if the number of faulty samples detected is higher than some threshold, the entire manufactured batch might be discarded. Delays may be costly, so identifying manufacturing anomalies in real time is important to minimize disruption to the productivity of the factory.

These are examples from just two sectors. Real-time anomaly detection is applied across practically every sector that we will cover in this chapter. With any use case involving real-time anomaly detection, the emphasis is on minimizing damages, since anomalies will inevitably arise.

Real-World Use Cases of Anomaly Detection

We will look at several industry verticals and businesses and how anomaly detection can be used.

Telecom

In the telecom sector, some of the use cases for anomaly detection are to detect roaming abuse, revenue fraud, and service disruptions.

What is **roaming abuse**? Roaming refers to the ability to use mobile services like calling and texting outside of your coverage area. Roaming abuse refers to the fraudulent use of roaming services without the intention to pay for them. So how do we detect **roaming abuse** in telecom sector? By looking at the location of the cellular devices, we can categorize the kind of behavior at any particular moment of the cellular device as normal or abnormal. This will help us to detect the cellular devices usage at that period of time. By looking at all the other information we know in general about roaming activity, we could also detect how this cellular device is being used and whether there is any roaming abuse taking place.

More specifically, with respect to roaming abuse detection, we can model time-series location data to detect abnormal travel patterns. Since these are sequences, we can model them using LSTMs, TCNs, or transformers. One application is to model what normal roaming history looks like for a person and then use the model to detect breaks in roaming patterns. This is a good example of semi-supervised anomaly detection. In a similar vein, autoencoders may be implemented as well, keeping in mind that the encoder and decoder portions of the autoencoder can be sequence models themselves like an LSTM. The idea is to flag anomalies by reconstruction loss.

Figure 11-3 shows how roaming works for your phone as you travel around the world.

Figure 11-3. *Roaming. Cell phone location history can be tracked throughout one's travels. Any time the cell phone connects to a cell network, connects to Wi-Fi, etc., location can be tracked*

Service disruption is another very high-impact use case for anomaly detection. Cellular devices are connected to cellular providers' networks via cell towers, which are located all over the place. Your cell phone connects to the nearest tower to participate in the cellular network. In case of events involving large crowds, such as a concert or a football game, the cellular towers that typically perform quite well become heavily overloaded, causing serious service disruptions and very bad customer experience for the duration of the overload.

Figure 11-4 shows service disruption of cell phone service in several areas in the northwestern United States.

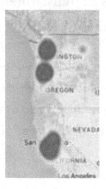

Figure 11-4. *Cell phone service disruptions. The darker zone in the middle indicates a higher frequency of service disruption*

If we know the various metrics of the cell phone towers and the associated devices at some period of time and for a long duration, along with any kind of information we have on the typical nature of activity around the towers in terms of whether concerts or games occurred in the vicinity or some major event is being expected in the vicinity of the cellular towers, then we can use time Series as a basis of how we represent all such activity and can subsequently use TCN, LSTM, or transformer algorithms to detect anomalies pertaining to the major events because they have a temporal dependency. This information can be helpful to evaluate how these services are being used and how effective the service is for particular cell phone towers.

The cell phone companies now have a way of understanding whether certain hours need to be upgraded or more towers need to be built. For instance, if major office buildings are being built near a particular tower, by using data on the time Series of all the towers owned by the cellular network, it is possible to detect the anomalies in other parts of the network and apply the principles to the tower that is going to be impacted by the newly constructed office buildings (which would add thousands of cell phone connections and could cause overloading on the tower and affect how the tower will be used in the near future).

Banking

In the banking sector, some of the use cases for anomaly detection are to flag abnormally high transactions, fraudulent activity, phishing attacks, etc. Credit cards are used by almost everyone in the world, and typically every individual has a certain pattern of using their credit card that is different from everyone else's pattern of use. This pattern creates an implicit profile of the individual using the credit card in terms of where they use it, when they use it, why they use it, and what they use it for. Every credit card company has such information about the credit card usage of a very large number of consumers, so they can use anomaly detection to detect when a specific credit card transaction might be fraudulent.

Autoencoders are very useful in such an anomaly detection use case. With such a case, we could take all the credit card transactions by individual consumers and capture and convert the features into numerical features such that we could assign certain scores to every credit card based on various factors along with a kind of indicator whether the transaction could be normal or abnormal. Then, using autoencoders, we could build an anomaly detection model that can quickly determine whether a specific transaction is normal or abnormal given everything that we know about all the other transactions of a customer. The autoencoder does not need to be extremely complicated. It can be built with just a couple of hidden layers for the encoder and couple of hidden layers for the decoder and still have pretty decent detection of abnormal activity (otherwise known as fraudulent activity) on the credit cards.

Similarly, we can use generative adversarial networks (GANs) to model what normal transaction history looks like for consumers and use the discriminator to flag anomalies. Here, the generator aims to generate synthetic data that is reminiscent of the normal transaction data distribution, and the discriminator learns to predict any abnormal-looking data point as fake. Fraudulent transactions would be labeled as fake.

Transaction history also has a temporal component to it. Transactions tend to exhibit patterns over time. For example, seasonal spending is a common pattern that often involves a higher volume of spending than might be usual for the consumer. A time-series model would learn these types of consumer-specific transactional patterns. For example, perhaps they are always a big birthday spender. Then, should any transactional patterns appear that deviate from the norm, these models would immediately pick it up and flag them as anomalies.

Figure 11-5 is a depiction of credit card fraud.

Figure 11-5. *Depiction of credit card fraud, where credit card details are obtained by someone (like a hacker) so they can make fraudulent purchases*

Environmental

When it comes to environmental aspects, anomaly detection has several applicable use cases. Whether it is deforestation or melting of glaciers, air quality or water quality, anomaly detection can help to identify abnormal activities.

Figure 11-6 is a photo of deforestation.

Figure 11-6. *Mass deforestation is harmful to the environment, especially in the Amazon Rainforest. (Source: commons.wikimedia.org)*

Let's look at an example of an air quality index. An air quality index provides some kind of measurement of breathable air quality, which can be measured by using various sensors placed at various locations in the region that is being monitored. These sensors measure and send periodic data to a centralized system that collects data from all the sensors. This data becomes a time Series, with each measurement consisting of several attributes or features. With each point in time having a certain number of features, which can then be input into a neural network such as an autoencoder, we can build an anomaly detector. Of course, we could use an LSTM, TCN, or transformer to do the same.

Figure 11-7 shows the air quality index in Seoul in 2015.

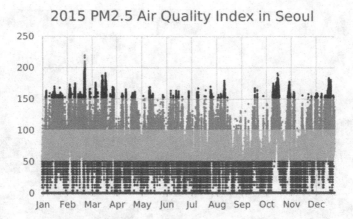

Figure 11-7. *Air quality index. Green indicates good quality, while yellow is moderate quality. Orange presents some issues for sensitive groups (like those with allergies), while red is unhealthy. Anything above red, like purple, is hazardous and very unhealthy. (Source: commons.wikimedia.org)*

Another environmental use case for anomaly detection is monitoring of air temperatures throughout the year. Since global temperatures are continuing to trend upward on average, it is helpful to train time-series algorithms on historical data, evaluate them on a daily basis, and have anomalous temperatures be flagged as a result of deviating significantly from what was predicted. These anomalies can then be collected and further studied, revealing specific trends or serving to be automatically compared to climate change forecasting models to find out how bad the warming is progressing in reality vs. what was predicted.

Satellite imagery is another domain where anomaly detection algorithms may be used. Detecting abnormal events such as water level drops in lakes, deforestation in the Amazon, wildfires, and so on can all be modeled using various deep learning models. As images are the primary medium of data, convolutional neural networks are utilized quite heavily.

CNNs can perform feature extraction to output vector embeddings that are efficient representations of the original image. These embeddings can be fed into autoencoders or GANs to determine if anomalous events are happening in the imagery in an unsupervised manner, allowing it to be scaled up easily as there is no labeling required.

Health Care

Health care is one of the domains that can benefit a lot from anomaly detection, whether it is to prevent fraud, detect cancer or chronic illness, improve ambulatory services, etc.

One of the biggest use cases for anomaly detection in health care is to detect cancer from various diagnostic reports even before any significant symptoms might indicate the presence of cancer. This is extremely important given the serious consequences of cancer for any person. Some of the techniques in anomaly detection that can be used in this context involve convolutional neural networks combined with autoencoders.

CNNs use the concept of dimensionality reduction to reduce the large number of features/pixels with colors into much lower dimensionality points using the neural network layers. By combining this CNN with autoencoders, we can use autoencoders to look at images such as MRI images, mammograms, and other images from diagnostic technologies in the health care industry.

Figure 11-8 shows a set of images from a CT scan.

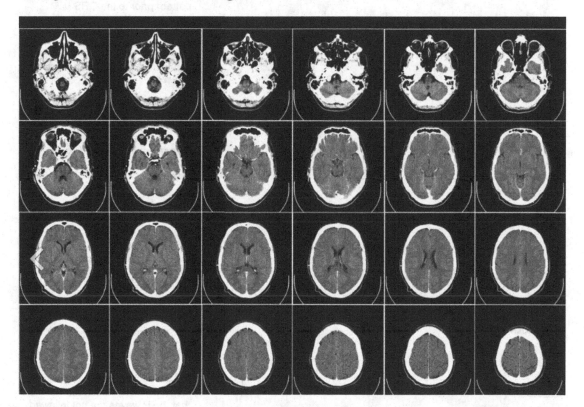

Figure 11-8. *CT scan images of someone's head. (Source: commons.wikimedia.org)*

Let's look at another use case of detecting abnormal health conditions of residents of a particular neighborhood. Typically, a local hospital serves residents of a specific neighborhood. The local hospital collects and stores various kinds of health metrics from all the residents in the neighborhood that visit the hospital. Some of the metrics are blood test results, lipid profiles, glycemic values, blood pressure readings, ECG results, etc. When combined with demographic data such as age, sex, health conditions, etc., this health data can be used to build a sophisticated AI-based anomaly detection model.

Figure 11-9 shows different health issues observed by looking at ECG results.

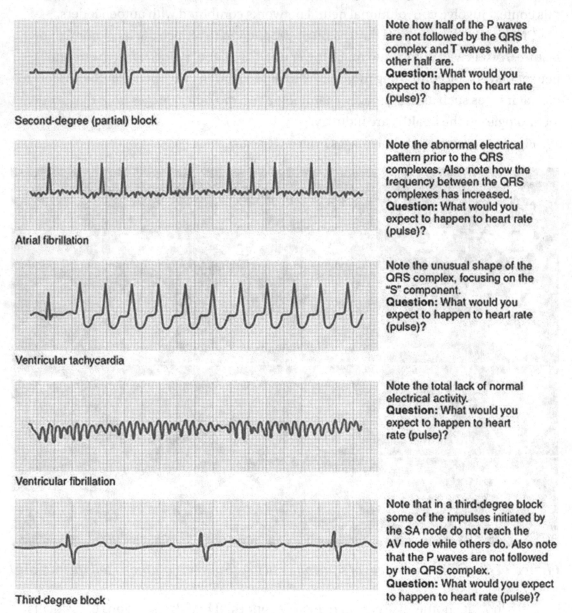

Second-degree (partial) block

Note how half of the P waves are not followed by the QRS complex and T waves while the other half are.
Question: What would you expect to happen to heart rate (pulse)?

Atrial fibrillation

Note the abnormal electrical pattern prior to the QRS complexes. Also note how the frequency between the QRS complexes has increased.
Question: What would you expect to happen to heart rate (pulse)?

Ventricular tachycardia

Note the unusual shape of the QRS complex, focusing on the "S" component.
Question: What would you expect to happen to heart rate (pulse)?

Ventricular fibrillation

Note the total lack of normal electrical activity.
Question: What would you expect to happen to heart rate (pulse)?

Third-degree block

Note that in a third-degree block some of the impulses initiated by the SA node do not reach the AV node while others do. Also note that the P waves are not followed by the QRS complex.
Question: What would you expect to happen to heart rate (pulse)?

Figure 11-9. *ECG results measuring someone's heart. (Source: commons. wikimedia.org)*

There are a lot of different use cases in health care for which different anomaly detection algorithms can be used to implement preventative measures. Just about any of the models that we have covered throughout this book (and beyond) can be applied in the health care setting due to the sheer diversity of data available.

Transportation

In the transportation sector, anomaly detection can be used to ensure proper functioning of the roadways and vehicles. By collecting different types of events from all the sensors that are operational on the roadways, such as toll booths, traffic lights, security cameras, and GPS signals, we could build an anomaly detection engine that detects abnormal traffic patterns.

Anomaly detection can also be used to look at times in schedules of public transportation and the related traffic conditions in this similar area of transportation. We could also look for abnormal activity in terms of fuel consumption, number of passengers the public transportation is supporting, seasonal trends, etc.

Furthermore, traffic simulations can be created to depict various people as they move throughout space and time via location sampling on a regular time-series interval. Sequence-based anomaly detection methods can be run to detect anomalies in both real life and in simulations by training on normal traffic pattern data for these individuals. The higher-fidelity the simulation, the better and more accurate the anomaly detection on real-life data. One such possibility is to combine anomaly detection algorithms and these high-fidelity simulations to allow for planning road constructions to detect if traffic jams still occur in a variety of situations.

Figure 11-10 show an image of a traffic jam due to peak time unexpected traffic.

Figure 11-10. *Traffic jam*

Social Media

In social media platforms such as Twitter, Facebook, and Instagram, anomaly detection can be used to detect hacked accounts spamming everyone, false advertisements, fake reviews, etc. Billions of people use social media platforms extensively, so the amount of activity on social media platforms is extremely high and is ever growing. To ensure the privacy and quality user experience of the individuals who are using the social media platforms, social media companies use many techniques to enhance the capabilities of their systems. They use anomaly detection to examine the activity of every individual for normal and abnormal behavior.

Similarly, any advertising platforms ads, any personalized friend recommendations, any news articles that the individual might have been interested in such as elections, can all be processed for abnormal or anomalous activity. It would be a great use case for anomaly detection if anomaly detection could detect troll activity on your tweets, propagandized bots, fake news, and so on. Anomaly detection can also be used to detect if your account has been taken over, because all of a sudden your account might be posting immense amount of tweets, pause tweets and commence, or troll other accounts and spamming everyone else.

User engagement levels can also be modeled, allowing for the automation of identification of trends simply by monitoring what content seems to be drawing the most attention in an abstracted, generalized manner. Once this anomalous content is discovered, it can be further boosted by recommendation algorithms to attract even more user engagement (thus boost profits earned from targeted advertising), or addressed and hidden if it is something deemed to violate community policies.

Figure 11-11 shows a fake news article posted on Facebook.

Figure 11-11. *Fake news article about Facebook*

Finance and Insurance

In the finance and insurance industries, anomaly detection can be used to detect fraudulent claims, fraudulent transactions, fraudulent travel expenses, the risk associated with a specific policy or individual, etc. The finance and insurance industries depend on the ability to target the right consumers and take the right amount of risk when dealing with finance and insurance. For instance, if they already know that a specific area is prone to forest fires or earthquakes or very frequent flooding, the insurance company which is insuring your home needs to have all the tools that they can get their hands on to quantify the amount of risk involved when writing the policy for the home homeowner insurance.

An example use of anomaly detection in finance is to detect wire fraud where a large amount of money is transferred in and out of the country using several different accounts—activity that is extremely difficult for humans to detect manually, considering the massive volume of transactions that take place every minute. AI techniques can be trained on very large amounts of data to detect very new and innovative ways to commit wire fraud, detection this is beyond the capabilities of any human or many of the statistical techniques that have been in place for decades.

Deep learning does help solve this very big problem of difficult anomaly detection in the financial and insurance industries due to the sheer size of data and the number of variables involved. With the advent of graphical processing units (GPUs), deep learning is now helping address many of the hard-to-crack use cases. Anomaly detection and deep learning can be used together to serve the needs of the finance and insurance industries.

Figure 11-12 shows reported mortgage loan frauds over time. The anomaly here is the sudden drop in reported loan frauds from what was projected.

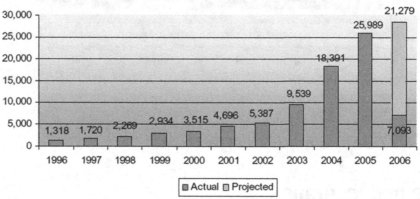

Figure 11-12. *Graph of mortgage loan reported frauds over time*

Cybersecurity

Another use case for anomaly detection is in cybersecurity or networking. In fact, one of the very first use cases for anomaly detection, decades ago, was the use of statistical models to try to detect any intrusion attempts into networks. As an example, anomaly detection is used to detect the very prevalent denial of service (DoS) attack. When a DoS attack is launched against a company's website or portal to disrupt service to customers, the attackers typically mobilize a large number of machines to run simultaneous connections and random useless transactions against the website or portal (a common target is some kind of a payment service for customers). As a result of this attack, the portal is not responsive to the customers, eventually leading to very poor customer experience and potentially loss of their business.

By training an anomaly detection system on data that has been collected for a long period of time, the system can detect the anomalous activity. This data is comprised of the typical use behavior, patterns in payment, how many users are active, how much the payment is at this particular time, as well as seasonal behaviors and other trends that exist for the payment portal exist for the payment portal, such as paying all the bills on the first of every month or doing all the shopping at the end of the month. When a DOS attack is suddenly launched against a company's payment portal, an anomaly detection algorithm can quicky detect such activity and notify the infrastructure or operational teams who can take corrective action such as setting up different firewall rules or better routing rules that attempt to block the bad actors from launching the attack or prolonging their attack against the portal.

Figure 11-13 shows an example of anomaly monitoring network flows.

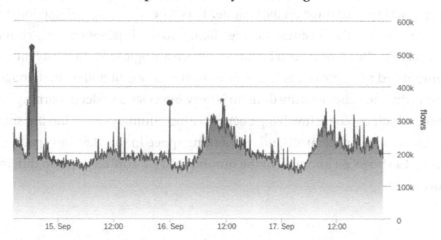

Figure 11-13. *Example anomaly monitoring network flows. The spikes are very high volumes of network flow, which can be considered as anomalies*

Another example would be when hackers are trying to get into the system given that they were somehow able to set up a Trojan to get into the network in the first place. Typically, detecting this anomalous activity is a process that involves a lot of scanning, such as port or IP scanning, to see what machines exist in the network when the services are being run. Suppose the machines are running SSH and telnet, the latter of which is easier to crack. The hacker can try to launch several different types of attacks that exploit the vulnerabilities of the telnet or asset service. Eventually, if one of the targeted machines responds, the hacker can get into the system and continue the penetration of the internal network until they accomplish what they came for.

Typically, networks have a pattern of usage—there are database servers with consistent read/write patterns, web servers with some traffic patterns, payroll systems, QA systems, end user-facing systems, and so on that all have some kind of expected usage pattern. It is usually well known expected behavior that is seen are for a long period of time in spite of the fact that there is a definitely a lot of change that is observed and expected over a long period of time as to how the machines are used how the networks are used in which machine talks to which other machine over what service and so on and so forth.

Anomaly detection can be used to detect if a specific port or service on a specific machine or machines is being connected to or transacted with at an abnormal rate, indicating some type of intrusion activity is taking place. This is extremely valuable information to the operations team, who can quickly pull in the cybersecurity experts and try to drill down into what is going on and take preventive or proactive action rather than simply reacting after the damage is done. This type of anomaly detection could be the difference between the business staying afloat and the business getting shut down (at least temporarily). There have been instances where a single cybersecurity intrusion has almost bankrupted a business, costing hundreds of millions of dollars in damages. This is the reason why the cybersecurity domain is very interested in deep learning, and all the use cases which could be involving deep learning anomaly detection are some of the top use cases in the cyber security and networking space in this day and age.

Figure 11-14 shows an anomaly in the number of TCP connections on different service ports

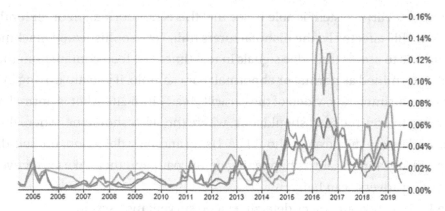

Figure 11-14. *TCP connections over service ports*

Not all the use cases in cybersecurity or networking are defensive in nature. We can also leverage anomaly detection to determine whether we need to upgrade systems, whether current systems are able to sustain the traffic for now and in the future, whether any node capacity planning needs to take place to bring everything back to normal, and so on. This is again very important for operations team to understand if there are trends which have not been foreseen a year ago that are now affecting the normal to abnormal behavior of the network. The sooner we know the better, so that we can start proactively planning to deal with this issue.

Video Surveillance

Another domain where anomaly detection is becoming extremely important is video surveillance. Nowadays, it is very common to see security cameras and video surveillance systems located in many places, such as in businesses, local schools, local parks, downtown streets, and houses. The point is, video surveillance is here to stay, especially given all the new technological advancements in smart apps and smartphones. In fact, we should expect much more video surveillance in the future.

In the very near future, we will see a lot more smart cars and self-driving cars, which depend on continuous processing of video using real-time analysis and detect various objects. At the same time, they can also detect any kind of anomaly. In a strictly security video surveillance sense, anomaly detection can be used to detect what is the normal for this specific camera which is looking at your backyard and when a specific anomaly is detected because of some kind of motion that is happening within the vicinity of the house. For example, some kind of animal or even an intruder is walking on the lawn.

Your home security system is able to see that this is not normal. In order for the cameras to do this effectively, the manufacturers train very sophisticated machine learning models to assess the video signals in real time. The feed coming from the cameras is determined as normal or abnormal. For example, if you are driving in a self-driving car on the interstate, video of the car will clearly indicate what is normal right now according to how the road should look, where the signs should be, where the trees should be, and where the next car should be. Using anomaly detection, the self-driving cars can avoid any abnormalities happening on the path and then take corrective action before anything adverse can happen.

Figure 11-15 shows an object-detecting video surveillance system.

Figure 11-15. *Object-detecting video surveillance system*

Manufacturing

Anomaly detection is also being used heavily in manufacturing sector. Specifically, since most of the manufacturing nowadays involves robots and a lot of automation, anomaly detection can be used to detect malfunctions or impending failure of parts of the manufacturing system.

In the manufacturing industry, because of all the automation that is happening, there is a lot of emphasis on various kinds of sensors and other types of metrics being collected in real time or near real time. This data can be used to build a sophisticated anomaly detection model to try to detect if there is any impending problem that will be seen very soon in the plant or the manufacturing cycle.

Another example of how anomaly detection can be used is the case of oil and natural gas platforms. An oil and natural gas platform typically has thousands of components all interconnected in various ways to make the plant functional. All these components can be monitored using sensors that do specific measurements of the various parameters of the components to which the sensors are attached. All these sensors can be part of an Internet of Things (IoT) platform. Collecting all the sensor output from all the thousands of sensors that are attached to the thousands of components over a long period of time makes it possible to train sophisticated anomaly detection models such as autoencoders, LSTMs, TCNs, and transformers.

Figure 11-16 shows a manufacturing plant with sensor readings

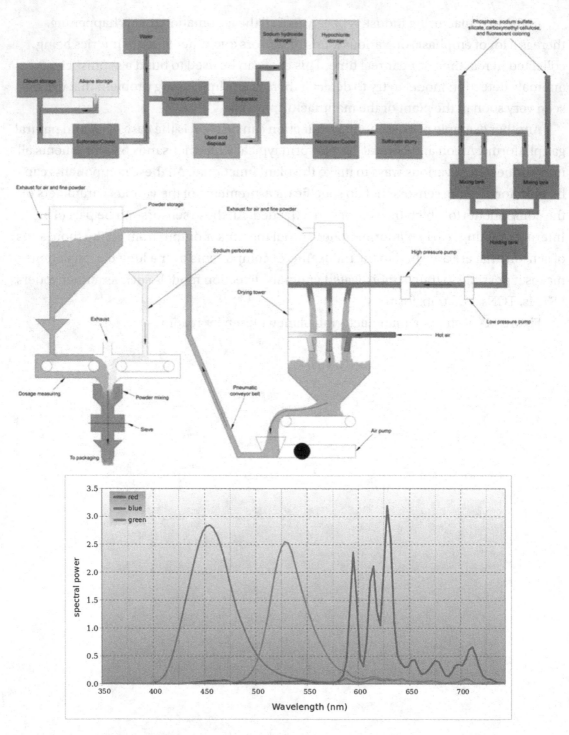

Figure 11-16. *Manufacturing plant with sensor readings*

Smart Home

Anomaly detection is also into smart home systems. Smart homes have lots of integrated components, such as smart thermostats, refrigerators, and interconnected devices, that can talk to each other. For example, Amazon Alexa can talk to your smart lights, which use smart bulbs. All these components can communicate with your smartphone via apps on your smartphone. Even thermostats are interconnected. So, how can we use anomaly detection in this use case? A simple example would be to monitor how you set your thermostat for the optimal temperature during all weather conditions and follow some sort of recommendation or recommended behavior. Because thermostats are personalized to some extent in every household, a very good deep learning algorithm could continuously look for all the usages for the thermostats across all the houses, including yours, and then be able to detect how you use it normally when you use it. Detectable anomalies could be something like a sudden rise / dip in temperatures, which could indicate that the AC or heater is not functioning. Nowadays, fridges have computer interfaces running on them and even come with usage monitoring capabilities. One example of anomaly detection would be your fridge reminding you to go grocery shopping since you didn't restock in a couple weeks - a definite anomaly if you do so weekly.

Figure 11-17 show an illustration of a smart home.

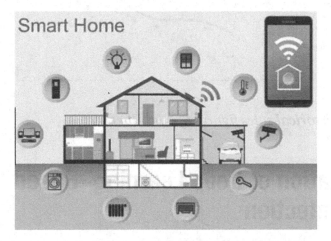

Figure 11-17. *Smart home with a depiction of all connected smart devices*

Retail

The retail industry uses anomaly detection algorithms for various use cases, such as to detect anomalies in the efficiency of the supply chain in terms of distribution of goods and services. Also interesting are the returns that happen all the time from customers because returned goods are tricky sometimes and it costs less to sell them in a clearance sale than to restock.

Looking at customer sales is also critical both in terms of revenue generated by sales and in terms of planning future products and sales strategies, especially when it comes to targeting consumers better. If incoming sales do not match up with predicted sales (by modeling normal sales data, which may or may not be accounting for an upward growth trend), then it is important to know so that product orders can be adjusted to avoid over- or undersupply.

Figure 11-18 show the historical sales figures of a product.

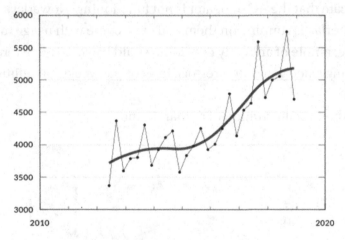

Figure 11-18. *Historical sales figures of a product*

Implementation of Deep Learning–Based Anomaly Detection

Given all the use cases in all the different industries, a small sampling of which were presented in the previous sections, the key steps in establishing an anomaly detection practice in your organization or business are as follows:

- Identifying business use case and getting aligned on the expectations

- Defining what data is available and understanding data and the nature of the data itself

- Establishing the processes to consume the data in order to process it

- Establishing the type of models to use

- Discussing strategy of how the models will be used and executed

- Investigating the results and feedback analysis as it effects the business

- Operationalizing the model used in the day-to-day activity of the business

In particular, the focus should be on how the models are built and what type of models to use. The type of anomaly detection algorithm used affects pretty much everything that you are trying to get out of an anomaly detection strategy. This in turn depends on the type of data available, as well as whether the data is already labeled or identified. One of the factors that affects the decision as to what type of anomaly detection will work best for the specific use case is whether it is a point anomaly, a contextual anomaly, or a collective anomaly. Another important factor is whether the data is an instantaneous snapshot at some point in time or continuously evolving or ever changing real-time time-series data. Also important is whether the specific features or attributes of the data are categorical or numerical, nominal, ordinal, binary, discrete, or continuous. It is also very important to know if the data is being labeled already or if some sort of a hint is provided as to what this data is, since it could steer you in the direction of supervised, semi-supervised, or unsupervised algorithms.

Although the technologies and algorithms are available to be used, there are several key challenges to implementing an anomaly detection approach based on deep learning:

- Integrating AI into existing processes and systems is difficult.

- The technologies and the expertise needed are expensive.

- Leadership needs to be educated on what AI can and cannot do.

- AI algorithms are not natively intelligent; rather, they learn by analyzing "good" data.

- There is a need for change in "culture," especially in large companies, who might be hesitant to adopt newer technologies or spend the resources needed to experiment, develop and integrate anomaly detection models using deep learning.

Future Trends

The field of anomaly detection is developing at a rapid pace thanks to the rapid progress of deep learning technology. Furthermore, the world shows no signs of stopping in terms of interconnecting everything possible. The rate and scope of data collection and storage are higher than ever before, and deep learning models are becoming far more capable than ever before. Here are some state-of-the-art trends for anomaly detection:

- **Multimodal anomaly detection**: Anomaly detection algorithms are being trained on multiple data source types simultaneously, such as training on images and text data, training on audio-visual data, training on human posture and voice, and so on. A good example in finance is using the combination of stock price values, news sentiment, and online discussion to train anomaly detection algorithms for price fluctuations. The idea is that the model could gain a deeper, richer understanding of what normal price fluctuations look like. For example, assume that bad news comes out about a top company. Immediately, the stock price might drop due to user panic. The model can learn to understand that this is normal behavior based on the extra insights it gains from training on news articles, adding another layer of predictive capability. When deviations from the predictions occur, they are sure to be true, unexpected anomalies.

- **Edge computing**: Processing power continually grows. Some smartphones nowadays are more powerful than computers from a decade ago. There are already deep learning models that can be run on mobile devices. Anomaly detection might then take place in real time at the source itself. Examples could be **malware detection** (by identifying unusual app behavior), **user detection** (the phone can detect who is using it by analyzing various behavioral and usage patterns and personalize the experience), **health monitoring** (for people who hook up a medical device to their phone, the phone can continuously monitor health signals and detect possible abnormalities), **decentralized content recommendation**, and so on.

Mobile phones are extremely common and are getting more and more powerful. Edge computing might become more prevalent, especially as computing costs get higher and data privacy concerns regarding centralized data collection increase. It might also be possible to see advanced algorithms like sequence modeling algorithms be trained on mobile devices too, allowing for personalized time-series forecasting for spending habits, budgeting, device usage, and so on that allow for detection of anomalies. For example, warning the user for using the phone way more than usual, or warning the user that they are spending far more than usual.

- **Few/zero-shot anomaly detection**: Some of the anomaly detection algorithms presented in this book are quite data hungry, such as the GAN and the transformer. However, what if algorithms could learn to detect anomalies with very little data, via few-shot or zero-shot learning? This is an ongoing field of research in deep learning and anomaly detection.

- **Large language models (LLMs) for anomaly detection**: LLMs can be tuned for a wide variety of language modeling tasks, such as summarizing works, generating text, coding, or discovering other patterns in text. As LLMs get more and more powerful, so does their potential to detect anomalies.

  Examples include automated and intelligent content moderation (word filters tend to be a scorched-earth method, with high recall but terrible precision and harming more than they help), fake user review detection, and even a coding bug detector. Nearly every NLP task is now possible to perform at a superhuman level thanks to the power and scalability of LLMs. Although they are lacking in some aspects (such as logical reasoning), LLMs are sure to make tremendous progress in the coming years. If it's not LLMs that dominate the language modeling space, a new type of model will succeed it.

- **Generative AI**: Deepfakes are getting extremely good and are almost in the realm of indistinguishability, at least to most human eyes. We need tools that detect deepfakes, in terms of both speech and video, so that we can protect people and their identities and fight misinformation. Unfortunately, this is an arms race that may eventually result in the generator winning, because once you get to the point where the generator perfectly replicates reality or at least becomes impossible to detect by a discriminator without too many false positives as in, the discriminator cannot detect generated deepfakes anymore without flagging too many real images. That being said, that is still a theoretical possibility and it's not certain how easily achievable it is, so there is still some hope for now.

Another potential use case is the detection of copyrighted art styles in AI-generated art. A point of recent debate and legal action is that the data used to train some of the AI art generator models was obtained without the permission of the creators. Basically, these AI art generators are creating art using the style of the original creators and it is being distributed. While it's too late now and there is no going back, at least these art pieces can be detected for potential legal action to take place.

It is important to keep in mind that speculation on the future is almost always a fool's gambit. There are some scenarios that are more likely than the others to occur, but the future is always uncertain. For all we know, someone might discover a new class of deep learning algorithms that that eclipses all cutting-edge models. Current modeling techniques may still be state-of-the-art for a while longer. Regardless, deep learning and the field of anomaly detection are continually evolving at an ever-increasing pace, and it is exciting to see these developments as they happen.

Summary

This chapter discussed practical use cases of anomaly detection in the business landscape. You have seen how anomaly detection can be used to address real-life problems in many businesses. As mentioned in the chapter introduction, every business and use case is different, and we cannot copy and paste code and build a successful model to detect anomalies in any dataset, so this chapter covered many use cases to give you an idea of the possibilities and concepts behind the thought process.

Remember that this is an evolving field with continuous inventions of new algorithms and enhancements to the current algorithms, which means that in the future the algorithms will not look the same. Just a few years ago, the RNN was the best algorithm for time Series, but it was succeeded by the GRU and the LSTM (Chapter 8) as well as the TCN in some cases (Chapter 9). And in recent years, the transformer (Chapter 10) has far outclassed LSTMs in various NLP tasks. Who knows what the future might hold? Even autoencoders have changed quite a bit; the traditional autoencoders have evolved into variational autoencoders (Chapter 6). The restricted Boltzmann machine, which was covered in the previous edition of this book, is no longer used much.

Index

511

P, Q